ENDOMYCORRHIZAS

Proceedings of a Symposium held at the
University of Leeds, 22–25 July 1974

ENDOMYCORRHIZAS

Proceedings of a Symposium held at the
University of Leeds, 22–25 July 1974

Edited by

F. E. SANDERS
University of Leeds, England

BARBARA MOSSE
Rothamsted Experimental Station, England

P. B. TINKER
University of Leeds, England

1975

ACADEMIC PRESS
London   New York   San Francisco

A Subsidiary of Harcourt Brace Jovanovich, Publishers

ACADEMIC PRESS INC. (LONDON) LTD.
24/28 Oval Road
London NW1

United States Edition published by
ACADEMIC PRESS INC.
111 Fifth Avenue
New York, New York 10003

Library of Congress Catalog Card Number: 75-34561
ISBN 0-12-618350-3

PRINTED IN GREAT BRITAIN BY
J. W. ARROWSMITH LTD., WINTERSTOKE ROAD, BRISTOL BS3 2NT

LIST OF PARTICIPANTS

ANDERSON, Mr. G.	ADAS, Leeds
ARNOLD, Mr. D. J.	ICI Plant Protection Ltd., Bracknell
ADEGEYE, Mrs. A. O.	University of Leeds
ATKINSON, Mr. M. A.	Department of Forestry, Oxford
BAREA, Dr. J. M.	Experimental Station of Zaidin, Granada, Spain
BAYLIS, Professor G.T.S.	University of Otago, New Zealand
BENJAMIN, Mr. L. R.	University of Leeds
BONFANTE, Dr. P. F.	Istituto Botanico dell'Universita di Torino, Italy
BOWEN, Mr. G. D.	Adelaide, South Australia
BROCKLEHURST, Mr. P. M.	University of Leeds
BROCKWAY, Mr. C. W.	Animal Industry and Agriculture Branch, Darwin, Australia
CALLOW, Dr. J. A.	University of Leeds
CALLOW, Dr. M. E.	University of Leeds
CHADWICK, Dr. M. J.	University of York
COOPER, Dr. K. M.	University of Otago, New Zealand
COX, Dr. G. C.	University of Leeds
DAFT, Dr. M. J.	University of Dundee, Scotland
EDMONDS, Dr. A. S.	University of Waikato, New Zealand
EVANS, Dr. L. V.	University of Leeds
FAIZY, Mr. S. A.	University of Leeds
FITTER, Dr. A. H.	University of York
GERDEMANN, Professor J.W.	University of Illinois, U.S.A.

GREENWOOD, Miss A.	University of Sheffield
GURR, Mr. K.	University of Reading
HADLEY, Dr. G.	University of Aberdeen, Scotland
HALLSWORTH, Dr. E. G.	CSIRO, Division of Soils, Glen Osmond, Australia
HARBERD, Dr. D. J.	University of Leeds
HARLEY, Professor J. L.	University of Oxford
HARRIS, Dr. P. J.	University of Reading
HATTINGH, Dr. M. J.	University of Stellenbosch, South Africa
HAYMAN, Dr. D. S.	Rothamsted Experimental Station, Harpenden
HEPPER, Miss C. M.	Rothamsted Experimental Station, Harpenden
HODGSON, Dr. D. R.	University of Leeds
JALALI, Dr. B. L.	Institut fur Bodenbiologie, Forschungsanstalt fur Landwirtschaft, West Germany
JANOS, Mr. D. P.	Organizacion Estudios Tropicales, Costa Rica
KASPARI, Dr. H.	Institut fur Mikrobiologie der Universitat, Bonn, West Germany
KHAFAGI, Mr. M.	University of Leeds
KHAIRI, Mr. S. M.	University of Leeds
KHAN, Professor A. G.	University of Islamabad, Pakistan
LEWIS, Dr. D. H.	University of Sheffield
MASON, Dr. P.	University of Edinburgh
MOSSE, Dr. B.	Rothamsted Experimental Station, Harpenden
NEHEMIAH, Mr. J.	Institut fur Pflanzenkrankheiten, Bonn, West Germany
NICOLSON, Dr. T. H.	University of Dundee
PAGET, Mr. D. K.	Rothamsted Experimental Station, Harpenden
PALMERLEY, Mrs. S.	Univeristy of Leeds

PATTISON, Mr. A. C.	Rothamsted Experimental Station, Harpenden
PEARSON, Dr. V.	University of Leeds
PHILLIPS, Mr. J. M.	Farnborough, Hants.
POSNER, Professor A. M.	University of West Australia, Perth
POWELL, Dr. C. LI.	Ruakura Soil Research Station, Hamilton, New Zealand
PREECE, Mr. T. F.	University of Leeds
PURVES, Miss S.	University of Aberdeen
READ, Dr. D. J.	University of Sheffield
SANDERS, Dr. F. E. T.	University of Leeds
SCANNERINI, Professor S.	Istituto Botanico dell'Universita di Torino, Italy
SCHENCK, Professor N. C.	University of Florida, U.S.A.
SCHÖNBECK, Professor F.	Institut fur Pflanzenkrankheiten, Bonn, West Germany
SPARLING, Mr. G. P.	University of Leeds
STRIBLEY, Dr. D. P.	University of Sheffield
TINKER, Professor P. B.	University of Leeds
TRINICK, Dr. M. J.	Rothamsted Experimental Station, Harpenden
VAN DIJK, Dr. C.	Weevers Duin Biological Station, The Netherlands
WAI, Dr. T. K.	University of York
WARCUP, Dr. J. H.	Waite Institute, Adelaide, South Australia
WEBSTER, Professor J.	University of Exeter
WILD, Miss J.	University of Leeds
WILLIAMS, Dr. P. J.	University of York
WONG, Mr. W. C.	University of Leeds
WOOLHOUSE, Professor H.W.	University of Leeds

PREFACE

The idea of this Symposium arose from informal discussions between the editors of this volume and a few others over three years ago. The most striking feature of recent research in mycorrhizas is the enormous interest in the vesicular-arbuscular mycorrhizas, With many new workers in this field, producing ever increasing amounts of new work, it seemed an excellent idea to hold a meeting to take stock of, and to summarise, the present state of the subject. It was essential to include work with other kinds of endomycorrhizas also, since there might be interesting analogies between the different forms; on the other hand, it was not practicable to include ectomycorrhizas for reasons of length and time, and also because these have been dealt with in detail in recent publications.

In the event we were pleasantly surprised, almost embarrassed, by the number of papers offered, and the number of people who attended. Some papers were invited, and these tend to be slightly longer than the rest in the published versions: their purpose was to give the background for the more detailed and specialised papers which follow. The arrangement of the sessions has been largely retained in this volume, with a few minor changes where sessions could not be arranged in the most logical sequence because of time pressures.

No record of the discussions has been given here though questions, answers and arguments flowed freely in the meeting. The recorded discussion was so long that it was quite impossible to include all of it in this volume, and it was felt best therefore to omit it, rather than to give a highly condensed or partial

record. Authors had the opportunity to revise manus-
cripts after the meeting, and most of the major
points in the discussion will have been incorporated
in this way.

One of the most pleasing and encouraging aspects
of the meeting was the variety of disciplines represented there: mycologists, agronomists, chemists,
foresters and ecologists. This leads us to believe
that the Symposium met a real need, and to hope that
the proceedings will be useful to everyone interested
in this topic.

We are greatly indebted to many people who
assisted in arranging the meeting and preparing the
proceedings. In particular, we thank Professor Harley
and Dr. Nicolson for their support and advice, Mr.
Gleave of the Special Courses Division, Department
of Adult Education, University of Leeds, and Mrs. M.
Hall and Mrs. S. Palmerley for clerical assistance.
We are also grateful to the Royal Society and the
Leverhulme Trust for grants in aid.

                              F. E. Sanders
                              B. Mosse
                              P. B. Tinker

CONTENTS

Physiology—phosphorus

Fine structure

Effects on plant growth

Ecology

Biological Interactions

PROBLEMS OF MYCOTROPHY

J. L. HARLEY

Department of Forestry, University of Oxford, U.K.

INTRODUCTION

I propose in this introductory paper to take
stock of the position that we have arrived at in
certain aspects of the study of mycorrhiza, and to
try to see in what direction research might go to
enter important new fields of ecology and physiology.
It has been said that mycorrhizasts spend too much
time reviewing their subject. So I shall not do
that, but assuming many of the past achievements pick
out a few of the potentially important growing points.
It is to be hoped that at least some of the problems
on which I comment will be elaborated or solved by
later speakers.
 Mycorrhizal symbioses are variable in several
respects; in the kind of fungus involved, the kind
of higher plant involved, and in their morphological
pattern and histology. They are part of a great
array of symbioses between heterotrophic and autotrophic organisms - from corals to beech trees, from
lichens to legumes. These grade from clearly mutualistic symbioses to biotrophic parasites on the one
hand and to casual root-surface and rhizosphere
associations on the other. In addition there are
derived types of association where both primary
partners, fungus and host, are each carbon heterotrophs, but in such cases the source of carbon may be
a second and autotrophic host. Three membered assoc-

iations of mycorrhizal fungus, nitrogen-fixing pro-
caryote and green autotrophic host are also common as,
for example, alder, most leguminosae, etc.

In all but a few of these symbiotic systems one
of the partners is carbon autotrophic and provides
fixed carbon compounds for the system. That is, the
heterotrophic partners are adapted to receiving a
supply of carbon direct from the photosynthetic
products of the autotrophic partners rather than
primarily or solely from the humus or dead tissues
indirectly derived from photosynthesis after death of
the autotroph. There is a short-circuiting of carbon direct from photosynthesis into the heterotroph
(Harley 1971).

It is easy to believe that this kind of short-
circuiting of carbon compounds should occur commonly
in those ecosystems, such as the sea, where there is
no large mass of soluble or insoluble abiotic carbon
compounds in the cycle. It is at first sight more
difficult to explain the prevalence of root-surface
organisms, mycorrhizal symbionts, nitrogen fixing
symbionts, lichens, etc. in terrestrial, especially
temperate woodland, ecosystems where abiotic carbon
compounds seem to be present in large quantities
especially in the surface soil horizons.

A very valuable recent paper by Gray and
Williams (1971) calls into question the availability
of carbon in such situations. It concludes that the
estimated yearly accretion of carbon, from the
obvious sources of leaf fall and so on, to the soil
is often inadequate to explain both the rate of CO_2
production from the soil and the maintenance of the
estimated biomass of the soil organisms. These
conclusions help us to understand the universal
existence both of very active populations of microorganisms around the root-systems of plants and the
extremely common development of mycorrhizal associations of all sorts.

If these associated microorganisms are in considerable measure powered by current photosynthesis,

an entirely different picture is gained of the
location and balance of microbial activities in the
soil than is usually accepted. The soil is not by
any means a uniform region of intense microbial activity but a complex pattern of activity determined by
availability of photosynthetic carbon. This conclusion confirms the ecological continuity between
rhizosphere and mycorrhizal and bacterial symbioses
(Harley 1948); it explains the prevalence of nitrogen fixation in the root region which has recently
been emphasized (Dommergues et al., 1973; Richards,
1973; Döbereiner et al., 1972, 1973; Silvester &
Bennett, 1973); it re-emphasises the probability of
a relationship between mycorrhizal and other infections and light intensity (Björkman 1942); and it
requires consideration in attempts to explain the
evolution of biotrophy and its early appearance in
the fossil record; for competition for available
carbon compounds produced a central selective force
which resulted in the evolution of diverse symbiotic
systems.

QUANTITIES OF CARBON COMPOUNDS INVOLVED

From both an ecological and a physiological
point of view it is important to gain some estimate
of the quantities of photosynthetic products which
are directly diverted to the mycorrhizal and other
symbiotic and associated heterotrophs in an ecosystem.
This is an outstanding problem for which only approximate solutions are as yet possible and there is room
for much further work upon it. Recent work on root
exudates from plants in culture has shown that as much
as 5-10% of the carbohydrates and amino acids (after
allowing for respiration) available to the roots of
seedlings are exuded into the medium; the actual
quantity varies according to the cultural conditions.
Barber and Gunn (1974) believed that even these
quantities, although they appear large, would by no

means explain the discrepancies in the carbon balance and soil respiration observed by Gray and
Williams (1971).   But, of course, it is difficult to
extrapolate from seedlings in culture to ecosystems.

On the other hand, calculations have also been
made of the order of magnitude of carbohydrate consumption by ectotrophic mycorrhizal fungi in ecosystems assuming them to be supplied by photosynthesis of the hosts (Harley, 1971, 1973).   Using estimates of weights of fruitbodies formed by mycorrhizal
fungi (Romell, 1939) and of respiration by mycorrhizal sheath (Harley *et al.*, 1956), two additive
estimates of carbon required may be obtained.   From
these, even if respiration of the sheath is ignored
to allow for overestimation, a conservative value
equivalent to about 10% of that used in timber production, 500 Kg per hectare per year, of a temperate
forest is estimated.   This does not include carbohydrate used by mycelium in the soil.

These estimates are only a preliminary cockshy
and they ought to be repeated, refined and extended
to other types of mycorrhiza, for they may well put a
new dimension in the study of the place of mycotrophy
in carbon cycling in ecosystems.

If it is true that all or a large part of the
carbon for mycorrhizal fungi comes from the host, a
further important consequence follows.   The fungi
are able, by use of this carbon source, to remove
inorganic nutrients selectively from the F and H soil
layers.   The effect of this will inevitably be to
slow up the rate of humus breakdown which is in most
soils limited by nutrients such as nitrogen and
phosphate (see Gadgil and Gadgil, 1971).   It is
possible that the activities of mycorrhizal fungi are
an important factor in the determination of soil
structure.

CARBON TRAFFIC IN THE SYMBIOTIC SYSTEM

Recent research on a variety of symbioses (ectotrophic mycorrhizas, lichens, biotrophic fungal parasites, corals and other plant-animal symbioses) has examined the actual traffic in carbon compounds from autotroph to heterotroph (Smith *et al.*, 1969) . A considerable degree of common behaviour involving similar compounds and related processes has been discovered.

Similar studies are now being made with endotrophic mycorrhizal associations.   Pearson and Reed (1973) have studied carbon movement to the endophyte in Ericaceae and Ho and Trappe (1973), Hayman (1974) and others, to Endogone, while S. E. Smith (1967) studied carbohydrate movement from endophytes to host in Orchidaceae.

Many of the results on carbohydrate movement in symbioses seem to indicate a directed flow from the site of photosynthesis in the autotroph to the site of accumulation or use in the heterotroph.   In many symbioses the movement of carbohydrate involves a change from a mobile compound in the autotroph to stable storage compounds in the heterotroph.   In the ectotrophic mycorrhiza of *Fagus* for example (Lewis & Harley, 1965 abc) sucrose seems to move through the host to the fungus where fungal carbohydrates, not readily reabsorbed or used by the host, are synthesised.   In this way a one-way source to sink mechanism seems to be set up.   Similar systems seem to occur commonly in other symbiotic systems.   However, in endogonaceous mycorrhiza, Hayman (1974) did not find any typically fungal carbohydrate in the endophyte which might be interpreted as a sink.   The absence of mannitol is not surprising for it has not been found in Phycomycetes; but trehalose has been identified in some biotrophic and free-living mycelial Phycomycetes including some Mucorales.

It is important, however, to see the concept of source to sink movement in its proper perspective.

In a single plant body, say of an angiosperm, the
source may well be the photosynthesising leaves.
The sinks will be of various kinds.   (1) Developing
organs with rapid metabolism may act as sinks by
keeping their floating concentration of carbohydrate
low by rapid use.   Such sinks are controlled by
hormonal signals which stimulate and control active
growth.   (2) The sink may be a storage organ, a seed
or a tuber, in which starch or some other insoluble
product with low active mass is synthesised.   There
the balance, condensation versus hydrolysis, is again
hormonally controlled (Kumar & Wareing, 1974).
(3) There is the storage organ, like the stem of
sugar cane, where the sink contains a soluble storage
substance of high active mass whose sequestration in
a vacuole is presumably dependent on hormones controlling membrane permeability or enzyme synthesis or
activity.

The beech mycorrhizal system is one where the
active masses of the soluble trehalose and mannitol
(but not of glycogen which is insoluble) are significant but they are retained in the fungus because
inter alia the host membranes are relatively impermeable to them.   There may be in addition a hormonal
signal produced by the fungus affecting the host, but
we have as yet no evidence of it in mycorrhizas.

Of course, in any plant body or symbiotic
system more than one sink, e.g. metabolic and storage,
may at any one time be operating from a single source.

Some of the experiments with *Fagus* mycorrhiza
(Lewis & Harley, 1965 b & c) where very low concentrations of highly radioactive sucrose were used,
illustrate some of these points.   Sucrose was translocated into the mycorrhizal tip, but there was no
evidence of directed transport to or preferential
accumulation in the fungal sheath.   The greater part
of the carbon in the tip region was in organic acids
(and other charged substances) and these were roughly
equally divided between host core and sheath.

Table 1. Translocation from high (10%)
and low (0.001%) 14$_C$-labelled, sucrose
concentrations in agar blocks to apices
of mycorrhizas of *Fagus* in 22 hours
(Lewis & Harley, 1965c)

| Region | % counts per minute | | | |
| | High sugar | | Low sugar | |
	A	B	A	B
Base	90.4	85.2	97.7	96.8
Apical sheath	7.4	10.0	1.1	1.6
Apical core	2.2	4.8	1.2	1.6

A   Block in contact with core only
B   Block in contact with core and sheath

This would indeed be expected since such substances
enter into the metabolic pathways of both symbionts
and these would be expected to have priority over
storage.   Their distribution might consequently reflect the metabolic states of the two kinds of tissue,
given that the membranes of both components were
permeable to them.   Indeed, in other circumstances,
e.g. in the uptake of ammonium, charged ionic carbon
compounds appear to pass from fungus to host
(Carrodus, 1967) as a normal part of uptake and
translocation.

Hayman's report (1974) that in the endogonaceous
mycorrhiza he studied there were no carbohydrates
peculiar to the fungus that might constitute a sink
is extremely interesting.   Deposition of fats has
been observed in many of the endogonaceous fungi, and
they and their synthesis might well be worth investigation as potential destinations of translocated
carbohydrates from the host.   On the other hand,
storage of carbon compounds might not be such a
feature of endogonaceous mycorrhizas as it is in the
sheaths of ectotrophic forms.   The active growth and

Harley, J. L.

metabolism of the fungus may well be the operative
sink, and the persistence and growth of the mycelium
and differentiated organs of the fungi would be
worthy of examination.

The experimental analyses of S. E. Smith (1967)
on *Rhizoctonia solani* and *Orchis purpurella* seedlings
illustrate active growth of the host acting as a
sink.   These showed translocation of carbohydrate
through the hyphae to the seedling over a period of
160 hours.   During the whole period about 50% of the
carbon going into the seedling was incorporated into
insoluble material.   No doubt much of this, since
active growth was occurring, was in cell-walls,
proteins and the like which constituted one set of
sinks.   The destination of the remainder only

Table 2.   Translocation of carbohydrate
through hyphae of *Rhizoctonia solani* (RS 10)
into seedlings of *Dactylorchis purpurella*
Calculated from graphs of S. E. Smith (1967)

Relative quantities of ^{14}C derived from
sucrose absorbed by hyphae and passed into
orchid

Category	Relative Quantity
Total translocated into seedling	100
Insoluble material	50
Soluble components:-	
Anionic (organic acid)	5
Cationic (Amino acid)	15
Sucrose	15
Glucose) Fructose)	5
Trehalose	6
Mannitol	4
Total soluble	50

changed a little over the whole experimental period
and after 160 hours (see Table 2) 20% was in charged
compounds (amino acids and organic acids), presumably
in the metabolic system, 20% was in sucrose, glucose
and fructose and 10% was still in fungal sugars
(trehalose and mannitol), possibly still in the
fungal hyphae within the tissues.   In this case the
sink seems to be the actively growing orchid seedling, and carbohydrate moves towards it through the
hyphae.
    The outstanding question is, of course, the
nature of the factors which allow the carbohydrates
to be released across the fungal membrane to the
host.   I will consider this intractable problem
later along with similar problems of permeability.

REVERSAL OF NORMAL MOVEMENT OF
CARBON COMPOUNDS

    This is a good opportunity to consider the
problem of the reversal of movement of carbohydrates in symbiotic systems.   Certain substances,
like the disaccharides sucrose and trehalose or the
polyhydric alcohols mannitol, sorbitol, etc., are
known to be transported in the phloem of angiosperms
or in fungal hyphae.   It is in such compounds that
considerable quantities of carbohydrate, the bulk
carbon and energy currency, move from sources where
they are formed to regions where they are metabolized
or converted to storage substances.   Short distance
movement of carbon takes place in minor quantities in
other forms.   In addition, carbon movement is also
associated with movement of other essential nutrients
(e.g. of nitrogen, as the carbon of amino acids) both
over long and short distances.   The setting up of a
source to sink system involving say sucrose, will not
affect a reverse movement of amino acids from the
site of absorption of nitrogen to sites of storage or

use.   But this is not a reverse flow of bulk carbon
or energy currency.   Indeed, the demonstration of
relatively small reverse flows of ^{14}C as by Reid and
Woods (1969) in ectotrophic mycorrhiza from fungus to
host is exceedingly interesting.   It deserves
detailed study, for it is likely to be an indicator
of movement of substances other than carbohydrates
from fungus to host.

FUNGAL SHEATHS AND THEIR FUNCTION

A compact fungal tissue, the sheath, mantle, or
mycochlaena is an organ of the mycorrhizas of many
forest trees and shrubs and a few smaller life-forms
(e.g. in Ericales).   In most mycorrhizas there is no
compact sheath.   Two whole sets of problems are
posed by these facts.   First, what are the factors
which promote sheath formation? (See Read &
Armstrong, 1972).   What peculiar properties do the
hosts or the fungi have in ectotrophic or sheathing
mycorrhizas which are not possessed by non-sheathing
mycorrhizas?   I do not propose to consider these
problems in any detail for I have not yet formed very
clear ideas about them;   except (i) that a study of
those host plants which may at different times form
either sheathing or non-sheathing mycorrhizas might
be worth undertaking, (ii) that factors hormonal and
otherwise leading to tissue formation in Basidiomycetes and other fungi should receive attention.
The second set of problems those about functioning of the sheath, have been considered experimentally
to some degree.
In forest trees the sheath is a conspicuous
feature of the mycorrhiza.   It comprises (in species
of *Fagus, Pinus, Nothofagus* and other genera) some
35-40% of the dry weight of the mycorrhiza, and is
responsible in *Fagus* for at least half of its CO_2
emission.   The renaming of ectotrophic mycorrhiza as
sheathing mycorrhiza (by Lewis, 1973) has therefore

merit in emphasising this.   The consistent presence
of a sheath in so many genera of gymnosperms and
angiosperms cannot be ignored in hypotheses of
function for it is an expensive structure in upkeep.
It must, you would think, have a selective value
although other kinds of plant with non-sheathing
mycorrhizas get on without it.

We have already seen that the sheath is a storage organ for carbohydrates.   Indeed, in *Fagus*
mycorrhiza whether in experiments it is fed exogen-
ously or through the host tissues with sugars, the
lion's share, some two thirds of the carbohydrates
accumulated, is in the sheath tissue (Lewis & Harley,
1965 a b & c).   One can see that this could be of
selective advantage to the fungi which are Basidiomycetes or Ascomycetes possessing perennial mycelium,
seasonally or periodically producing large fruit
bodies.   But this periodicity of fruiting is itself
usually also a reflection of periodic climate, where
only certain seasons of the year are suitable for
active vegetative and reproductive activity.   Meyer
(1973) and Moser (1967), Singer and Morrello (1960)
have all commented on the dominance of the ectotrophic forms in colder boreal forests, in montane
regions and in other zones where climate limits growth
for considerable periods.   By their storage of
reserves the fungi are enabled to exploit rapidly the
optimum periods.   In so far as this enables them to
absorb nutrients which are in part passed to the host,
this is of selective advantage to the whole symbiotic
system.   But there is also another side of the coin.
The sheath appears also to act as a storage tissue
for nutrients absorbed from the soil, that is by the
fungus.   Early in our work on phosphate uptake we
(Harley & McCready, 1952) demonstrated the accumulation of phosphate in the sheath of beech mycorrhiza.
There has been some misunderstanding of this in the
sense that many have assumed it only to occur in
excised mycorrhizas.   This is not so, as Table 3
shows.

Table 3. Accumulation of phosphate in the sheath of attached and excised mycorrhizas of adult beech trees at three seasons from Harley & McCready, 1952).

Condition of roots	Attached				Excised			
Date	31 Mar	1 May	23 June		31 Mar	1 May	23 June	
Condition of leaves	in bud	expanding	fully expanded		in bud	expanding	fully expanded	
Conc. H_2PO_4 supplied mM.	0.074	0.32	0.16	1.6	0.074	0.32	0.16	1.6
Mean percentage phosphate in sheath	88	88	89.8	86.8	91	89	84.9	91.3

Moreover, it has since been observed by many of those who have worked with whole young plants (see, for instance, Lobanov, 1960; Clode, 1956; Morrison, 1962). Table 4 gives estimates by various people on accumulation of substances in *Fagus* mycorrhiza and, except for chloride, each is accumulated in the sheath to some extent.

Table 4.   Accumulation in the fungal sheath of Fagus mycorrhiza.

Percentage of material absorbed found in the sheath.

Substance	Concentration	% in Sheath	Authority
Cl^-	lmM.	46 ± 3)	F.A.Smith (1972)
H_2PO_4	lmM.	83 ± 3)	
H_2PO_4	0.01 - 1.0 mM.	83 - 95	J.L.Harley & C.C.McCready (1952)
NH_4^+	10mM NH_4Cl	65 Total)	B.B.Carrodus (1967)
	+ $14_{CO_3}^-$	73 Amino)	
NH_4^+	ditto	74 Total	J.L.Harley (Unpubl.)
Rb^+	0.04 - 0.1mM	50 - 62	J.M.Wilson (Unpubl.)
Glucose	27.7 mM	70	D.H.Lewis & J.L.Harley (1965 a,b)

The sheath, therefore, seems to act as a storage organ both for carbohydrates and for soil derived nutrients. The ectotrophic mycorrhizal system appears in this way to be adapted to sharply seasonal climates and perhaps short growing-seasons. The feeding roots are converted into sheathed mycorrhizas of several years duration which have storage and absorptive functions.

Such a system would appear to be less well adapted to herbaceous plants, especially of small life-forms - annuals, cryptophytes, geophytes, and hemicryptophytes - where whole new shoots as well as root systems are rebuilt each year mainly from stored reserves in seed or other organs. Moreover, in regions of equable climate and long growing-season the formation of an expensive sheath even in arborescent species might have no selective value for there would seem to be less demand for a storage organ of that kind. On the other hand, the essential feature of all mycorrhizal systems is the fungal mycelium connecting the plant to the soil and its selective advantage requires particular consideration.

THE FUNCTIONING OF THE MYCORRHIZAL SYSTEM

If we ignore the sheath as a specialization peculiar to large life-forms in particular kinds of seasonal or periodic habitat, we can picture the essential mycorrhizal system as having three phases. (1) A soil-inhabiting mycelium. (2) A fungal phase within the tissues of the host. (3) The host tissues themselves.

In carbon autotrophic mycorrhizal systems carbon compounds synthesised in the green tissues of the host not only nourish the host itself but also pass into the fungus in phase (2) and so into the external mycelium as its source of carbon. Soil-derived nutrients absorbed by the mycelium in the soil pass into the fungal phase (2) and so into the host tissue.

For such a system to be of selective advantage in ecological situations, a factor or factors limiting growth of each partner must be supplied by the other. We have concluded that the fungus acquires carbon from the host, but why is it selectively advantageous that the host be dependent on the fungus for soil derived nutrients? Sanders and Tinker (1973) have crystallized our thoughts in this matter by emphasizing the part played by the fungus in exploiting the sources of non-mobile nutrients such as phosphate. It is clearly less costly in carbon compounds, and more flexible in operation, that the absorbing surface should be of extensive hyphae rather than consisting of a ramifying root system of much larger biomass (and hence carbon demand) to surface ratio.

So much seems reasonable, and thanks to Sanders' and Tinker's elegant work, almost simplistic. The question has been posed, however, of possible differences between the absorptive properties of the fungi and their hosts. There are several experimental results, especially on phosphate absorption, by different workers on both ectotrophic and endotrophic mycorrhiza which might indicate the possibility that fungi might be able to exploit different sources from the host. In addition the work of Bartlett and Lewis (1973) indicates that mycorrhizas may possess surface phosphatases which could allow them to exploit organic soil phosphates more efficiently and rapidly than uninfected roots. By contrast neither the experiments of Sanders and Tinker (1973) nor those of Hayman and Mosse (1972) with *Endogone* mycorrhizas showed any such effect. This is clearly still an important problem, and with it goes a second problem with respect to phosphate absorption. To what extent do mycorrhizal fungi secrete into the soil organic acids - hydroxyacids - such as malic or citric, which might be effective in bringing into solution phosphates from iron and aluminium phosphate complexes by chelating iron and aluminium (Johnson, 1952-1960).

The great problems arise, however, in phase (2) the interface between fungus and host. Here there is a bidirectional metabolically dependent polar movement of substances of an apparently selective kind.

It is bidirectional because movement into and out of the host, from and to the fungus, appears to be occurring simultaneously. It is selective because only certain substances, not the whole soluble contents of cells and hyphae, seem to be moving. It is polar because carbon comes from the host to the fungus and soil-derived nutrients in the reverse direction.

It is, of course, possible to suggest that in those endotrophic mycorrhizas where digestion of the fungus is evident, there are actually two different routes. Digestion might result in the release of materials from the fungus, which itself absorbs carbon compounds by its active intact hyphae. This, of course, has been queried, as by Hadley and Purves (1974). Moreover, there is no such evidence of two possible routes in ectotrophic mycorrhizas where a two-way movement across the Hartig net and contact zone appears to be inevitable. Indeed, it is exceptional in symbiotic systems in general for digestion to occur. Hence the general case would appear to be bidirectional movement of material across a single interface.

The problem of leakage of metabolites and ions from plant tissues has been the subject of a very recent review by Simon (1974). He has collected together evidence about factors which render membranes leaky and emphasized their possible effects on membrane structure. One of the factors he mentions is dessication and remoistening. Although this may have a particular relevance to certain kinds of symbiosis, e.g. lichens, nothing yet arises which helps with the general case. In any event most of the examples of leakage on which he comments are non-selective leakage.

The factors which were observed to be important in beech mycorrhizas in the initiation of leakage, from the sheath especially, were low oxygen concentration and other factors which might affect respiration rate. Their effects did not operate equally on all substances. For instance, whereas excised mycorrhizas aerated in 3% O_2 or less, readily lost potassium by a metabolically dependent system (Harley & Wilson, 1959), it required more prolonged anaerobiosis to cause a leak of phosphate. It is possible, therefore, to consider a gradient of oxygen availability as a possible factor in this exchange of material.

In normally growing mycorrhizas it seems credible that the external mycelium and sheath surface might be adequately aerated to be able to absorb potassium whilst the Hartig net might be deficiently aerated and permit its release. This would be a mechanism of throughput for potassium similar to that suggested by Crafts and Broyer (1938) for the movement of nutrients from the cortex to the stele of uninfected roots. Our observations with phosphate (Harley et al., 1953) were equally suggestive of a similar mechanism of movement. The uptake of phosphate from solutions of low oxygen concentration resulted in a greater proportional movement of phosphate to the host than uptake from highly aerated solutions. Further inspection showed that low oxygen supply reduced uptake rate of phosphate by the host less than that by the fungal sheath. Here also, on the analogy with Crafts and Broyer one might construct a mechanism of throughput of phosphate depending upon gradient of oxygen in the tissues and differential effect on the components. However, this is far from satisfying and must be rejected because at the very same sites and in the reverse sense along the gradient, carbohydrate is being passed from host to fungus, and carbohydrate uptake

by cells of hyphae is oxygen dependent.   I give this
example to show how halting and tentative are our
steps towards our explanation of this central
problem.

These problems of movement and interchange of
substances between symbionts have been particularly
studied in symbiosis between fungi and algae in
lichens, and between invertebrate animals and algae.
D. C. Smith (in press) pointed out that release from
the autotroph is selective (i.e. in few kinds of
compound) and represents a substantial proportion of
the carbon fixed in photosynthesis.   Moreover,
whereas the algae leaked photosynthates in symbiosis
and for a short time after isolation in culture, they
soon became self-contained after a few hours of
living free.   It was clear, therefore, that their
leakiness arose from some factor in the symbiotic
system.   He concluded that it was not a general
relaxation of permeability barriers in the membrane
for that would have been unselective.   Nor was it
likely to arise from the diversion of materials
destined for cell wall or external slime production.
Both of these would not have been adequately specific.   He concluded that there must be a direct
effect on selective transport systems in the membranes
of the algae.   There was some evidence of the production of thermolabile factors promoting leakage in
some animal symbioses, but apart from these no very
convincing stimuli to leakage which might operate in
natural conditions were put forward.

Smith's most recent experiments have confirmed
pretty clearly that movement of carbohydrate in
lichens does depend on leakage from the algae and is
not due directly to an absorptive activity of the
fungus.   Selective inhibition of carbohydrate uptake
by the fungus does not by itself affect loss of
carbohydrate by the algae in whole lichens.   By
contrast, factors which affect the ability of algal
membranes to transport specific sugars, and factors
which affect the internal metabolic use of these

sugars, and hence their concentration, do jointly affect leakage rate.

Clearly the problems of transport and interchange of material are general to all symbiotic systems and central to hypotheses of function. But they have a much wider importance too, in the movement between cell and cell in all organisms, individual or dual.

CONCLUSION

The outstanding problems in the study of mycorrhiza are common to mycotrophy in general and to the whole subject of symbiosis between autotroph and heterotroph. Mycorrhizasts have mainly concentrated up to now on describing the effects of infection on the growth and mineral nutrition of the host plant. The problems of carbon nutrition of symbiotic systems are equally important and deserve increased attention. The importance of symbiosis in ecosystems has been much underestimated and the extent and effects of diversion and short-circuiting of carbon and nutrients through symbiotic systems requires further evaluation. The problem of the effects of each symbiont on the selective permeability of the membranes of the other and of the movement of nutrients within the symbiotic system is a central and essential problem for the future.

REFERENCES

BARBER, D. A. and GUNN, K. B. (1974). The effect of mechanical forces on the exudation of organic substances by the roots of cereal plants grown under sterile conditions. *New Phytol.*, 73, 39–45.

BARTLETT, E. M. and LEWIS, D. H. (1973). Surface phosphatase activity of mycorrhizal roots of beech. *Soil Biol. Biochem.*, 5, 249–257.

BJÖRKMAN, E. (1942). Über die Bedingungen der Mykorrhizabildung bei Kiefer und Fichte. *Symb. botan. upsal.*, 6 (2), 1-191.

CARRODUS, B. B. (1967). Absorption of nitrogen by mycorrhizal roots of beech. II Ammonium and nitrate as sources of nitrogen. *New Phytol.*, 66, 1-4.

CLODE, J. J. E. (1956). As micorrizas na nigracao do fosforo - estudo com o ^{32}P. *Publicões, Dir. ger. servs. flor. aquic.*, 23, 67-206.

CRAFTS, A. S. & BROYER, T. C. (1938). Migration of salts and water into xylem of the roots of higher plants. *Am.J. Bot.*, 25, 525-535.

DOMMERGUES, Y., BALANDREAU, J., RINANDO, G. & WEINHARD, P. (1973). Non-symbiotic nitrogen fixation in the rhizosphere of rice, maize and different tropical grasses. *Soil Biol. Biochem.*, 5, 83-89.

DÖBEREINER, J., DAY, J. M. & DART, P. J. (1972). Nitrogenase activity and oxygen sensitivity of the *Paspalum notatum-Azotobacter* association. *J. gen. Microbiol.*, 71, 113-116.

DÖBEREINER, J., DAY, J. M. & DART, P. J. (1973). Rhizosphere associations between grasses and nitrogen-fixing bacteria: effect of O_2 on nitrogenase activity in the rhizosphere of *Paspalum notatum*. *Soil Biol. Biochem.*, 5, 157-159.

GADGIL, R. L. & GADGIL, P. W. (1971). Mycorrhiza and litter decomposition. *Nature, London,* 233, 133.

GRAY, T. R. G. & WILLIAMS, S. T. (1971). Microbial productivity in the soil. In: *Microbes and Biological Productivity* eds. Hughes A. H. & Rose A. A., C.U.P., pp. 255-286.

HADLEY, G. & PURVES, S. (1974). Movement of ^{14}C carbon from host to fungus in orchid mycorrhiza. *New Phytol.*, 73, 475-482.

HARLEY, J. L. (1948). Mycorrhiza and soil ecology. *Biol. Rev.*, 23, 127-158.

HARLEY, J. L. (1971). Fungi in ecosystems. *J. Ecol.*, 59, 653-680.

HARLEY, J. L. (1973). Symbiosis in ecosystems. *J. Nat. Sci. Counc. Sri Lanka*, 1, 31-48.

HARLEY, J. L. & McCREADY, C. C. (1952). The uptake of phosphate by excised mycorrhizal roots of the beech II. Distribution of phosphorus between host and fungus. *New Phytol.*, 51, 56-64.

HARLEY, J. L., McCREADY, C. C. & BRIERLEY, J. K. (1953). The uptake of phosphate by excised mycorrhizal roots of the beech IV. The effect of oxygen concentration upon host and fungus. *New Phytol.*, 52, 124-132.

HARLEY, J. L., McCREADY, C. C., BRIERLEY, J. K. & JENNINGS, D. H. (1956). The salt respiration of excised beech mycorrhizas II. The relationship between oxygen consumption and phosphate absorption. *New Phytol.*, 55, 1-28.

HARLEY, J. L. & WILSON, J. M. (1959). Absorption of potassium by beech mycorrhizas. *New Phytol.*, 58, 281-298.

HAYMAN, D. S. (1974). Plant growth responses to vesicular-arbuscular mycorrhizas. VI. Effect of light and temperature. *New Phytol.*, 73, 71-80.

HAYMAN, D. S. and MOSSE, B. (1972). Plant growth responses to vesicular-arbuscular mycorrhizas. III. Increased uptake of labile P from the soil. *New Phytol.*, 71, 41-47.

HO, I. & TRAPPE, J. M. (1973). Translocation of ^{14}C from *Festuca* plants to their endomycorrhizal fungi. *Nature (New. Biol.)*, 244, 30-31.

JOHNSON, H. W. (1952). The solubilization of phosphate, I. N.Z. *Jl Sci. and Technol. B.*, 33, 437-446.

JOHNSON, H. W. (1954). The solubilization of "insoluble" phosphate, II. N.Z. *Jl Sci. and Technol, B.*, 36, 49-55.

JOHNSON, H. W. (1954). The solubilization of "insoluble" phosphate, III. N.Z. *Jl Sci. and Technol. B.*, 36, 281-284.

JOHNSON, H. W. (1956). Chelation between calcium and organic anions. N.Z. *Jl Sci. and Technol. B.*, 37, 522-537.

JOHNSON, H. W. (1959). The solubilization of "insoluble" phosphate 4. N.Z. *Jl Sci.*, 2, 109-120.

JOHNSON, H. W. (1959). The solubization of "insoluble" phosphate V. N.Z. *Jl Sci.*, 2, 215-218.

JOHNSON, H. W. (1960). The production of organic acids by mycorrhizal fungi and the possible importance of the phenomenon in phosphate metabolism of trees supporting a mycorrhiza. Summary in 8th New Zealand Sci. Congress, Auckland.

KUMAR, D. & WAREING, P. F. (1974). Studies on tuberization of *Solanum andigena*. II. Growth hormones and tuberization. *New Phytol.*, 73, 833-840.

LEWIS, D. H. (1973). Concepts in fungal nutrition and the origin of biotrophy. *Biol. Rev.*, 48, 261-278.

LEWIS, D. H. & HARLEY, J. L. (1965 a, b & c). Carbohydrate physiology of mycorrhizal roots of beech. I. The identity of endogenous sugars and utilization of exogenous sugars. *New Phytol.*, 64, 224-237. II. Utilization of exogenous sugars by uninfected and mycorrhizal roots. *New Phytol.*, 64, 238-255. III. Movement of sugar between host and fungus. *New Phytol.*, 64, 256-269.

LOBANOW, N. W. (1960). *Mykotrophie der Holzpflanzen.* V.E.B. Deutscher Verlag der Wissenschaften, Berlin. pp. 350.

MEYER, F. H. (1973). Distribution of ectomycorrhizae in native and man-made forests. In *Ectomycorrhizae,* (eds. Marks, G. C. & Kozlowski, T. T.) Academic Press, New York, pp. 79-105.

MORRISON, T. M. (1962). Die ectotrophe Emarhungsweize der Waldgrenze. *Mitt. forstl. Bund Vers Anst.* Wein, 75, 357-380.

PEARSON, V. & READ, D. J. (1973). The biology of mycorrhiza in the Ericaceae. II. Transport of carbohydrate and phosphorus by the endophyte and the mycorrhiza. *New Phytol.*, 72, 1325-1331.

READ, D. J. & ARMSTRONG, W. (1972). A relationship between oxygen transport and the formation of the ectotrophic mycorrhizal sheath in conifer seedlings. *New Phytol.*, 71, 49-53.

REID, C. P. P. & WOODS, F. W. (1969). Translocation of C^{14}- labelled compounds in mycorrhizae and its implications in interplant nutrient cycling. *Ecology,* 50, 179-187.

RICHARDS, B. N. (1973). Nitrogen fixation in the rhizosphere of conifers. *Soil Biol. Biochem.*, 5, 149–152.

ROMMELL, L. G. (1939). The ecological problem of mycotrophy. *Ecology*, 20, 163–167.

SANDERS, F. E. & TINKER, P. B. (1973). Phosphate flow in mycorrhizal roots. *Pestic. Sci.*, 4, 388–395.

SILVESTER, W. B. & BENNETT, K. J. (1973). Acetylene reduction of roots and associated soil of New Zealand conifers. *Soil Biol. Biochem.*, 5, 171–179.

SIMON, E. W. (1974). Phospholipids and plant membrane permeability. *New Phytol.*, 73, 377–420.

SINGER, R. & MORELLO, J. H. (1970). Ectotrophic forest tree mycorrhizae and forest communities. *Ecology*, 41, 549–551.

SMITH, D. C. (In press). Transport from symbiotic algae and symbiotic chloroplasts to host cells. *Symp. Soc. exp. Biol.*, 28.

SMITH, D. C., MUSCATINE, L. & LEWIS, D. H. (1969). Carbohydrate movement from autotrophs to heterotrophs in parasitic and mutualistic symbiosis. *Biol. Rev.* 44, 17–40.

SMITH, F. A. (1972). A comparison of the uptake of nitrate, chloride and phosphate by excised beech mycorrhiza. *New Phytol.*, 71, 875–882.

SMITH, S. E. (1967). Carbohydrate translocation in orchid mycorrhiza. *New Phytol.*, 66, 371–378.

EVOLUTION OF VESICULAR-ARBUSCULAR MYCORRHIZAS

T. H. NICOLSON

Department of Biological Sciences,
University of Dundee, U.K.

INTRODUCTION - THE PROBLEMS

Consideration of vesicular-arbuscular (VA) mycorrhizas presents a number of interesting problems which are difficult to explain within the usual principles of mycology. Fungi of the *Endogone* genus are the most prevalent causing these infections and they are universally present in practically all soils in association with a great variety of plants of different taxonomic groups (Nicolson, 1967; Gerdemann, 1968). However, the relationships are obligate as far as the endophytes are concerned. Despite this they display a remarkable lack of host/ endophyte specificity when compared with other groups of obligate parasites such as the rusts (Uredinales) or powdery mildews (Erysiphales).

The endophytes involved in VA mycorrhizas produce propagules such as chlamydospores or azygospores which are frequently large and poorly adapted for dissemination. Despite this they are distributed in all soils from virgin to climax ecological habitats.

Within the Angiospermae a few families display ectomycorrhizal infections but as a group they largely form V A mycorrhizas. There are, however, certain families such as the Cruciferae and Cyperaceae which appear to consistently escape infection by any types

Nicolson, T. H.

of mycorrhizal fungi (Gerdemann, 1968).   It is
strange indeed that such families should be non-
mycorrhizal considering that practically all other
plants develop these associations.

PHOSPHORUS AND VA MYCORRHIZAS

At present the main interest in VA mycorrhizas
is in the phosphorus nutrition of the host.   The
importance of VA mycorrhizas for the absorption of
this element was initially indicated by the work of
Baylis (1959) and Gerdemann (1964) and further
advanced by a number of papers published over a two
year period (Baylis, 1967; Daft & Nicolson, 1966;
Holevas, 1966; Murdoch, Jackobs & Gerdemann, 1967).
This work showed how closely involved were these
mycorrhizas in the phosphorus nutrition of the host.
Few papers now appearing do not allude to phosphate
relations in some form and in the recent review by
Mosse (1973) over one quarter of the text is devoted
to this.   Indeed, one has heard the remark that
workers involved with VA mycorrhizas are more or less
obsessed with phosphorus.   Be that as it may the
relationship is of obvious importance and the earlier
suggestion that the mycorrhizas may act as an auxil-
iary absorbing system (Nicolson, 1967; Daft &
Nicolson, 1966) has been confirmed (Sanders & Tinker,
1971; Hayman & Mosse, 1972; Hattingh *et al.*, 1973).
These workers have shown that the external mycelium
of the endophytes acts as an absorbing network
exploring the soil beyond the roots' own exploration
regions and "the value of these mycorrhizas for the
phosphate nutrition of plants in deficient environments may rival that of *Rhizobium* in nitrogen
nutrition" (Sanders & Tinker, 1973).

Possible co-evolution of hosts and endophytes

It has been suggested that the explanation for

many of the unusual features of V A mycorrhiza may
be that there has been collateral evolution between
hosts and endophytes from a time very early in the
evolution of a land flora (Nicolson, 1967).   While
eukaryotic microorganisms, including fungi, evolved
far back in Precambrian time (Cloud, 1972; Schopf,
1970), land plant evolution did not take place until
some 400 million years ago on the borderline between
the Silurian/Devonian eras (Chaloner, 1970).   The
earliest of land plants such as the simple *Cooksonia*
have not been examined for the possible presence of
mycorrhizal infections and this would be difficult
as the fossil remains are mainly found in the form
of compression material.   The famous Rhynie fossil
plants, preserved in a petrified form, were investigated by Kidston and Lang over a period of years in
the early part of this century.   These were for long
taken as the oldest of vascular plants and the starting point of land plant evolution.   The position
has been reassessed and the Rhynie material is now
dated to a more recent time of 370 million years ago,
some tens of millions years after the possible
origin of a land flora (Chaloner, 1970).   However,
by that time there existed relationships between the
underground parts of plants and fungi which are
remarkably like modern V A mycorrhizas* and the
structures produced convincingly resemble *Endogone*
fungi.   In a paper by Kidston and Lang (1921) which
examined the "thallophyte" flora associated with the
genera *Rhynia* and *Asteroxylon* some ten plates with
over one hundred microphotographs are produced and
one cannot fail to be impressed by the resemblance
between some of the fungi figured and *Endogone*
species.   One representative plate is reproduced
here as Plate I.   In this plate spores are shown
which are of a size and with a wall complexity

---

* Use of the term 'mycorrhiza' is not strictly
  correct here as these plants did not have true
  roots.

similar to that seen in some present-day endophytes
(Mosse, 1970; Old, Nicolson & Redhead, 1973).   In
photographs where hyphae and vesicles are shown
these could equally well have been taken from the
roots of Angiosperms with V A mycorrhiza.   Kidston
and Lang considered that in certain cases the associations were of a mycorrhizal nature.   Further
similar material has been re-examined recently by
Boullard and Lemoigne (1971) and they also consider
a mycorrhizal relationship was present.   Hence, it
would appear that associations resembling modern VA
mycorrhizas were present very early in plant evolution.   As new plant types were evolved and
occupied new habitats mycorrhizas could have evolved
collaterally so that they became widespread and
relatively non-specific.   Hence, there may have
been co-evolution between the endophytes and the
majority of the vascular flora similar to what may
have occurred with rusts on a more restricted scale
(Savile, 1971).   That the endophytes became obligate symbionts with no free saprophytic ability would
not have been a particular disadvantage since, being
distributed in all habitats, host plants would have
been more or less always available.

EVOLUTION OF ROLE IN PHOSPHORUS NUTRITION

Considering the importance of phosphorus in
metabolic processes it is not surprising that some
symbiotic associations have evolved which are concerned in obtaining it from the environment.   The sole
source of phosphorus is the parent rock material and
from this it is gradually incorporated into the
soils which support plant growth.

It is unlikely that the first plants which
evolved and established a terrestrial ecosystem en-
countered a completely virgin "soil".   It would
seem likely that a microbial population would have
occupied such "soils" previously as one finds at

present with virgin materials prior to higher colonisation.   It has been convincingly argued that
blue-green algae would have been prominent under
such conditions as they are in present day virgin
areas (Fischer, 1965).   These, together with certain heterotrophic prokaryotes would have been
involved in the fixing of atmospheric nitrogen so
that this element would not have been in short supply
when the first vascular plants appeared.   The
limiting nutrient would have been phosphorus as
indeed it is in present-day pioneer habitats
(Harley, 1970; Walker, 1965).   Under such conditions any relationship between plant and microorganism which ameliorated phosphorus extraction from the
environment would have been selected and perpetuated.
It is suggested here that VA mycorrhizas may have
originated in this way.   Subsequently these plant/
endophyte associations evolved collaterally so that
by the Carboniferous era relationships closely
resembling modern VA mycorrhizas were widespread in
Gymnosperms (Osborn, 1909; Halket, 1930; Butler,
1939).   The origins and early evolutionary history
of the Angiosperms remains "an abominable mystery"
as it was in Darwin's day but practically all
modern families, with the exception of the few which
show ectotrophic mycorrhizas or have no mycorrhizas
(Gerdemann, 1968), display the phenomenon of V A
mycorrhiza.   These associations may have evolved
during the course of land plant evolution as an
adaption for the more efficient absorption of phosphorus and "As nodulation has been evolved as a
mechanism for nitrogen fixation, vesicular-arbuscular
mycorrhiza may have been evolved as a means for the
more efficient extraction of phosphorus from the
pedosphere" (Daft & Nicolson, 1969).
    There is no obvious reason why certain families
do not form these mycorrhizas.   One never ceases to
be surprised when examining Crucifers or members of
the Cyperaceae and they never show any sign of VA
mycorrhiza even when their roots intermingle with

infected roots of other plants and may have endophyte
mycelium entangled with their root hairs.   Presumably some factor in root physiology confers resistance.   Or the answer may lie in Professor Baylis's
(1970, 1972a & b) theories regarding the evolution of
root systems and root hairs in relation to VA mycorrhizas and phosphorus nutrition.   However, recent
reports indicate that this resistance is not absolute
and infections can occur in these families
(Gerdemann, 1974).

TAXONOMY OF THE ENDOPHYTES

The taxonomy of the endophytes has also proved
to be a baffling subject.   Again it is hoped that
work presented at the conference may elucidate the
position within the presently-constructed *Endogone*
group.   But their taxonomic affinities with other
fungal groups will doubtless remain a problem for
some time to come.   Fine structure investigations
(Mosse, 1970) and analysis of cell wall constituents
(Bartnicki-Garcia, 1970) may assist in this direction.

CONCLUSIONS

Over the past few decades considerable progress
has been made in studies on V A mycorrhizas.   We now
know much more of their ecology and significance in
plant nutrition and may soon be able to utilise this
knowledge for increasing crop production and for
other practical purposes.   The picture regarding the
taxonomy of the endophytes is becoming clearer.   But
many problems still remain and it is hoped that the
momentum of present research will continue, so that
these will also be clarified over the next decades.
It is appropriate that an association which may have
had a basic influence on the evolution of vascular
plants should receive such attention.

REFERENCES

BARTNICKI-GARCIA, S. (1970). Cell wall composition
and other bio-chemical markers in fungal phylogeny.
In: *Phytochemical phylogeny*.   Ed. HARBORNE, J. B.
Sym. Phytochemical Society, Univ. Bristol 1969,
pp. 81-103.   Academic Press.

BAYLIS, G. T. S. (1959). The effect of vesicular-
arbuscular mycorrhizas on growth of *Griselinia
littoralis* (Cornaceae).   *New Phytol.*, 58, 274-280.

BAYLIS, G. T. S. (1967). Experiments on the ecological significance of phycomycetous mycorrhizas.
New Phytol., 66, 231-243.

BAYLIS, G. T. S. (1970). Root hairs and phycomycetous
mycorrhizas in phosphorus-deficient soil.
Pl. Soil, 33, 713-716.

BAYLIS, G. T. S. (1972a). Minimum levels of available
phosphorus for non-mycorrhizal plants.   *Pl. Soil*,
36, 233-234.

BAYLIS, G. T. S. (1972b).   Fungi, phosphorus and the
evolution of root systems.   *Search*, 3, 257-258.

BOULLARD, B. & LEMOIGNE, Y. (1971). Les champignons
endophytes du *Rhynia Gwynne - Vaughnii* K. et L.
Botaniste, 54, 49-89.

BUTLER, E. J. (1939). The occurrences and systematic
position of the vesicular-arbuscular type of
mycorrhizal fungi.   *Trans. Br. mycol. Soc.*,
22, 274-301.

CHALONER, W. G. (1970). The rise of the first land
plants.   *Biol. Rev.*, 45, 353-377.

CLOUD, P. (1972). A working model of the primitive
     earth.   *Am. J. Sci.*, 272, 537-548.

DAFT, M. J. & NICOLSON, T. H. (1966).   Effect of
     Endogone mycorrhiza on plant growth.   I.   *New
     Phytol.*, 65, 343-350.

DAFT, M. J. & NICOLSON, T. H. (1969). Effect of
     Endogone mycorrhiza on plant growth.   II. Influence
     of soluble phosphate on endophyte and host in maize.
     New Phytol., 68, 945-952.

FISCHER, A. G. (1965). Fossils, early life and
     atmospheric history.   *Proc. Natn. Acad. Sci.*
     USA. 53, 1205-1213.

GERDEMANN, J. W. (1964). The effect of mycorrhiza on
     the growth of maize.   *Mycologia,* 56, 342-349.

GERDEMANN, J. W. (1968).   Vesicular-arbuscular
     mycorrhiza and plant growth.   *A. Rev. Phytopath.,*
     6, 397-418.

GERDEMANN, J. W. (1974). Vesicular-arbuscular mycor-
     rhizae.   In: *Structure and function of roots.*
     Cabot Foundation Symposium - Harvard (1974)
     (In press).

HARLEY, J. L. (1970). The importance of microorgan-
     isms to colonizing plants.   *Trans. Proc. bot.*
     Soc. Edinb., 41, 65-70.

HALKET, A. C. (1930). The rootlets of *Amyelon*
     radicans, Will., their anatomy, their apices and
     their endophytic fungus.   *Ann. Bot.,* (Lond.),
     44, 865-905.

HATTINGH, M. J., GRAY, L. E. & GERDEMANN, J. W.(1973).
     Uptake and translocation of ^{32}P - labeled phosphate
     to onion roots by endomycorrhizal fungi.   *Soil*

Sci., 116, 383-387.

HAYMAN, D. S. & MOSSE, B. (1972). Plant growth
responses to vesicular-arbuscular mycorrhiza.
III. Increased uptake of labile P from soil.
New Phytol, 71, 41-47.

HOLEVAS, C. D. (1966). The effect of vesicular-
arbuscular mycorrhiza on the uptake of soil phosphorus by strawberry (*Fragaria* sp. var. Cambridge
Favourite). *J. hort. Sci.*, 41, 57-64.

KIDSTON, R. & LANG, W. H. (1921). On Old Red Sandstone plants showing structure from the Rhynie
chert bed, Aberdeenshire. Part V. The Thallophyta
occurring in the peat bed; the succession of the
plants throughout a vertical section of the bed,
and the conditions of accumulation and preservation
of the deposit. *Trans. R. Soc. Edin.*, 52, 855-902.

MOSSE, B. (1970). Honey-coloured sessile *Endogone*
spores. III. Wall structure. *Arch. Mikrobiol.*,
74, 146-159.

MOSSE, B. (1973). Advances in the study of vesicular-
arbuscular mycorrhiza. *A. Rev. Phytopath.*,
11, 171-196.

MURDOCH, C. L., JACKOBS, J. A. & GERDEMANN, J. W.
(1967). Utilization of phosphorus sources of different availability by mycorrhizal and non-mycorrhizal maize. *Pl. Soil*, 27, 329-334.

NICOLSON, T. H. (1967). Vesicular-arbuscular
mycorrhiza - a universal plant symbiosis. *Sci.
Prog.*, *Oxf.*, 55, 561-581.

OLD, K. M., NICOLSON, T. H. & REDHEAD, J. F. (1973).
A new species of mycorrhizal *Endogone* from Nigeria
with a distinctive spore wall. *New Phytol*, 72,
817-823.

Nicolson, T. H.

OSBORN, T. G. B. (1909). The lateral roots of
   Amyelon radicans Will., and their mycorrhiza.
   Ann. Bot., (Lond.), <u>23</u>, 603-611.

SANDERS, F. E. & TINKER, P. B. (1971). Mechanism of
   absorption of phosphate from soil by *Endogone*
   mycorrhizas. *Nature,* London, <u>233</u>, 278-279.

SANDERS, F. E. & TINKER, P. B. (1973). Phosphate flow
   into mycorrhizal roots. *Pestic. Sci.,* <u>4</u>, 383-395.

SAVILLE, D. B. O. (1971). Coevolution of the rust
   fungi and their hosts. *Q.Rev. Biol.,* <u>46</u>, 211-218.

SCHOPF, J. W. (1970). Precambrian microorganisms and
   evolutionary events prior to the origin of
   vascular plants. *Biol. Rev.* <u>45</u>, 319-352.

WALKER, T. W. (1965). The significance of phosphorus
   in pedogenesis. In: *Experimental Pedology.*
   (Ed. Hallsworth and Crawford).   11th Easter School
   in Agricultural Sciences, Univ. Nottingham.
   pp. 295-315.

TAXONOMY OF THE ENDOGONACEAE

J. W. GERDEMANN

Department of Plant Pathology,
University of Illinois, Urbana, USA

AND

JAMES M. TRAPPE

Pacific Northwest Forest and Range Experiment Station,
Forestry Sciences Laboratory,
Corvallis, Oregon, USA

INTRODUCTION

Taxonomy, as well as the experimental sciences, provides an understanding that allows one to predict from a hypothesis events that are unobserved (Rogers, 1958). Whereas progress in the experimental sciences is generally welcomed, progress in taxonomy often causes irritation to the nontaxonomist who must have a name for the organism with which he is working. When names change he may regard it more as an outrage rather than progress toward a more reasonable system.

The Endogonaceae, until quite recently, were rarely collected and little was known concerning their life cycles and taxonomic relationships. We now realize that they are among the most common of all soil-borne fungi. Much new information about them has been obtained. Yet much mystery still remains and we are still in the process of developing

hypotheses which we hope will provide greater predictability.   In order to achieve predictability,
similar individuals must be grouped into species,
and species with similar morphology must be placed
in genera.   "Splitters" and "lumpers" will always
differ as to where the lines are to be drawn;
however, lines must be drawn if we are to avert
chaos.

In order to understand why we have proposed
certain changes, one must examine the history of the
Endogonaceae and then apply the "International Rules
of Botanical Nomenclature."

HISTORY

Link (1809) described the genus *Endogone* and
indicated a relationship to the genus *Tuber*.   In
Endogone he placed one new species, *Endogone
pisiformis* Link ex Fries.   Since it represents the
type of the genus, it is of great importance to
establish the species that best corresponds to his
description.   Link's description, although reasonably good for its time, does not state whether the
"sporangia" (spores) form as the result of a sexual
process.   Thaxter (1922) discussed this problem in
detail (p. 296-298) and concluded that Link's
description best corresponds with a well known
sporocarpic species in which zygospores bud out from
the point of union of two gametangia.   We find
Thaxter's reasoning logical and compelling and
accept his conclusions.

Fries (1849) established the family Endogonei
which he placed in the order Tuberacei along with
Hymenogastrei and Tuberei.

Tulasne and Tulasne (1845) described the genus
Glomus for two   species, *G. microcarpus* Tul.& Tul.
and *G. macrocarpus* Tul.& Tul.   They did not place
them in *Endogone* because they failed to observe the

"sporidia minuta, globosa" of Link. The "sporidia"
of Link were oil droplets, which are small and of
uniform size in zygospores. Illustrations of
spores of *G. microcarpus* and *G. macrocarpus* (Tulasne
& Tulasne, 1851) clearly show that the spores are
terminal, borne on vegetative hyphae and that the
oil droplets are highly variable in size, thus
clearly establishing that they are chlamydosporic
species.

Berkely and Broome (1846) described *Endogone
lactiflua* Berk. & Broome. Their description is
very brief, but fortunately the type specimen is
still available and the species can be clearly
defined (Trappe & Gerdemann, 1972). It is a zygosporic species, the zygospore of which buds from the
larger of two gametangia, and the spore is enclosed
in a mantle of interwoven hyphae.

Tulasne and Tulasne (1851) transferred *Glomus
microcarpus* and *G. macrocarpus* to the genus *Endogone*.
They recognized two other species, *E. pisiformis* and
E. lactiflua. They examined the Berkeley and Broome
collection but apparently failed to notice that the
"sporangia" (zygospores) originated in a way
distinctly different from that of the two species
they originally placed in *Glomus*.

Bresadola (1896) gave the name *Endogone reniformis* Bres. to a sporocarpic species, the sporocarps
of which consisted of hyphae and "asci" which contain
globose spores. The "asci" of *E. reniformis* are
true sporangia which lack columellae and contain
sporangiospores.

Thus, three types of species, zygosporic,
chlamydosporic, and sporangial, were included in the
genus *Endogone*. They all shared one character of
producing sporocarps that were somewhat similar in
appearance. Also, the zygospores and chlamydospores superficially resembled each other. It
remained for Bucholtz (1912) to show the true nature
of these structures. He described sexual origin of

zygospores, and as a result of his studies, the
Endogonaceae was placed in the Mucorales. It has
often been assumed that the chlamydosporic and
sporangial species are the asexual stages of zygosporic species. However, at present there is little
evidence for this point of view.

Following Thaxter's monograph of the Endogonaceae, the concept of the genus *Endogone* was further
broadened. The genus *Sphaerocreas* was recognized
by Thaxter (1922) because it differed from other
chlamydosporic species in having sporocarps, from
which bundles of hyphae radiate, and acrogenous
chlamydospore development. The one species in the
genus, *Sphaerocreas pubescens* Sacc. & Ellis, was
transferred to *Endogone* by Zycha (1935).

Kanouse (1936) described the genus *Modicella*
for the sporangial species included in *Endogone*.
However, she failed formally to transfer any species
into the genus. Zycha *et al.* (1969) retained
sporangial species in *Endogone*.

Peyronel (1924, 1937) found sporocarps of
several chlamydosporic species closely associated
with plants having vesicular-arbuscular mycorrhizae
and thus was the first to identify correctly these
endophytes. Butler (1939) noted the resemblance of
vesicles in roots and in soil to chlamydospores of
Endogone (Glomus) and concluded that they probably
represented *Endogone* species that had lost the
ability to form sporocarps. Godfrey (1957a) and
Mosse (1956, 1959) discovered that certain species
form chlamydospores free in the soil and in sporocarps. By use of wet-sieving and decanting of soil
such "free-borne" chlamydospores were shown to be
very common (Gerdemann & Nicolson, 1963), and the
genus *Endogone* was emended to include species that
produce chlamydospores singly in soil or in root
tissue (Nicolson & Gerdemann, 1968).

Two other groups of fungi with large "*Endogone*-
like" spores were also discovered by wet-sieving and

decanting and were shown to produce endomycorrhizae.
The first of these (Gerdemann, 1955a) produces a
large globose to subglobose spore on a "bulbous"
suspensor-like structure with a fine hypha generally
projecting from the bulbous base to the spore
surface.   This species produces endomycorrhizae with
arbuscules, but vesicles in roots have not been
observed.   Distinctive vesicles are borne on coiled
hyphae in soil.   Spores produced by this species,
and others similar to it, differ in many respects
from chlamydospores.   Since they most closely
resemble zygospores of previously described *Endogone*
species, they were placed in the genus *Endogone*
(Nicolson & Gerdemann, 1968).   The second group
produces an "*Endogone*-like" spore laterally on a
hypha which terminates in a large thin-walled
vesicle (Gerdemann & Nicolson, 1963).   Such species
have been shown to form vesicular-arbuscular mycorrhizae.   The germination and fine structure of spores
of a member of this group were described in detail by
Mosse (1970, a, b, c).   Although such spores were
generally referred to as belonging to *Endogone,* they
were never formally described and placed in that
genus.

   Thus the genus *Endogone* grew into an assemblage
of diverse species, about which few generalizations
could be made.   All except the sporangial species
produced large globose to sub-globose spores;
however, species within the genus differed in origin
of spores, spore morphology, type of germination,
mycorrhizal relationships, and habitat.   Clearly
one could predict little about a species included in
such a genus, and in discussions or writing it
became necessary to define the group to which each
individual *Endogone* species belonged.
   We (Gerdemann & Trappe, 1974) have revised the
genus *Endogone* sensu Zycha *et al.* (1969) by recog-
nizing three genera that were previously described
(*Endogone, Glomus, Modicella)* and by describing a

new genus *(Gigaspora)*.   One additional new genus

described and the basis for identification of the
isolate as *Endogone* was not stated.   We attempted to
obtain isolates of *E. pisiformis* from freshly collected sporocarps (Gerdemann and Trappe, 1974).   Slow
growing yellow cultures were obtained.   The hyphae
were nonseptate, varied greatly in diameter, and were
irregular in shape.   Thin-walled inflated cells up
to 83μm diameter were abundant.   Cells contained
yellow oil similar in appearance to that in sporocarps.   Zygospores, chlamydospores, and sporangia
did not form.   These reports of successful cultur-
ing of *Endogone* need to be verified.

    In the Pacific Northwest most *Endogone* species
are relatively rare, and sporocarps have been found
only in association with ectomycorrhizal hosts.   The
sporocarps are generally hypogeous and are found more
or less by chance by digging or raking with a truffle
fork.   They have not been found in cultivated
fields, orchards, or grasslands.   *Endogone* species
are known from cold or temperate regions and we are
not aware of any records of their collection from
areas with a tropical climate.   One species,
Endogone lactiflua (E. flammicorona, Trappe & Gerd.)
has been shown to form ectomycorrhizae experimentally
(Fassi & Palenzona, 1969).   Evidence from association suggests that many other *Endogone* species are
ectomycorrhizal.

B.   *Gigaspora* Gerdemann & Trappe (1974)

    In *Gigaspora,* azygospores are borne singly in
soil.   They are large, generally globose or subglobose, and borne terminally on a bulbous suspensor-like
cell.   A slender hypha usually extends from the bulbous suspensor to the spore.   The spore wall consists
of a number of layers that are continuous except for
a small occluded pore.   The spores contain oil globules more or less uniform in size and suggestive of
the contents of *Endogone* zygospores.   The spores of
Gigaspora have previously been called zygospores

(Nicolson & Gerdemann, 1968); however, there is no
obvious sexual fusion and they are perhaps best
regarded as azygospores (Gerdemann & Trappe, 1974).
Germ tubes are produced directly through the wall
near the base of the spore.   In one species,
G. gilmorei Trappe & Gerd., the spore wall is transparent and the wall layers thicker than in most other
Gigaspora species.   In this species the wall structure and method of germination is very similar to that
described by Mosse (1970 a, b, c) for "honey-coloured
sessile *Endogone* spores" *(Acaulospora laevis* Gerd. &
Trappe).   *Gigaspora* species produce distinctive
vesicles on coiled hyphae in soil.   They may represent sporangial stages; however, spores have not
been observed in them, and their function is apparently temporary storage of food.   The hyphae of
Gigaspora species have a unique type of wound healing
that has not been reported for any other fungi
(Gerdemann, 1955b).

   Gigaspora species are world-wide in their distribution.   Sporocarps have not been observed, and
single spores are generally collected by wet-sieving
and decanting.   "Pot culture techniques" (Gerdemann
& Trappe, 1974) are useful when few spores are
present, or if the soil contains too much organic
matter for wet-sieving and decanting.   *Gigaspora*
species are likely to be present wherever endomycorrhizal hosts occur, i.e. cultivated fields, native
grasslands, and forests.   They produce endomycorrhizae with arbuscules.   Vesicles within roots have
been noted for one species.

C.   *Acaulospora* Gerdemann & Trappe (1974)

   The azygospores of *Acaulospora* are similar to
those of *Gigaspora* species.   They form, however, in
a distinctly different way.   A large thin-walled
vesicle with dense contents is produced terminally on
a broad funnel-shaped hypha.   The azygospore buds
laterally from the stalk, and the contents of the

vesicle are transferred to the spore. As the spore reaches its mature size, the emptied vesicle collapses. The vesicles frequently have disappeared on spores wet-sieved from soil, leaving no evidence of the spores' origin. The thin-walled vesicle may represent a gametangial stage. However, conjugation of gametangia has not been observed, and the only apparent function is temporary storage of food.

Mosse (1970 a, b, c) has shown that the spore wall structure is very complex, consisting of many layers. She also described spore germination in detail. Peripheral compartments are formed near the base of the spore from which germ tubes are produced. The compartments are formed by the separation of spore wall membranes, which bulge into the spore. Wall material is deposited upon the split membranes, and walls form which divide the large compartments into smaller ones. Germ tubes arise from these compartments and penetrate the wall.

Azygospores occur singly in soil and are likely to be found wherever endomycorrhizal hosts occur. They have been collected by wet-sieving and decanting and pot-culture. *Acaulospora* produces endomycorrhizae with arbuscules and vesicles.

D. *Glomus* Tulasne & Tulasne (1845)

In *Glomus*, chlamydospores are borne on undifferentiated, nongametangial hyphae. Spores are generally terminal; however, intercalary spores and spores with more than one basal attachment sometimes occur. The spore walls may be single or double. Double walls are distinguished from laminate walls which are composed of an indefinite number of less clearly defined tightly fused layers. Mature spores contain oil droplets that are highly variable in size. At maturity, spore contents are separated from attached hyphae by a septum or occluded by spore wall thickening. Chlamydospores are formed in sporocarps, roots or free in the soil. They are generally

hypogeous and can be collected by raking with a
truffle fork, wet-sieving and decanting, or "inoculated pot cultures" (Gerdemann & Trappe, 1974).
Chlamydospores germinate by renewed growth from
subtending hyphae (Mosse, 1956, 1959; Godfrey,
1957b). A number of species have been maintained in
pot culture for many years.

We have not found zygospores or sporangia
associated with *Glomus* species in any of our collections. Chlamydospores occasionally contain smaller
spores, but this phenomenon is of unknown significance. It has generally been assumed that chlamydospores are asexual stages of zygosporic species.
Thaxter (1922) found zygospores associated with
chlamydospores in sporocarps of *Endogone fasciculata*
Thaxter *(Glomus fasciculatus* (Thaxter) Gerd. &
Trappe). These sporocarps are loose structures that
contain sphagnum and other foreign matter. The
hyphae attached to chlamydospores are thick-walled,
while the hyphae associated with zygospores are
thin-walled. These sporocarps quite possibly
consist of a chance mixture of two species
(Gerdemann, 1965). Godfrey (1957a) described a sporocarp of *Endogone microcarpa* (Tul.& Tul.) Tul. & Tul.
(Glomus microcarpus) that contained chlamydospores
and zygospores. However, the hyphae associated
with the two kinds of spores appear to differ, and
the possibility of a freak combination of two
species, or hyperparasitism, cannot be dismissed.
Zygosporic *(Endogone)* species are apparently absent
from grasslands and cultivated fields where chlamydosporic *(Glomus)* species are particularly common.
In the Pacific Northwest *Endogone* species are associated with ectomycorrhizal plants, and *Glomus* species
are associated with endomycorrhizal hosts. This
does not preclude the possibility that sexual stages
occur in *Glomus*. It is, however, very unlikely that
chlamydosporic species are merely asexual forms of
the presently known zygosporic species.

It is questionable whether *Glomus* species have

been obtained in axenic culture.   Barrett's (1961)
"*Rhizophagus* cultures" may have been members of this
genus.   An isolate obtained from *G. mosseae* (Nicol.
& Gerd.) Gerd. & Trappe, using Barrett's method of
isolation closely resembled his "*Rhizophagus* cultures" (Gerdemann, 1968).   However, the culture did
not produce spores, and attempts to synthesize
mycorrhiza with it consistently failed.   Therefore,
it is not certain whether the isolate was *G. mosseae*
or some other fungus.

We consider the genus *Rhizophagus* as a likely
synonym of *Glomus*.   Dangeard's (1896, 1900) description and illustrations leave little doubt that he
observed a member of this genus.

Glomus species are generally associated with
endomycorrhizal plants, and there is good experimental evidence that a number of species produce endomycorrhizae with vesicles and arbuscules.   They are
world-wide in their distribution and are the most
common endomycorrhizal fungi.   They may, in fact, be
the most common of all soil-borne fungi.

E.   *Sclerocystis* Berkeley and Broome (1875)

Sclerocystis is closely related to *Glomus*,
differing from it only in that the chlamydospores are
arranged in a single orderly layer around a central
plexus of hyphae.   Three species are known only from
tropical or subtropical regions and greenhouses.
The one other described species, *S. rubiformis* Gerd.
& Trappe, is widely distributed in temperate regions.
The three tropical species produce small sporocarps
that are fused together in mats on the soil surface.
One of them, *S. coremioides* Berk & Broome, also forms
single sporocarps in the soil.   *S. rubiformis* is
known only from sporocarps wet-sieved from soil.

S. coremioides and *S. rubiformis* have been
shown to produce vesicular-arbuscular mycorrhizae
similar to those formed by *Glomus* species.

F. *Glaziella* Berkeley (1879-80)

Glaziella contains only one species and it has been found only in the tropics near beaches.   It produces hollow, orange to scarlet sporocarps that may be as much as 5 cm broad.   Chlamydospores, similar to those produced by *Glomus*, up to 415μm diameter are scattered in the sporocarp walls. Mycorrhizal relationships are unknown.

G. *Modicella* Kanouse (1936)

In *Modicella* thin-walled sporangia are produced in small hypogeous or epigeous sporocarps.   Sporangia lack columellae and contain sporangiospores. No other spore stages are known, and the genus resembles other genera of Endogonaceae only in that it produces sporocarps.   A relationship with the Mortierellaceae has been suggested (Thaxter, 1922; Walker, 1923), and Kanouse (1936) placed the genus in that family.   We have retained the sporangial species in the Endogonaceae (Gerdemann & Trappe, 1974), but we lack any great conviction as to their natural relationships.   Walker (1923) isolated *M. malleola* (Harkn.) Gerd. & Trappe in axenic culture but failed to obtain any type of fruiting.   It is not known if *Modicella* species produce mycorrhizae.

CONCLUSIONS

Our revision of the Endogonaceae (Gerdemann & Trappe, 1974) can be regarded as a temporary solution to a difficult taxonomic problem.   As more information becomes available, natural relationships, or the lack of them, will become more apparent.   Some genera may even be removed from the Endogonaceae when their life cycles are more completely known.   Other genera are still relatively broad and contain elem-

ents so diverse that they may eventually be sub-
divided.   No doubt there are many species that are
still unnamed.   We hope that our revision represents progress toward a more natural, and predictive,
system of classification.

REFERENCES

BAKERSPIGEL, A. (1958).   The structure and mode of
    division of the nuclei in the vegetative spores
    and hyphae of *Endogone sphagnophila* Atk.
    Am. J. Bot., 45, 404-410.

BARRETT, J. T. (1961).   Isolation, culture, and
    host relation of the phycomycetoid vesicular
    arbuscular mycorrhizal endophyte *Rhizophagus*.
    In; *Recent Adv. Bot.* Univ. Toronto Press, Toronto,
    2, 1725-1727.

BERKELEY, M. J. (1879/1880).   Fungi Brasilienses.
    Vidensk. Meddr. dansk naturh. Foren., pp. 31-34.

BERKELEY, M. J. & BROOME, C. E. (1846).   Notices of
    British Hypogaeous fungi.   *Ann. Mag. nat. Hist.*,
    18, 73-82.

BERKELEY, M. J. & BROOME, C. E. (1875).   Enumera-
    tion of the fungi of Ceylon.   *J. Linn. Soc.*
    14, 29-140.

BRESADOLA, J. (1896).   Fungi Brazilienses.
    Hedwigia, 35, 276-302.

BUCHOLTZ, F. (1912).   Beiträge zur Kenntnis der
    Gattung *Endogone* Link.   *Beih. bot. Zbl.*,
    29, 147-225.

BUTLER, E. J. (1939).   The occurrences and systemat-
    ic position of the vesicular-arbuscular type of

mycorrhizal fungi.   *Trans. Br. mycol. Soc.,*
22, 274-301.

DANGEARD, P. A. (1896).   Une maladie du peuplier
dans l'ouest de la France.   *Botaniste,* 5, 38-43.

DANGEARD, P. A. (1900).   Le *"Rhizophagus populinus"*
Dangeard.   *Botaniste,* 7, 285-291.

FASSI, B. & PALENZONA, M. (1969).   Sintesi micorrizica tra *"Pinus strobus"*, *"Pseudotsuga douglastii"*
ed *"Endogone lactiflua"*.   *Allionia,* 15, 105-114.

FRIES, E. M. (1849).   *Summa Vegetabilium Scandin-
aviae,* 2, 261-572.

FURLAN, V., & FORTIN, J. A., (1973). Formation of
endomycorrhizae by *Endogone calospora* on *Allium
cepa* under three temperature regimes.   Natur.
Canadien, 100, 467-477.

GERDEMANN, J. W. (1955a).   Relation of a large soil-
borne spore to phycomycetous mycorrhizal infections.   *Mycologia,* 47, 619-632.

GERDEMANN, J. W. (1955b).   Wound-healing of hyphae
in a phycomycetous mycorrhizal fungus.   *Mycologia,*
47, 916-918.

GERDEMANN, J. W. (1965).   Vesicular-arbuscular
mycorrhizae formed on maize and tuliptree by
Endogone fasciculata.   *Mycologia,* 57, 562-575.

GERDEMANN, J. W. (1968).   Vesicular-arbuscular
mycorrhiza and plant growth.   *A. Rev. Phytopath.,*
6, 397-418.

GERDEMANN, J. W., & NICOLSON, T. H. (1963).   Spores
of mycorrhizal *Endogone* species extracted from soil
by wet sieving and decanting.   *Trans. Br. mycol.
Soc.,* 46, 235-244.

GERDEMANN, J. W. & TRAPPE, J. M. (1974).   The
Endogonaceae in the Pacific Northwest.   *Mycologia
Memoir,* 5. 76 pp.

GODFREY, R. M. (1957a).   Studies of British species
   of *Endogone*.   I. Morphology and taxonomy.
   Trans. Br. mycol. Soc., 40, 117-135.

GODFREY, R. M. (1957b).   Studies on British species
   of *Endogone*.   III. Germination of spores.
   Trans. Br. mycol. Soc., 40, 203-210.

KANOUSE, B. B. (1936).   Studies of two species of
   Endogone in culture.   *Mycologia*, 28, 47-62.

LINK, H. F. (1809).   Observatione in Ordine
   Plantarum naturales.   *Ges. Naturforsch. Freunde*,
   Berlin, 3, 3-42.

MOSSE, B. (1956).   Fructifications of an *Endogone*
   species causing endotrophic mycorrhiza in fruit
   plants.   *Ann. Bot.*, 20, 349-362.

MOSSE, B. (1959).   The regular germination of resting spores and some observations on the growth
   requirements of an *Endogone* sp. causing vesicular-
   arbuscular mycorrhiza.   *Trans. Br. mycol. Soc.*,
   42, 273-286.

MOSSE, B. (1970a).   Honey-coloured, sessile
   Endogone spores.   I. Life history.   *Arch.
   Mikrobiol.*, 70, 167-175.

MOSSE, B. (1970b).   Honey-coloured, sessile
   Endogone spores.   II. Changes in fine structure
   during spore development.   *Arch. Mikrobiol.*,
   74, 129-145.

MOSSE, B. (1970c).   Honey-coloured, sessile
   Endogone spores.   III. Wall structure.   *Arch.
   Mikrobiol.*, 74, 146-159.

NICOLSON, T. H., & GERDEMANN, J. W. (1968).   Mycorrhizal *Endogone* species.   *Mycologia*, 60, 313-325.

PEYRONEL, B. (1924).   Specie di *"Endogone"* produt-
trici di micorize endotrofiche.   *Boll. Staz.
Patol. veg. Rome.*, 5, 73-75.

PEYRONEL, B. (1937).   Le *"Endogone"* quali produt-
trici di micorize endotrofiche nelle fanerogame
alpestri.   *Nuovo. G. bot. ital.*, 44, 584-586.

ROGERS, D. P. (1958).   The philosophy of taxonomy.
Mycologia, 50, 326-332.

THAXTER, R. (1922).   A revision of the Endogoneae.
Proc. Am. Acad. Arts Sci., 57, 291-351.

TRAPPE, J. M. & GERDEMANN, J. W. (1972).   *Endogone
flammicorona* sp. nov., a distinctive segregate
from *Endogone lactiflua*.   *Trans. Br. mycol. Soc.*,
59, 403-407.

TULASNE, L. R. & TULASNE, C. (1845).   Fungi
nonnulli hypogaei, novi v. minus cogniti auct.
G. bot. ital., I. 2, 35-63.

TULASNE, L. R. & TULASNE, C. (1851).   Fungi
Hypogaei.   Friedrich Klincksieck, Paris, 222pp..

WALKER, L. B. (1923).   Some observations on the
development of *Endogone malleola* Hark. *Mycologia*,
15, 245-257.

ZYCHA, H. (1935).   Mucorineae.   *Kryptogamenflora
Mark Brandenburg*, 6a, 264 pp.

ZYCHA, H., SIEPMANN, R. & LINNEMANN, G. (1969).
Mucorales. J. Cramer, Lehre, 355 pp.

A CULTURABLE ENDOGONE ASSOCIATED WITH EUCALYPTS

J. H. WARCUP

*Waite Agricultural Research Institute,
The University of Adelaide,
South Australia*

INTRODUCTION

During experiments on the growth of eucalypt seedlings in natural and heat-treated soil, some seedlings in certain batches of untreated soil grew better than usual. Roots of some of these larger plants were examined, but obvious ectomycorrhizas were not always present. Secondary roots of the plants, however, generally lacked root hairs and hyphae were present on and around the roots. Closer examination of stained roots showed that at least two fungi were present forming sheaths on the root surface: one had septate hyphae and structure of the septa suggested that the fungus was an Ascomycete; the other was basically aseptate and had hyphae of the general appearance of an *Endogone* but no spores.

Being interested in Ascomycetes that might be mycorrhizal I decided to culture from the sheathed roots. On isolation plates of segmented washed roots plated on a weak agar medium, young hyphae developed resembling those of the presumed Ascomycete. Besides these and several other types of fungal hyphae, an unusual aseptate mycelium appeared, especially from sheaths near root tips. Hyphae of this aseptate fungus, removed in blocks of agar and placed

on fresh medium, continued growth and sometimes could
be further subcultured.

Some months later small fructifications associated with similar sheaths and *Endogone*-like mycelium
were found near older eucalypt seedlings in the same
soil. Attempts to culture from young fructifications were successful and gave a mycelium similar to
that obtained from root sheaths.

THE FUNGUS AND ITS EFFECT ON <u>EUCALYPTUS</u>

Endogone eucalypti (ined.) has small irregularly
globose fructifications often arising from a single
stout hypha and containing 2-18 (78) subglobose
zygospores (50-60 x 55-65 μm) enclosed in a definite
peridium of 1-3 layers of rather thick-walled hyphae.
Vegetative hyphae are variable in diameter (to 16 μm)
and may become thick-walled (to 5 μm) when old. The
original isolates (Plate II, 1) were from seedling
roots of *Eucalyptus obliqua* L'Herit grown in pots in
soil from Kuitpo forest some 50 km. south of
Adelaide, South Australia. Subsequently the fungus
has been obtained from naturally occurring seedlings
of *E. obliqua* from the same area and from the Gramp-
ian Mountains, Victoria.

While *Endogone eucalypti* was found on *Eucalyptus
regnans* F. Muell. and *E. obliqua* grown as seedlings
in Kuitpo soil, it was not found on any of the weed
grasses and herbs, such as *Briza maxima* L., *B. minor*
L., *Danthonia* sp., *Anagallis arvensis* L., *Geranium*
sp., *Gnaphalium* sp. and *Senecio* sp., that germinated
in the pots. These plants, however, were all infected by various VA endophytes. Conversely these
vesicular-arbuscular (VA) *Endogone* species did not
infect the eucalypt seedlings.

Roots of seedlings of *E. regnans* grown in sterile conditions and inoculated with *Endogone eucalypti*
were initially colonized by "runner" hyphae which
grew along the root, often in the groove between

epidermal cells, and in many cases formed a net-like
growth between cells. From these runner hyphae
finger-like secondary hyphae often meshed together to
form a sheath across the root surface, usually one
but occasionally 2-3 cells deep. If a complete
sheath formed on lateral roots they were often devoid
of root hairs, but on longer roots sheath and root
hairs may occur together, and the fungus may then
also form a sheath (Plate II, 2) around the base or
along the sides of some of the root hairs. In
lateral roots epidermal cells may also be slightly
elongated radially as is common with ectomycorrhizas
of *Eucalyptus* (Chilvers & Pryor, 1965). In general,
the fungus seems a sheath-former rather than a typical ectomycorrhizal fungus with sheath and Hartig
net.

Whether *E. eucalypti* aids nutrition of its host
is of interest: the original plants from which the
fungus was obtained were larger than average though
other mycorrhizal fungi were also present. Seedlings of *Eucalyptus regnans* were grown for 2 months
in a growth room at 20°C, 22000 lx. and a 12 h day,
in an autoclaved (30 min/15 lb) alluvial loam from
Kuitpo Forest. The soil was inoculated with *E. eucalypti* (isolate N 191) or given a dressing of superphosphate, equivalent to 125 kg/ha, applied 1 cm from
the base of the seedling when the first leaves were
starting to show. The inoculum consisted of hyphae
in an agar block applied to the base of the seedling
about 0.5 cm below soil level. The autoclaved soil
in pots quickly became recolonised by fungi such as
Trichoderma, Mortierella, Mucor, Rhizopus and *Penicillium* and contained a high population of bacteria.

Table 1 shows that seedlings had a low % P and
remained very small unless they were inoculated or
given a dressing of superphosphate. As the level of
other nutrients in this soil is relatively high the
responses to phosphate or inoculation were very
marked. Growth responses to superphosphate were
visible after 8 days, those to N 191 after 15 days.

Table 1.   Mean dry weight and % mineral
composition of *Eucalyptus regnans* grown in
autoclaved Kuitpo soil with or without
E. eucalypti isolate N191 or with superphosphate.

Treatment	Mean dry wt (g)	% dry matter						
		P	K	Mg	S	Ca	Na	Cl
Super-phosphate[†]	0.74	0.105	0.82	0.45	0.27	0.7	0.89	1.8
N191	0.55	0.195	1.22	0.45	0.33	0.6	0.58	1.6
Nil*	0.03	0.07	0.57	0.55	0.30	0.9	0.46	1.3
Nil**	0.03	0.07	0.55	0.39	0.28	0.4	0.37	0.6

Mean of three[†], eight* or thirty five** plants.

Thus, in part differences in dry weight between inoculated plants and those given superphosphate reflect
differences in response rates.   In Kuitpo Forest
soil inoculation with N 191 also increased % K
(Table 1), but S and Mg (and also Cu and Zn) concentrations were not greatly affected.
    Significant growth responses to inoculation with
several isolates of *E. eucalypti* were also obtained
in an autoclaved sandy soil.

GROWTH OF THE FUNGUS IN CULTURE

    Work on the growth of *Endogone eucalypti* in
culture was undertaken in the hope of obtaining
information that might lead to the isolation and
growth of further species of *Endogone*.   *E. eucalypti*

was initially isolated on a weak Czapek-Dox + yeast
agar (sucrose, 0.5 g; NaNO$_3$, 0.33 g; KH$_2$PO$_4$,
0.2 g; MgSO$_4$.7H$_2$O, 0.1 g; KCl, 0.1 g; yeast extract 0.1 g; distilled water, 1,000 ml; Davis agar,
15 g) but further subculturing on this medium gave
poor and variable results with some isolates soon
dying.   Growth was better on a weak cornmeal agar
(Difco dehydrated cornmeal agar, 2 g; distilled
water, 1,000 ml; + 1.2 % agar) and, in terms of
radial spread, was best on water agar.   On this
medium some isolates covered a 9 cm plate and a few
produced fructifications.   Even on dilute cornmeal
and water agars growth tended to be variable.   As
inhibiting substances in the agar were suspected, the
fungus was grown in water, from hyphae in small
blocks of agar;   growth, however, was poor.

On all media used the fungus formed swellings
(Plate II, 3) occasionally in primary but more
usually in second order hyphae.   Most swellings were
small and irregular in size and shape, though occasionally in older well-grown colonies on water agar
larger swellings resembling vesicles were formed;
however, most were intercalary or lateral rather than
terminal.   Another feature on all media was that
some hyphae tended to "leak" contents so that lens-
shaped aggregations of cytoplasmic materials occurred
in the agar.   On the other hand, if freshly boiled
half-seeds of hemp (Barrett, 1961) or cucumber were
placed cut surface downwards on a colony in agar, the
fungus was markedly invigorated, colonized the seed
and produced some aerial hyphae around it;   on media
alone, the fungus grew almost exclusively in the
agar.   When growing well, the fungus anastomoses
extensively (Plate II, 4) and marked cytoplasmic
streaming (Mosse, 1959) can be observed.

These observations led me to suspect that poor
growth of the fungus in culture might be due to both
the presence in media of inhibiting substances and in
simple semi-defined media to the lack of some
"growth factor(s)" using that term in a wide sense.

The effects of different sources of water and agars were tested (Table 2).

Table 2.   Effect of source of water in water agar on colony diameter of isolate N191.

Source of water	Colony diameter (mm)
Tap	0
Rain	0
Glass distilled	7
Double glass distilled	19
deionized	21

Apart from tap water, the other grades of water were prepared from rainwater, the deionized water from double glass-distilled rainwater.   Table 2 shows that the source of water markedly affected the growth of the fungus in water agar.   Different sources of agar (Davis F, Davis bacteriological, Oxoid No. 3, Difco bacto, Ionagar No. 2, Purified agar) had only minor effects on the growth of isolate N 191 though growth was always better on Davis agars.   Further experiments showed that over a 5-day period the fungus grew over the temperature range 5-35°C with an optimum about 25°C.   The fungus grew over a tested range of pH 4-8 with best growth at pH 5-6.

    The ability of *E. eucalypti* to utilize various compounds was then investigated.   As the fungus tends to grow faster in water agar than in agars prepared with nutrients much subsequent work has been done in still liquid culture in 150 ml pyrex flasks containing 20 ml of medium.   In all early experiments the fungus grew better with low nutrient levels;   a simple basic medium (KH_2PO_4, 0.1 g; $MgSO_4.7H_2O$, 0.05 g;   KCl, 0.05 g;   double distilled water 1,000 ml) was therefore used.   Initial experiments showed that with this basic solution and yeast

extract (0.1 g/l, BBL) as nitrogen source, isolate
N191 utilized glucose, sucrose, starch, ethanol, and
to a lesser extent glycerol as carbon sources.
Growth was better with starch than glucose or
sucrose, but since growth was also better on starch
in the absence of added nitrogen, this better growth
may not be due solely to starch as a carbon source.
It was also noted that the amount of growth depended
markedly on the source of the material used; one lot
of glucose was inhibitory, one batch of yeast extract
gave double the growth of another, thus the above
results may not be entirely representative of the
ability of the fungus to use the material examined.
With the same base solution and glucose (1 g /l) as
carbon source, isolate N191 did not utilize nitrate,
ammonium and bacto peptone but did use yeast extract,
soytone and proteose peptone.   Ability to use
nitrate or ammonium in the presence of vitamins has
not been tested.   Most nitrogen sources were used at
a level of 0.1 - 0.2 g/litre.   Maximum levels of
each have not been investigated, but in yeast
extract-starch solutions growth with 0.4 g/l yeast
extract was either no better or worse, than that with
0.2 g/litre (Table 3).

Table 3.   Growth of Isolate N191 in still
culture with different levels of starch
and 2 levels of yeast extract (dry wt.
10^{-4} g).

Yeast extract (g/l)	0.2	0.4
Starch (g/l)		
0	21	20
1	70	72
5	120	86
10	120	80
20	191	86

Autoclaving the carbohydrate separately from the
basal medium made no difference to growth when the
carbohydrate was starch, and improved growth slightly
when it was glucose, but the difference was not large
enough to warrant the extra work involved.   Addition
of trace elements (Fe^{+++}, 0.02 mg; Zn^{++}, 0.02 mg;
Mn^{++}, 0.01 mg/l) also improved growth in the miner-
als-yeast extract-starch medium.

Since *E. eucalypti* was stimulated by hemp and
cucumber seed the effect of some plant extracts was
tested.   The most important reaction was with Difco
lima bean extract (Table 4) and further tests showed
that diluted (0.1-0.5 g/l) Difco lima bean agar (with
extra agar) is an excellent medium for the fungus.

An attempt has been made to isolate on starch-
yeast and on dilute lima bean agars prepared with
double distilled water some of the vesicular-arbuscular endophytes associated with weed species in the
Kuitpo soil.   Growth from washed root segments was
markedly improved - fresh growth commonly occurring
from external hyphae - compared with earlier attempts
using the Czapek-Dox + yeast medium.   However, no
Endogone was obtained in culture.   It would seem

Table 4.   Growth of Isolate N191 on media
with yeast extract, starch and lima bean
extract.

Medium basal solution + (g/l)	Dry wt (10^{-4} g)
0.2 g yeast extract	11
" + 10 g glucose	64
" + 10 g starch	71
" " + 0.1 g lima bean extract	188
" " + 0.5 g lima bean extract	190

that *E. eucalypti* is less exacting in its requirements than many other species of *Endogone*.   Nevertheless, the results suggest that further attempts to
culture species of *Endogone* would be worthwhile.
For instance, it is doubtful whether the procedures
used by bacteriologists when working with a nutrit-
ionally exacting bacterium have ever been tried with
Endogone.

DISCUSSION

The data presented suggest that the kinds of
association of species of *Endogone* with host plants
may be wider than usually envisaged.   Fassi (1965)
and Fassi *et al.,* (1969) have presented evidence
that *E. flammicorona* Trappe & Gerdemann forms ectomycorrhizas with species of *Pinus* and *Pseudotsuga,* a
result confirmed for *Pinus* by Tandy (pers.comm.), and
E. eucalypti has been found to form sheaths with
several species of *Eucalyptus*   The data also show
that *E. eucalypti* may assist its hosts in uptake of
P and K.

That *E. eucalypti* does not occur on a range of
grasses and herbs growing in the same soil as infected eucalypts suggests that while many species of
Endogonaceae have a wide host range (Gerdemann, 1968;
Mosse, 1973), others may be more restricted.

It is also interesting that *E. eucalypti,*
E. flammicorona (Tandy, pers. comm.) and at least one
other *Endogone*-like endophyte in Kuitpo forest soil
do not appear to form vesicles or spores.   Perhaps a
reason why spores are scarce in some soils where
endophytic mycorrhizas are common (Mosse & Bowen 1968;
Baylis, 1969) is that not all the endophytes form
them.

The fact that *E. eucalypti* grows in culture and
that simple experiments have markedly improved its
growth in media, holds hope for the isolation and
growth in pure culture of other species of *Endogone*.

Warcup, J. H.

ACKNOWLEDGMENTS

I am greatly indebted to Mr. J. T. Hutton, C.S.I.R.O., Division of Soils, Adelaide for the X-ray fluorescence data on element composition of the plants.

REFERENCES

BARRETT, J. T. (1961). Isolation, culture and host relation of the phycomycetoid, vesicular-arbuscular endophyte *Rhizophagus*. *Recent Adv. Bot.*, Univ. Toronto Press, Toronto, 2, 1725-1717.

BAYLIS, G. T. S. (1969). Host treatment and spore production by *Endogone*. *N.Z. Jl. Bot.*, 7, 173-4.

CHILVERS, G. A. & PRYOR, L. D. (1965). The structure of eucalypt mycorrhizas. *Aust. J. Bot.*, 13, 245-59.

FASSI, B. (1965). Micorrize ectotrofiche di *Pinus strobus* L. prodotte da un'endogone *(Endogone lactiflua* Berk.). *Allionia,* 11, 7-15.

FASSI, B., FONTANA, A. & TRAPPE, J. M. (1969). Ectomycorrhizae formed by *Endogone lactiflua* with species of *Pinus* and *Pseudotsuga*. *Mycologia,* 61, 412-414.

GERDEMANN, J. W. (1968). Vesicular-arbuscular mycorrhiza and plant growth. *A. Rev. Phytopath.*, 6, 397-418.

MOSSE, B. (1959). The regular germination of resting spores and some observations on the growth requirements of an *Endogone* sp. causing vesicular-arbuscular mycorrhiza. *Trans. Br. mycol. Soc.*, 42, 273-286.

MOSSE, B. (1973). Advances in the study of vesicular-
arbuscular mycorrhiza.   *A. Rev. Phytopath.*,
11, 171-196.

MOSSE, B. & BOWEN, G. D. (1968). The distribution of
Endogone spores in some Australian and New Zealand
soils, and in an experimental field soil at
Rothamsted.   *Trans. Br. mycol. Soc.*, 51, 485-492.

TECHNIQUES USED TO STUDY THE INTERACTION BETWEEN ENDOGONE AND PLANT ROOTS

C. M. HEPPER AND B. MOSSE

Rothamsted Experimental Station,
Harpenden, Herts., U.K.

INTRODUCTION

Inoculation with surface-sterilised *Endogone* spores can produce typical vesicular-arbuscular (VA) mycorrhiza in axenically grown plants on agar slopes in test tubes (Mosse, 1962). In certain media large amounts of external mycelium were attached to the roots (Mosse & Phillips, 1971). Unfortunately, this mycelium could not be isolated in a viable condition for use in further experiments. Typical VA infections have now been established in root organ cultures (Mosse & Hepper, 1975) which are interesting research tools providing better facilities for manipulation and for observation of the infection process. In spite of obvious advantages root organ cultures are highly artificial systems. We have therefore tried to supplement observations on germ tube development and infection by using agar coated slides buried in soil (Chinn, 1953; Jackson, 1957).

THE ESTABLISHMENT OF VESICULAR-ARBUSCULAR MYCORRHIZA IN ROOT ORGAN CULTURES

Method
Details of the methods used to establish mycorrhizal root organ cultures of *Trifolium pratense* have

already been described (Mosse & Hepper, 1975). A modified White's tissue culture medium was used containing, per litre of water: KCl, 65 mg; KNO_3, 80 mg; $Ca(NO_3)_2.4H_2O$, 300 mg; $MgSO_4.7H_2O$, 720 mg; KH_2PO_4 or $CaHPO_4.2H_2O$, 9.3 or 10.7 mg respectively; FeNaEDTA, 4.6 mg; $MnCl_2.4H_2O$, 4.9 mg; KI, 0.75 mg; H_3BO_3, 1.5 mg; $ZnSO_4.H_2O$, 1.92 mg, $CuSO_4.5H_2O$, 1 μg; $Na_2MoO_4. 2H_2O$, 0.168 μg; glycine, 3 mg; thiamine HCl, 0.1 mg; nicotinic acid, 0.5 mg; pyridoxine, 0.1 mg; sucrose 20 g. The pH of the medium was adjusted to 7.0 before sterilising for 10 minutes in a pressure cooker at $121^{\circ}C$.

Root organ cultures of *Trifolium pratense* have now been maintained over two years by the usual methods of subculturing in liquid medium, and cultures of *T. parviflorum* and *Lycopersicum esculentum* (var. Moneymaker) have also been used.

To establish mycorrhizal infections, root segments bearing several laterals were placed in the medium in petri dishes and germinated *Endogone* spores placed close to them. These spores ("yellow vacuolate", Mosse & Bowen, 1968; *Glomus mosseae*, Gerdemann & Trappe, 1974) were pre-germinated on distilled water agar because the trace elements in the tissue culture medium, in particular manganese and zinc, completely inhibit spore germination, although not subsequent hyphal growth (Hepper & Smith, unpublished). Pre-germination of spores also allows those which are contaminated to be discarded.

Observations

Typical VA infections were produced in root organ cultures of *Trifolium pratense* (Mosse and Hepper, 1975), *T. parviflorum* and *Lycopersicum esculentum*. Infections were established via typical entry points and showed normal arbuscules and vesicles but often intercellular hyphae predominated. Generally infections in agar were more successful because it was difficult to maintain roots and inoc-

ulum in close contact in the liquid medium, but even
in agar infection was not always predictable.   In
particular pH changes induced in the medium by the
growing roots seemed to be critical (Mosse & Hepper,
1975).

   The fungus frequently made extensive growth
before infection took place.   Fig. 1 illustrates
the amount of growth on agar from a single *Endogone*
spore before the roots became infected.

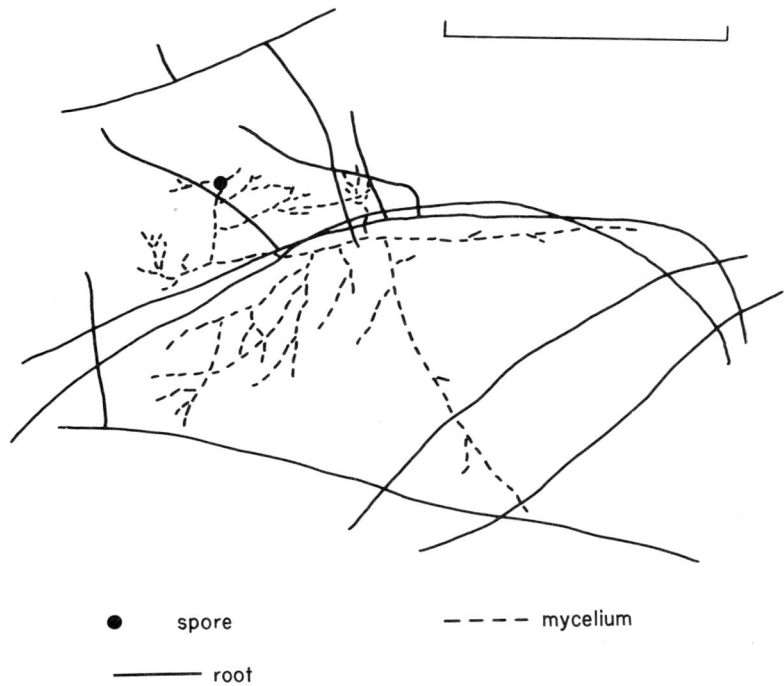

● spore   — — — — mycelium

———— root

Fig. 1.   Extent of mycelial growth from
a spore of *E. mosseae* before the roots have
become infected.   Line represents 1 cm.

Hepper, C. M. & Mosse, B.

Germinated spores placed in liquid medium were stimulated by exudates diffusing from growing roots even when there was no contact between the host and fungus (Fig. 2). No further hyphal growth was obtained, however, when the original spore was detached from the mycelium.

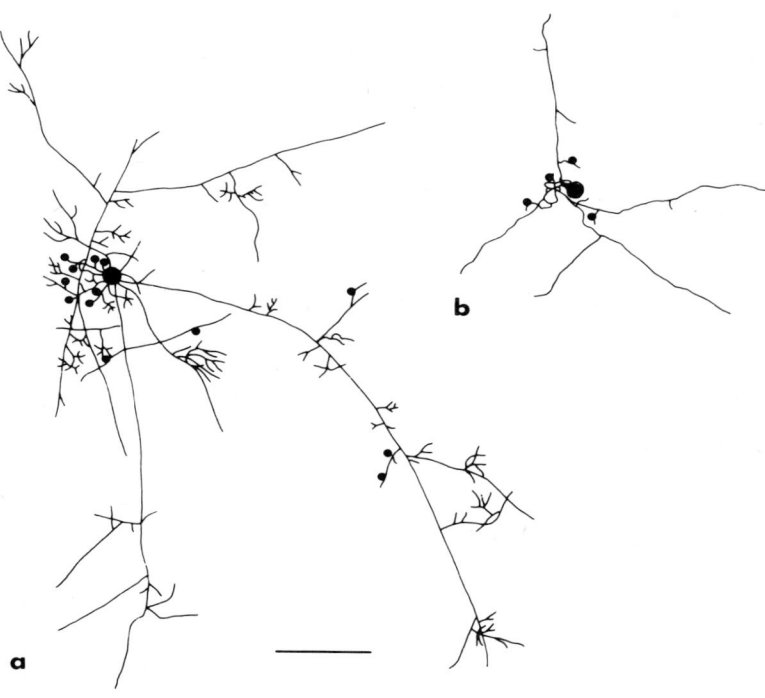

Fig. 2. Growth of mycelium from a spore of *E. mosseae* in the presence of a) and without b) *T. pratense* roots. Line represents 1 mm.

After infection there was usually vigorous
growth of the external mycelium, sometimes extending
for several centimetres into the medium.   The mycelium produced in the presence of growing roots char-
acteristically formed many small vegetative spores
and some lateral hyphae branched repeatedly producing
a "rhizoid-like" structure reminiscent of intra-cell-
ular arbuscules (Fig. 3a).   Such branches only
occurred in the presence of living roots.   Fig. 3b
shows these tufts of fine, thin-walled hyphae at a
late stage of development when septa had formed and
most of the cytoplasm had receded.

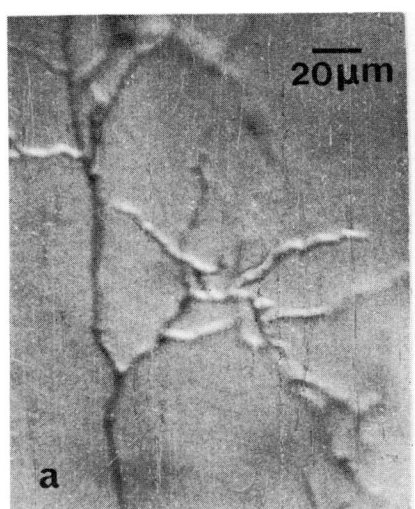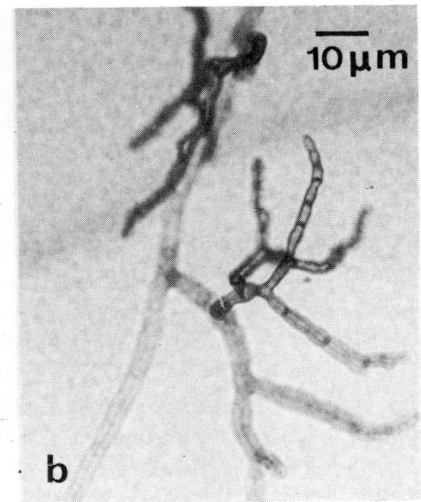

Fig. 3.   Rhizoid-like branches:
a) young, b) after retraction of cytoplasm and formation of septa.

Hepper, C. M. & Mosse, B.

Uses of root organ cultures

One advantage of this system is the possibility of providing a separate medium for the host and fungus by the use of divided plates or by feeding the root base from a separate vial (See Raggio & Raggio, 1956).   This would facilitate the study of the effects on infection of various nutrients or other compounds applied separately to the host or to the fungus.

By sub-culturing from infected root organ cultures it might be possible to maintain a continuous two-membered system of mycorrhizal roots, and supply reasonable amounts of external mycelium for simple studies of fungal metabolism.

BURIED SLIDE TECHNIQUE

Method

The method used was essentially as described by Brown and Hornby (1971).   Microscope slides or 5 cm square glass plates were coated with 1.5% distilled water agar.   Several spores were placed separately on the agar and the slide was buried, inclined at an angle, in a pot of soil or other rooting medium. Seeds of *T. pratense* were placed in the soil near the slide so that when they germinated the roots grew down in contact with the slide.   The pots were kept in a greenhouse on a damp sand tray.   After 2 weeks the slides were removed from the pots and allowed to dry out.   The tops of the plants and the remaining soil were carefully removed leaving a layer of dried agar holding the spores, mycelium and roots in position.   The slides were then stained with acetic-aniline blue (Jones & Mollison, 1948).   Even in non-sterile soils where other fungi grew across the slide, it was quite easy to identify the endogonaceous mycelium.

The two soils used in these studies were:
1. Arable (fallow) soil from Ashridge, pH 6.5, containing 11.6 mg/kg phosphorus soluble in NaHCO$_3$. This is soil No. 7 of Hayman and Mosse (1971).   2. Soil carrying a wheat crop from Little Knott field at Rothamsted, pH 6.5, containing 33.4 mg/kg phosphorus soluble in NaHCO$_3$.   Both soils were passed through a 3 mm sieve before use.

Results

The percentage germination of spores on slides buried in unsterile soil was similar to that of spores germinated axenically on water agar;   there was no evidence of a fungistatic factor inhibiting spore germination in either soil nor stimulation of germination as in soil agar plates (Mosse, 1959).
The mycelium on buried slides consisted of long, relatively unbranched hyphae (Fig.4) which frequently

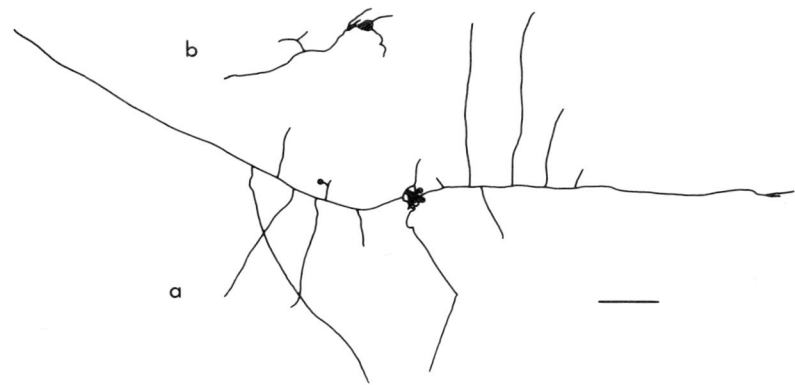

Fig. 4.   Type of mycelial growth from a spore of *E. mosseae* a) germinated on a slide in non-sterile soil, b) germinated axenically on distilled water agar.
Line represents 1 mm.

anastomosed with those from other spores, as do
hyphae from spores germinated on cellophane overlying a soil agar plate (Mosse, 1959).   Much more
extensive hyphal growth was obtained from spores
placed in non-sterile soil than from those germinated
axenically on distilled water agar (Fig. 4).   Growth
of the mycelium was not markedly stimulated by roots
growing down the slide and none of the "rhizoid-like"
branches induced by the living roots in root organ
cultures developed on the buried slides.   If such
branches are due to morphogenetic effects of root
exudates, in the soil these exudates may be adsorbed
or metabolised by other microorganisms.

In root organ cultures the first infections
often occurred some distance from the original inoculum (Mosse & Hepper, 1975) and hyphae frequently
grew past roots without any apparent reaction to them
(Fig. 1).   This also occurred on buried slides in
the soil (Fig. 5, hypha A).   If, on the other hand,
infection was going to occur the approaching hypha
often changed direction (Fig. 5, hypha B), so that
the tip approached the root at approximately 90°.
Generally it was then deflected along the root and
formed an appressorium nearby (Fig. 5a).   Following
infection fungal growth was often markedly stimulated
and many small vegetative spores developed on the
external mycelium (Fig. 5, C).   Entry places into
the root were quite unpredictable but once infection
had occurred, further entry points on the same roots
were likely.   If the physiological state of the root
determines points of entry the special conditions
required appear to be very localised.

Uses of the technique

By using pieces of infected root instead of
spores the technique might be employed to study re-
generation from infected roots, in particular to
assess rapidly the viability of freeze-dried material.
By comparing growth in non-sterile and irradiated

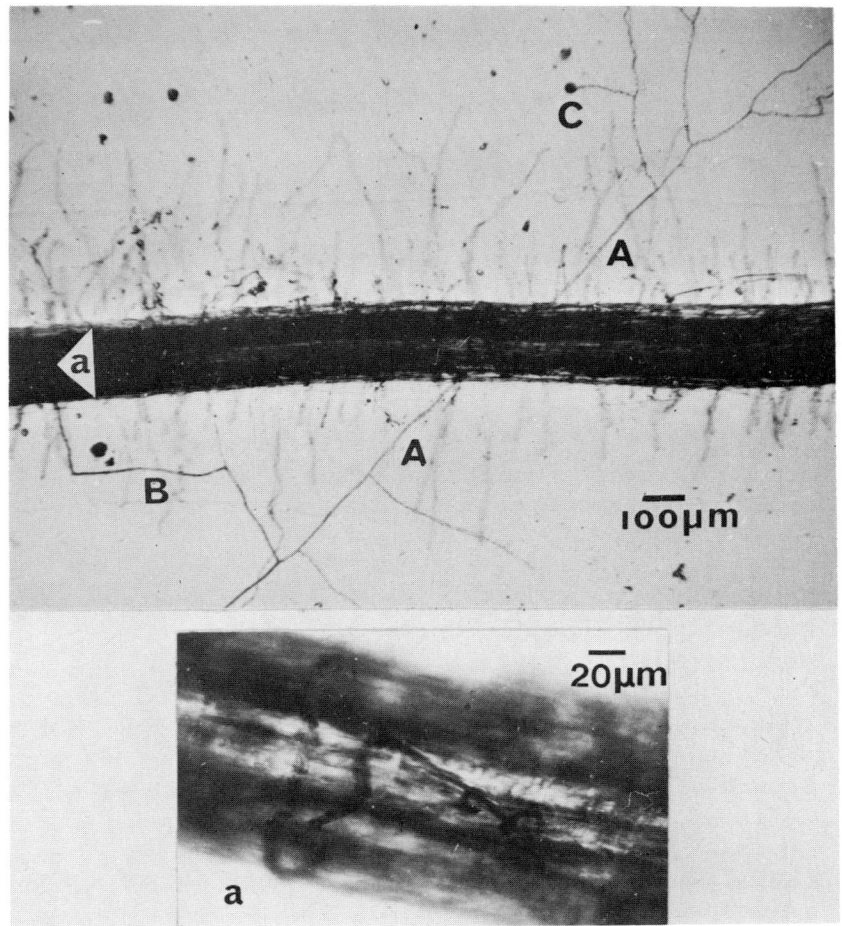

Fig. 5. Growth of mycelium on buried slide showing non-infective main hypha A passing underneath the root without any change in direction. Infective branch hypha B changes direction prior to entering the root. Fig. 5a shows detail of entry point. Note vegetative spore at C.

soils it will be possible to evaluate effects of other microorganisms on development of the endophyte. The technique could also be useful to study the extent of the endophyte's independent life in the soil.

REFERENCES

BROWN, M. E. & HORNBY, D. (1971). Behaviour of *Ophiobolus graminis* on slides buried in soil in the presence or absence of wheat seedlings. *Trans. Br. mycol. Soc.*, 56, 95-103.

CHINN, S. H. F. (1953). A slide technique for the study of fungi and actinomycetes in soil with special reference to *Helminthosporium sativum*. *Can. J. Bot.*, 31, 718-724.

GERDEMANN, J. W. & TRAPPE, J. M. (1974). The *Endogonaceae* in the Pacific Northwest. *Mycologia Memoir,* No. 5, 75 pp.

HAYMAN, D. S. & MOSSE, B. (1971). Plant growth responses to vesicular-arbuscular mycorrhiza. I. Growth of *Endogone*-inoculated plants in phosphate-deficient soils. *New Phytol.*, 70, 19-27.

JACKSON, R. M. (1957). Fungistasis as a factor in the rhizosphere phenomenon. *Nature, London*, 180, 96-97.

JONES, P. C. T. & MOLLISON, J. E. (1948). A technique for the quantitative estimation of soil microorganisms. *J. gen. Microbiol.*, 2, 54-69.

MOSSE, B. (1959). The regular germination of resting spores and some observations on the growth requirements of an *Endogone* sp. causing vesicular-arbuscular mycorrhiza. *Trans. Br. mycol. Soc.*, 42, 273-286.

MOSSE, B. (1962). The establishment of vesicular-
arbuscular mycorrhiza under aseptic conditions.
J. gen. Microbiol., 27, 509-520.

MOSSE, B. & BOWEN, G. D. (1968). A key to the recognition of some *Endogone* spore types.   *Trans. Br.
mycol. Soc.*, 51, 469-483.

MOSSE, B. & HEPPER, C. (1975). Vesicular-arbuscular
mycorrhizal infections in root organ cultures.
Physiol. Pl. Path., 5, 215-223.

MOSSE, B. & PHILLIPS, J. M. (1971). The influence of
phosphate and other nutrients on the development of
vesicular-arbuscular mycorrhiza in culture.
J. gen. Microbiol., 69, 157-166.

RAGGIO, M. & RAGGIO, N. (1956). A new method for the
cultivation of isolated roots.   *Physiol. Pl.*,
9, 466-469.

ENDOGONE STRAIN AND HOST PLANT DIFFERENCES IN DEVELOPMENT OF VESICULAR-ARBUSCULAR MYCORRHIZAS

D. I. BEVEGE AND G. D. BOWEN

CSIRO, Division of Soils, Glen Osmond, South Australia, 5064.

INTRODUCTION

Differential response of *Paspalum notatum* to various strains of *Endogone* was demonstrated by Mosse (1972) and it is therefore appropriate to examine the possible variation in the ontogeny of infections caused by different fungi. Furlan and Fortin (1973) recently studied the effect of temperature on mycorrhiza development in onion caused by *E. calospora;* infection started 4-8 weeks after inoculation depending on the temperature and was closely followed by arbuscule formation.

Additionally, the lack of VA infection in certain plant groups (notably Cruciferae and Chenopodiaceae) and limited infection of others (e.g. Pinaceae) is of more than passing interest; investigation of the barriers to infection of such plants may shed some light on infection processes in general.

This note describes studies on the time-course of infection in onion and subterranean clover *(Trifolium subterraneum* L. var. Bacchus Marsh) when inoculated with spores of three species of *Endogone: E. mosseae* (Nicolson & Gerdemann, 1968) from sporocarps provided by Dr. B. Mosse, (EM): *E. araucareae*

(Bevege, 1971) from spores produced in pot culture
with *Araucaria cunninghamii* Ait. (Coniferae) , (EA);
and "white-reticulate" spores (Mosse & Bowen, 1968)
collected from orchard soil near Adelaide by sieving
(WR.).   Subsidiary observations were made on maize,
Arabidopsis thaliana (L) Heynh (Cruciferae) and *Pinus
radiata* D. Don (Pinaceae), inoculated with one or
more of the *Endogone* species listed above.

METHODS

Seed of clover, onion, maize and radiata pine
was germinated aseptically on nutrient agar and
transplanted, when radicles were 1-2 cm long, into
free draining polythene tubes (25 mm diam), containing 150g of an infertile sand from Waneroo, Western
Australia, that had been sterilised by gamma irradiation (2.5 Mrads).   A layer of spores (EM, EA or
WR) was placed 8 cm deep in the tube.   Two seedlings
were planted in each tube which received 20 ml of $\frac{1}{4}$
strength complete nutrient solution (Hoagland & Arnon,
1938) and was thereafter watered daily.   The tubes
were placed in a controlled environment cabinet at
$20^{\circ}C$ with 8 hr night/16 hr day at 20,400 lumen m^{-2}.

Two tubes of clover and onions from each treatment were harvested progressively;   clover at 8, 13,
18, 19, 20, 21, 22, 24, 26, 28 and 32 days after
transplanting, onion at 8, 16, 17, 18, 20, 22, 24,
28, 30, 32 and 35 days.   The slightly different
time sequence was due to the slower development of
onion.   Maize plants were harvested at 28 and 32
days and radiata pine at 24, 44, and 91 days.

Polythene tubes were sliced transversely and all
roots harvested for 1 cm on each side of the inoculum
layer.   Roots were washed, cleared (Bevege, 1968),
stained (Phillips & Hayman, 1970) and examined by
microscope.

Seedlings of *Arabidopsis* were grown in funnels
containing soil kept moist by a wick dipping into

0.1 strength Hoagland and Arnon solution.   Spores
(EA or WR) were placed 2 cm below the soil surface
and seed was placed on the surface after 48 hours
stratification in 10^{-3} M KNO_3 at $2^{O}C$ with continuous
illumination (20,400 lumen m^{-2}).   Seed germinated
after 3 days and plants were harvested 24 days later
when they were flowering.

RESULTS

All times have been calculated from the estimated time the root passed through the inoculum layer;
i.e. 4 days with clover and 9 days with onion.

Development of infections in clover and onion

Data on spore germination and early stages of
infection are summarised in Table 1.
WR spores associated with clover germinated at
4 days (the earliest germination observed); EM
spores had not germinated by day 9 but had at day 14
and similarly EA spores germinated between days 4-14.
With onion WR and EM spores germinated by day 7 and
EA spores between day 7 and 8.
All endophyte strains made little pre-infection
growth in the rhizosphere of clover and no infection
occurred till day 14.   First infections occurred
earlier in onion and even WR which infected more
slowly than EA and EM had formed appressoria by day
7.   At day 7 the first trifoliate leaf of clover
was still expanding and the second was expanded at
day 14, whereas in onion the second leaf was already
expanding by day 7.   Possibly plants are not susceptible to infection till the second leaf is expanded but the experimental design was not sufficient to
elucidate this.
In both clover and onion, growth of EM was more
general with several infection points and longitudinal hyphae within the roots, while EA and WR produced

Bevege, D. I. and Bowen, G. D.

Table 1. Observations on spore germination and infection.

	Spore Type & Plant					
Criterion	White-reticulate		E. mosseae		E. araucareae	
	Clover	Onion	Clover	Onion	Clover	Onion
Spore Germination[1]	4	0-7	9-14	0-7	4-14	8
Infection[1]	14	13+	14	7	14	9
Degree of infection	local		general		local	

[1] Days after root passage through inoculum layer when this criterion was first observed.

+ Appresoria had formed at 7 days but no infection occurred until day 13.

more localised infections.

Data on later stages in the development of the
infection are summarised in Table 2. For every
criterion studied there were differences, often
quite marked, either between fungal strains or
between host species.

Infections spread rapidly in clover but with
all endophytes remained quite localised in onion
even after 6-10 days. This slower spread in onion
is consistent with later production of vesicles and
arbuscules in this host. Similarly infection in
maize remained localised with all three spore types.

Both the extent and rate of hyphal growth outside the root varied with spore type, being considerable for EM from the outset and much less for EA and
WR. By day 28 however, hyphae of WR had become
abundant in clover cultures and had formed chlamydospores similar to those used initially as inoculum.
EM formed sporocarps at this time. EA failed to
produce spores on clover by day 28 but numerous
zygospores (Bevege, 1971) were found in clover cultures retained to day 84 in the subsidiary radiata
pine trial. In association with maize, zygospores
were formed by day 28. None of the spore types
formed spores or sporocarps with onion.

Arabidopsis thaliana and *Pinus radiata*

WR spores did not germinate in association with
Arabidopsis. They were comparable to those used in
the clover and onion trials and we conclude that root
exudates of *Arabidopsis* either do not contain specific germination factors or contain germination
inhibitors toward WR. Spores of EA germinated with
resultant considerable hyphal entanglement with root
hairs; no growth occurred on the root surface, no
appressoria were formed nor infection achieved, but
vesicle clusters normally associated with the post-
infection phase (Bevege, 1971) developed on the
hyphae.

Table 2. Development of the vesicular-arbuscular infection.

Days after infection when particular structures were first observed

Structure	White-Reticulate		E. mosseae		E. araucareae	
	Clover	Onion	Clover	Onion	Clover	Onion
Arbuscules	2	6	1	4	0	2
Collapsed arbuscules*	4	7	7-9	10-15	14-15	14-15
Vesicles	0	6	1	14	Not Formed[1]	
External hyphae	1-2	6	1-2	0	1-2	6
Internal spread	Rapid	Slow	Rapid	Slow	Rapid	Slow

* Days after formation i.e. longevity.

[1] Vesicles did not form on clover, onion or maize but do so in Araucaria cunninghamii (Bevege, 1971).

Radiata pine was tested with all three strains of *Endogone*.   No infection was observed but hyphae grew around and along roots.   EM formed occasional appressoria and external vesicles by day 24 and day 44 respectively; EA formed external vesicles by day 88 but not appressoria.   Spores of WR germinated to produce some surface hyphae by day 91, but no vesicles or appressoria.

DISCUSSION

These studies have shown marked variation between different host endophyte pairs in most factors studied, e.g. time for germination of spores and infection, development of arbuscules and vesicles, spread of infection, and growth of hyphae outside of the root.   These differences may have implications for the establishment of particular endophytes and the rapidity and extent of plant response to the infection.   For example, one might expect quite different inoculum dose - response curves for associations with poor spread of the infection compared with those in which the infection spreads rapidly.   In view of the importance of hyphal growth into soil for uptake of poorly diffusible nutrients (Bowen & Rovira, 1969; Sanders & Tinker, 1971; Hattingh *et al.* 1973) differences between associations in production of extramatrical hyphae may be directly related to different plant responses observed by Mosse (1972).   Indeed later observations we have made show Mosse's type E3 produced far more extramatrical hyphae on onion than the types reported here under similar conditions and the growth of EA on *Araucaria* usually exceeds that found in the studies above on clover and onion.
The present observations indicate the potential of available material for experimental approaches to some aspects of mycorrhizal nutrition of plants e.g. the importance (if any) of arbuscule degeneration for

nutrient transfer.

Blocks to *Endogone* infection of a host can occur from failure to stimulate spore germination (*Arabidopsis* – WR combination) or to blocks to the infection process – all 3 types grew in the rhizosphere of *P. radiata* as did EA with *Arabidopsis* but did not infect.   This material and *E. lactiflua* (which Fassi *et al.* 1969, reported to form ectomycorrhizas on *Pinus* species) may be useful for more detailed studies on the processes of infection. Infection of legume nodules by endotrophic mycorrhizal fungi apparently occurs only with difficulty for we have been unable to infect nodules of subterranean clover despite great ease in infecting adjacent root tissue (Daft & El-Giahmi, 1975).   Such proximity of poorly and highly susceptible tissue is of further interest in the physiology of infection, quite apart from the obvious ecological advantage conferred on nitrogen fixation by the efficient phosphate (and possibly molybdenum) uptake of mycorrhizas.

ACKNOWLEDGMENTS

This study was supported by a Nuffield Foundation research grant to D.I.B.   We wish to acknowledge the assistance of Miss B. Arnott throughout.

REFERENCES

BEVEGE, D. I. (1968). A rapid technique for clearing tannins and staining intact roots for detection of mycorrhizas caused by *Endogone* spp., and some records of infection in Australasian plants. *Trans. Br. mycol. Soc.,* 51, 808-810.

BEVEGE, D. I. (1971). Vesicular arbuscular mycorrhizas of *Araucaria*: Aspects of their ecology and physiology and role in nitrogen fixation. Ph.D. Thesis, Univ. of New England, Armidale, Australia, 349 pp.

BOWEN, G. D. & ROVIRA, A. D. (1969). The influence of microorganisms on root growth and metabolism. In *Root Growth*, ed. Whittington, W. J., pp. 170-201. Butterworths Scientific Publications.

DAFT, M. J. & EL-GIAHMI, A. A. (1975). This symposium, p. 581.

FASSI, B., FONTANA, A. & TRAPPE, J. M. (1969). Ectomycorrhizae formed by *Endogone lactiflua* with species of *Pinus* and *Pseudotsuga*. *Mycologia*, 61, 412-414.

FURLAN, V. & FORTIN, J. A. (1973). Formation of endomycorrhizae by *Endogone calospora* on *Allium cepa* under three temperature regimes. *Naturaliste can.*, 100, 467-477.

HOAGLAND, D. R. & ARNON, D. I. (1938). The water culture method for growing plants without soil. *Circ. Univ. Calif. Agr. Exp. Stn.*, 347. (Revised 1950).

HATTINGH, M. H., GRAY, L. E. & GERDEMANN, J. W. (1973). Uptake and translocation of ^{32}P-labelled phosphate to onion roots by endomycorrhizal fungi. *Soil Sci.*, 116, 383-387.

MOSSE, B. (1972). Effects of different *Endogone* strains on the growth of *Paspalum notatum*. *Nature, London*, 239, 221-223.

MOSSE, B. & BOWEN, G.D. (1968). A key to the recognition of some *Endogone* spore types. *Trans. Br. mycol. Soc.*, 51, 469-483.

NICOLSON, T. H. & GERDEMANN, J. W. (1968). Mycorrhizal *Endogone* species. *Mycologia,* <u>60</u>, 313-325.

PHILLIPS, J. M. & HAYMAN, D. A. (1970). Improved
procedures for clearing roots and staining parasitic and vesicular-arbuscular mycorrhizal fungi
for rapid assessment of infection. *Trans. Br.
mycol. Soc.,* <u>55</u>, 158-161.

SANDERS, F. E. & TINKER, P. B. (1971). Mechanism of
absorption of phosphate from soil by *Endogone*
mycorrhizas. *Nature, London,* <u>233</u>, 278-279.

FACTORS AFFECTING SYMBIOTIC GERMINATION OF ORCHID SEED

J. H. WARCUP

Waite Agricultural Research Institute,
The University of Adelaide,
South Australia.

INTRODUCTION

Orchids differ from most other higher plants in that their seeds, which are minute, generally undifferentiated, and with little stored food, are incapable of germination solely in a dilute mineral salts solution but require an external source of organic carbon. Since 1922 (Knudson, 1922) it has been shown that in the laboratory many orchids may be germinated on media containing inorganic salts and sucrose, but it is generally agreed (Harley, 1969) that in nature orchids only germinate in the presence of endophytic fungi that infect germinating seed and later the roots or other absorbing organs of young and adult plants.

In general the endophytic fungi have been isolated from adult orchids and the method for confirming them to be endophytes has been that of back-inoculation into cultures of sterilized seed. This approach, however, has not been without difficulties (Arditti, 1967) and since Knudson's work most orchid seed has been germinated asymbiotically. Recently there has been renewal of interest in symbiotic germination but there are still many points which require clarification. This neglect is perhaps

Warcup, J. H.

surprising for orchids have one of the few mycorrhizal systems as yet available for detailed analysis in agar culture.

In this paper I wish to discuss some aspects of
symbiotic germination of orchid seed that have been
of interest to me; the contribution makes no pret-
ence of being a detailed review of literature.   Data
are restricted to work on green or autotrophic
orchids.   Since germination has been used in more
than one sense in relation to orchids (Harley, 1969)
the term is used here to denote those stages between
first swelling of the embryo and development of the
first leaf.

THE FUNGI

The fungi endophytic with green orchids are
commonly placed in the form-genus *Rhizoctonia*.
This genus, based as it is on mycelial states of
certain fungi, is ill-defined, and perfect states of
Rhizoctonias are known to occur in both the Basidiomycetes and the Ascomycetes (Whitney & Parmeter,
1964; Warcup & Talbot, 1966).   Basidiomycetes with
known *Rhizoctonia* states occur in the Auricularia-
ceae, Tulasnellales sensu Talbot (1973) and the
Corticiaceae.

Table 1 records the perfect states at present
known for orchid Rhizoctonias.   Species isolated
from orchids but not yet shown to be symbiotic with
orchid seed are marked with an asterisk.   The data
in Table 1 show that at present most identified
fungi from green orchids belong to the Tulasnellales.
Sebacina vermifera is, so far, the only member of
the Tremellaceae, while the position of *Corticium
catonii* in modern classification has not been examined.   Whether fungi belonging to other genera in
the Tulasnellales are likely to be orchid endophytes
is difficult to estimate.   *Oliveonia pauxilla*
(without a *Rhizoctonia* state) has been found to be

Table 1.   Known Perfect States of Orchid
Rhizoctonias.

Tulasnellales
(1) Tulasnellaceae
        Tulasnella allantospora Wakefield & Pearson
        T. asymmetrica Warcup & Talbot
        T. calospora (Boudier) Juel
        T. cruciata Warcup & Talbot
        T. violea (Quel.) Bourdot & Galzin
        T. sp. 0632
      [+]T. sp. 257
(2) Ceratobasidiaceae
        Thanatephorus cucumeris (Frank) Donk
      *T. orchidicola Warcup & Talbot
        T. sterigmaticus (Bourdot) Talbot
        T. sp. 0426

        Ceratobasidium cornigerum (Bourdot) Rogers
        C. obscurum Rogers
        C. sphaerosporum Warcup & Talbot
        C. sp. 0507
        C. sp. E12
        C. sp. 0615
      *C. sp. 0648

      [+]Oliveonia pauxilla (Jacks.) Donk

Tremellaceae	Sebacina vermifera Oberwinkler
?	Corticium catonii Burgeff

---

* Species isolated from orchids but not yet shown to
  be symbiotic with orchid seed.

[+] Species not yet known from orchids but symbiotic
  with orchid seed.

---

partially symbiotic with *Cymbidium findlaysonianum*
Lindl. but not with other orchids tested (Table 6).
This suggests that members of other genera of the
Tulasnellales may be found to be orchid symbionts if

tested against a range of orchids; however, irres-
pective of this, it seems probable that further
species of *Tulasnella* and *Ceratobasidium* will be
found to be orchid endophytes.

Whether any of the Rhizoctonias belonging to the
Corticiaceae are endophytes of orchids is not known.
Some experiments with a few members of the Cortic-
iaceae with *Rhizoctonia* states, *Waitea circinata*
Warcup & Talbot, *Botryohypochnus isabellinus* (Fr.)
Erikss., and a *Botryobasidium* species, and seed of
Spiranthes sinensis (Pers.) Ames, *Thelmitra aristata*
Lindl., *Cymbidium findlaysonianum* and *Dactylorhiza
purpurella* (T. & T.A. Steph.) Soó, orchids which
between them have wide symbiotic capabilities, have
been carried out but the results were all negative.
However, for reasons discussed later, such negative
results may not be conclusive.

Likewise, there are few data on whether Ascomycete Rhizoctonias are orchid symbionts. Three different Ascomycetes, all as yet unidentified, have been
isolated from within roots of species of *Pterostylis*
(Warcup, unpublished data). The data available
suggest that mature roots of some orchids such as
Pterostylis may carry a wider range of fungi intern-
ally than young roots. Although two of the Ascomycetes were isolated from fungal coils within the
orchid roots this is not necessarily conclusive evidence that they are symbionts. Coils formed by
orchid endophytes may occasionally be colonized by
other fungi (Warcup, 1971). In such cases if the
hyphae growing from a coil on an isolation plate are
different from those of the coil, this contamination
is easily recognized, but if the fungus is itself a
Rhizoctonia then contamination would be difficult to
detect. Ascomycetes are much more common fungi
within the roots of a wide range of plants than is
commonly appreciated (Warcup, unpublished data).
Symbiosis tests with these Ascomycetes and seed of
Pterostylis have so far been unsuccessful.

The question of the resupinate wood-rotting

fungus *Sistotrema brinkmannii* (Bres.) J. Erikss.
needs comment.   During crossing experiments with
Rhizoctonias from *Orchis mascula* (L.) L., Sprau
(1937) obtained fruiting areas which he isolated and
named *Corticium masculi*.   Rogers (1944) placed
C. masculi under *Sistotrema (Trechispora) brinkmannii* as a nomen nudum.   Downie (1959) expressed
doubts whether *C. masculi* is an orchid endophyte
since she found the fungus as a contaminant in one
of her *Rhizoctonia* cultures.   Derx (1937) isolated
a number of Basidiomycetes from his garden and found
that *Corticium octosporum* (= *S. brinkmannii*) aided
growth of protocorms of a *Laelia* X *Cattleya* hybrid,
but the medium used was a starch-gelatin one so that
it is possible that the fungus was providing soluble
nutrients (Knudson, 1925), not acting as an endophyte.   I have made symbiosis tests with three
strains of *S. brinkmannii* and seed of *Thelymitra
aristata, Spiranthes sinensis, Pterostylis nutans*
R.Br. and *Dactylorhiza purpurella* without success,
but since *S. brinkmannii* is an aggregate species
these negative results may not be conclusive.

While clamp-bearing Basidiomycetes are considered to be predominantly associated with saprophytic
orchids (Burgeff, 1959), I have found an unidentified
clamped fungus isolated from *Gastrodia sesamoides*
R.Br. to be fully symbiotic with seed of *Cymbidium
findlaysonianum*.   It has, however, no effect on seed
of a number of other orchids.   *C. findlaysonianum*
is an autotrophic species but the genus does contain
saprophytic species (Burgeff, 1936).

Whatever the taxonomic position of the fungi,
the data available - including the number of morphol-
ogically distinct isolates in my collection - suggest
that the list of perfect states in Table 1 is far
from complete.

Most orchid Rhizoctonias have been isolated from
orchids, but knowledge of their identity and taxonomic position allows more rational testing of these
and related fungi from any source (cf. Derx, 1937) to

see if they are able to stimulate germination of
orchid seed.   Such an approach was also used by
Downie (1959) when, having isolated strains of *Rhizoctonia solani* Kühn from *Dactylorhiza purpurella* she
carried out symbiosis tests, some successful, with
isolates of *R. solani* from cauliflower, wheat,
tomato and potato; as it happened, however, her
isolates of *R. solani* included both *Thanatephorus
cucumeris* and *Ceratobasidium cornigerum* (the isolate
from wheat).   In my work (Warcup, 1973, and unpublished) it has been found that isolates of *Thanatephorus cucumeris, Ceratobasidium cornigerum* and
Tulasnella calospora from non-orchid sources are as
efficient as isolates from orchids, and indeed there
is no reason why they should not be for in nature
orchid seed must come in contact with an appropriate
fungus in soil or decaying organic matter.

THE MEDIUM

Considering that Bernard (1909) emphasized that
a delicate balance has to be maintained between the
fungus and the orchid embryo, it is surprising that
the effect of composition of the medium as a major
factor of the environment has received little attention in symbiotic germination studies.   This is also
in direct contrast to asymbiotic studies where much
effort has been directed at improvement of seed
germination and seedling growth (Arditti, 1967).

My interest in media arose from the fact that I
took a fungus medium (herein called M1) that I had
used for years, substituted cellulose (Hadley, 1969)
for sucrose and used it as a medium for orchid seed
germination (Warcup, 1973).   Although a number of
orchids germinated excellently on this medium,
others did not, even when presented with what, from
isolation studies, should be the correct fungus.
This led to consideration of other orchid germination media (Harvais & Hadley, 1967b).   A small pre-

Table 2. Composition of some orchid culture media (g per litre)

	Knudson B	Knudson C	Burgeff Eg1	Burgeff Sb	Pfeffer*	M1
Sucrose	20.0	20.0	20.0	–	1.0**	–
Cellulose	–	–	–	–	9.0	10.0
Starch	–	–	–	5.0	–	–
$Ca(NO_3)_2$	1.0	1.0	1.0	–	0.8	–
KNO_3	–	–	–	–	0.2	–
$NaNO_3$	–	–	–	–	–	0.3
$(NH_4)_2SO_4$	0.5	0.5	0.25	–	(0.5)+	–
Na nucleinate	–	–	–	0.5	–	–
KH_2PO_4	0.25	0.25	0.25	1.0	0.2	0.2
K_2HPO_4	–	–	0.25	–	–	–
$MgSO_4 \cdot 7H_2O$	0.25	0.25	0.25	0.3	0.2	0.1
KCl	–	–	–	–	0.1	0.1
$FeSO_4 \cdot 7H_2O$	0.025	0.025	0.02	0.01	–	–
$MnSO_4 \cdot 4H_2O$	–	0.0075	–	–	–	–
$CaCl_2$	–	–	–	0.1	–	–
NaCl	–	–	–	0.1	–	–
Micronutrients	–	–	–	–	+	–
Yeast extract	–	–	–	–	–	0.1

* Harvais & Hadley (1967b)   ** Glucose
+ As required for certain species (Hadley, 1969)

liminary test of the germination of *Thelymitra luteocilium* R.D. Fitzg. and *Cattleya trianaei* Lindl. et
Reichb. f. on Knudson C, Burgeff Eg1 and M1, all with
1% cellulose as carbon source, showed that symbiotic
germination of these orchids was best on M1.   Examination of the composition of these and a few other
media (Table 2) showed that the main differences
between them are (a) the source and quantity of nitrogen, and (b) the presence of micronutrient solution
and/or iron in some media.   A series of media was
prepared based on the mineral solution of M1 but
varying the nitrogen source and the presence or absence of the micronutrient solution of Harrison and
Arditti (1970).   Data on symbiotic germination of
Pterostylis vittata Lindl., *Thelymitra aristata* and
Cattleya trianaei are presented in Table 3.   For
Pterostylis and *Thelymitra* the germination scale 0-6
(Warcup, 1973) was used;   for *C. trianaei,* where the
protocorm quickly becomes green, stages 5 and 6 cannot be distinguished hence a scale 0-5 is used.
The data show that there was little difference in
seed germination on calcium nitrate and sodium nitrate as source of nitrogen but wherever ammonium
sulphate was present growth was poor, though *Thelymitra aristata* was less affected than *Cattleya trianaei.*   There was also a suggestion that the micronutrient solution depressed growth slightly under
these conditions.   With *Cattleya trianaei* there was
a good correlation between poor germination of seed
and low final pH of the medium.   Further experiments
have shown that a number of orchids show poor symbiotic germination on media with ammonium sulphate.
That low pH alone is not the reason for the poor
germination on ammonium sulphate is suggested by the
data in Table 4 on comparative germination of *Thelymitra luteocilium* with two different isolates of
Tulasnella calospora on medium M1 containing different levels of ammonium sulphate, sodium nitrate and
Difco yeast extract as nitrogen source.   In all
media with ammonium sulphate as nitrogen source the

Table 3. Effect of nitrogen source and addition of micronutrient solution on symbiotic germination of seed of *Pterostylis vittata*, *Thelymitra aristata* and *Cattleya trianaei*.

Basal Solution (M1) plus	*Pterostylis vittata*	*Thelymitra aristata*	*Cattleya trianaei*	pH initial	final
$NaNO_3$	6*	6	5†	(5.2)	(5.0)
$Ca(NO_3)_2$	6	6	5	(4.9)	(4.5)
$NaNO_3$ + micronutrient	4	4	5	(5.0)	(4.6)
$CaNO_3$ + micronutrient	5	6	4	(5.1)	(4.5)
$NaNO_3$ + $(NH_4)_2SO_4$	1	4	0	(5.0)	(3.4)
$CaNO_3$ + $(NH_4)_2SO_4$	1	3	0	(4.8)	(3.2)
$NaNO_3$ + $(NH_4)_2SO_4$ + micronutrient	2	4	0	(5.0)	(3.2)
$CaNO_3$ + $(NH_4)_2SO_4$ + micronutrient	1	3	0	(4.8)	(3.3)
$(NH_4)_2SO_4$	2	3	1	-	(3.5)

* germination scale 0-6

† germination scale 0-5

Warcup, J. H.

Table 4. Comparison of germination of *Thelymitra luteocilium* with two isolates of *Tulasnella calospora* on medium M1 with different levels and sources of nitrogen; final pH's of media are also given.

N source	Isolate 0689			Isolate 0584		
	$NaNO_3$	Yeast	$(NH_4)_2SO_4$	$NaNO_3$	Yeast	$(NH_4)_2SO_4$
Level						
N*	5 (4.1)†	3 (6.1)	0 (3.5)	4 (4.3)	2 (6.7)	3 (3.3)
N/6	6 (4.4)	6 (4.4)	2 (3.5)	6 (4.2)	6 (4.8)	6 (3.5)
N/24	6 (4.3)	5 (4.5)	2 (3.5)	6 (4.3)	6 (4.8)	6 (3.4)
N/48	6 (4.3)	6 (4.6)	3 (3.5)	6 (4.5)	6 (4.8)	6 (3.4)
Nil	6 (-)			6 (-)		

* 2 gm per litre

† Final pH of medium

final pH was low, however, germination with isolate
0584 was good whereas that with 0689 was poor.   The
data in Table 4 also show that good germination
occurred without added nitrogen to the medium.

Other experiments have shown that *Thelymitra
luteocilium* is less sensitive than some orchids to
the level of nitrogen in the medium, and that for
symbiotic germination of these orchids the nitrogen
level of most orchid media (Table 2) is too high for
maximum germination.   Since several orchids showed
excellent germination on medium M1 without added
nitrogen, trials were conducted with water agar as
the medium.   On the water agar the source of carbon
was the filter paper on which the seed are planted in
the tube (Warcup, 1973):   in the absence of the paper
germination of most orchids was poor.   The results
of two such experiments are given in Table 5.   In
general, in such experiments, germination was better
on M1 than on water agar but in some cases growth on
water agar was good.   Further, as shown in Table 5
different isolates of fungi differed in their growth
stimulation on the two media.   The type of agar used
also influenced seed germination on water agar, presumably through variation in the traces of nutrients
present in the agar.

Table 5.   Comparative germination of
Thelymitra luteocilium and *Dendrobium form-
osum* Roxb. on medium M1 and water agar.

	Medium	*Thelymitra luteocilium*		*Dendrobium formosum*	
		M1	Water	M1	Water
Fungus isolate					
Tulasnella calospora	0584	5	6	4	2
" "	0689	5	5	6	4
" "	54	4	3	2	2
Nil		1	1	1	1

These results show that while orchids appear
adaptable to wide variations of inorganic salt combinations, the quantities of nitrogen in standard
media appear higher than necessary for good symbiotic
germination.   Further the concentration of nutrients,
including nitrogen, would appear to be a factor in
whether some fungal isolates appear "efficient"
(Warcup, 1973) or inefficient in symbiosis tests.
This means that results of such tests should always
specify the medium used.

SPECIFICITY OF MYCORRHIZAL ASSOCIATION

Bernard (1909) gained the impression that the
orchid-fungus association was highly specific.   This
question has since raised considerable, and at times
heated (and biased) discussion which is still not
completely resolved.   Harley (1969) summarized the
position then by saying "amongst the species of
Rhizoctonia isolated from orchids a few have always
been found to inhibit the growth of seedlings and the
remainder exhibit varying degrees of specificity in
respect of the group whose germination they stimulate".   This "degree of specificity" has been considered to have either an ecological or a taxonomic
basis.
Ecological groupings (Curtis, 1937; 1939), that
different species of orchids in one habitat have
similar endophytes which may be different from those
in the same orchids in a different habitat, seem a
logical possibility.   Whether they have any basis
requires more extensive data than are available.
Curtis' data were from comparatively few orchids
whose symbiotic capabilities were unknown.   Data
from southern Australia show that orchids in the same
habitat may not have the same endophyte.   Species of
Pterostylis, Caladenia and *Thelymitra* occur in dry
sclerophyll forest and at times grow within a few
centimetres of one another, yet isolation studies

have shown that each has its own endophyte with only rare "cross-overs" and these as double infections.

Since Bernard's work there have been numerous suggestions that related orchids may have similar endophytes, i.e. that there is some degree of specificity based on taxonomic groupings. For instance Warcup (1971) showed that many species of *Caladenia* and closely related orchids throughout southern Australia have the same endophyte, *Sebacina vermifera*. On the other hand the same fungus may occur in unrelated orchids and several different Rhizoctonias may be isolated from the same orchid (Curtis, 1939; Downie, 1959; Harvais & Hadley, 1967a; Warcup, 1971). Thus present data suggest that different orchids differ in the range of their symbiotic capabilities. But so little is known about the field occurrence of endophytes in orchids that speculation on the extent and nature of specificity in relation to taxonomic groups is not at present profitable.

The idea of specificity has also been applied to symbiotic germination and in fact arguments against any degree of specificity have come from consideration of the results of symbiotic germination tests, especially with starch as carbon source, where a wide range of organisms other than orchid Rhizoctonias have been reported to induce orchid seed germination (Curtis, 1939). It seems quite possible that a number of organisms may be able to liberate sufficient sugars from starch (Knudson, 1925) to allow seed to germinate; however, this information is not really germane to whether there is specificity in the association that occurs in nature between orchids and endophytic fungi.

I have made a number of experiments on medium M1 with cellulose as carbon source with *Thelymitra aristata* or *Pterostylis nutans* and diverse fungi isolated from wheat-field soil (Warcup, 1957) or pine nurseries. Some 80 different fungi were used including a number of soil Basidiomycetes, many sterile mycelia and representatives of such genera as

Mucor, Mortierella, Penicillium, Aspergillus, Clad-osporium, Alternaria, Humicola, Chaetomium and *Stysanus.* No isolate, other than a few recognized as Rhizoctonias, stimulated germination of seed, and not all Rhizoctonias caused stimulation.

Since, apart from the data of Hadley (1970), there is little information on the ability of different species of endophytes to stimulate germination of seed of a range of orchids, experiments have been carried out with many of the fungi listed in Table 1 with 14 species belonging to 10 genera of orchids. The medium used was M2-yeast (half strength M1 minerals, no yeast extract, 1% cellulose and 1.5% Davis agar) which has given better results than M1 with some orchid-fungus combinations. The data are presented in Table 6. Germination data are given in three categories: 0 = no germination beyond that of uninoculated seed on the medium; + = greater germination than uninoculated seed but no shoot differentiated; s = formation of a shoot; thus the tests do not consider the question of relative efficiency (Warcup, 1973). Germination was assessed up to 8 weeks.

The data in Table 6 show that both orchids and fungi differ markedly in the range of partners with which they form effective symbioses; whether many of the partial symbioses (+) would become fully effective (s) on different media or under different environmental conditions is not known. Further, while the isolates of *Tulasnella calospora, T. asymmetrica* and *Ceratobasidium cornigerum* were chosen because they had been found to be effective with a number of orchids, little is known about the effectiveness of the strains of some of the other fungi.

The data show that in these orchids there are degrees of specificity with some taxonomic basis. Certainly, if one wishes to germinate seed of a *Cattleya,* a *Thelymitra* or a *Diuris,* a species of *Ceratobasidium* would seem an unlikely symbiotic partner; conversely with *Pterostylis,* species of *Tulas-*

Table 6. Symbiotic associations of seed of fourteen orchids and strains of seventeen species of endophytic fungi.

Fungus	Isolate No.	Diuris pedunculata R.Br.	Thelymitra aristata	T. grandiflora R.D. FitzG.	Dendrobium dicuphum F. Muell	D. Nobile Lindl.	Cattleya trianaei	Laelia lobata Veitch	Spiranthes sinensis	S. cernua (L.) Richard	Orchis morio L.	Dactylorhiza purpurella	D. incarnata (L.) Soó	Cymbidium findlaysonianum	Pterostylis nutans
Tulasnella calospora	062	S	S	+	0	S	+	S	S	S	S	S	+	S	0
" "	0584	S	S	+	S	S	+	S	S	S	S	S	S	S	0
" "	0689	S	S	+	0	S	S	S	S	S	S	S	S	S	0
T. asymmetrica	0497	0	S	+	0	+	+	+	0	0	0	−	−	+	0
" "	0591	0	S	S	0	S	+	S	0	0	+	S	S	S	0
T. cruciata	0471	0	S	+	+	S	+	S	0	0	+	+	+	S	0
T. violea	0353	0	S	+	0	+	+	S	0	0	0	S	0	S	0
T. sp.	0632	0	S	+	+	S	+	S	0	0	Sp*	S	−	S	0
T. allantospora	0579	0	S	0	0	0	0	0	0	0	0	0	0	0	0
T. sp.	257	0	S	−x	0	0	0	0	0	0	0	0	0	+	0
Thanatephorus cucumeris	T35	0	0	0	0	0	0	0	S	S	S	S	S	−	S
Th. sterigmaticus	0708	0	0	+	0	0	0	−	S	+	−	S	+	+	S
Th. sp.	0426	0	0	0	0	0	0	0	0	0	0	+	0	Ɔ	0
Ceratobasidium cornigerum	0167	0	0	0	0	0	0	+	S	S	S	S	S	S	S
" "	AD14	0	0	0	0	0	0	+	S	S	S	S	+	+	S
C. sphaerosporum	0657	0	0	0	0	0	0	0	S	S	S	S	S	0	+
C. sp.	0507	0	0	+	0	0	0	0	+	S	S	S	S	+	S
C. sp.	0615	0	0	0	0	0	0	0	+	−	S	−	S	+p	S
C. sp.	E13	0	0	0	0	0	0	0	+	+	+	+	+	−	+
C. obscurum	08	0	0	0	0	0	0	0	0	0	+p	+p	0	0	0
Oliveonia pauxilla	T330	0	−	0	0	0	0	0	0	0	−	0	0	+	0

* pathogenic to some/all seed.

x no test.

Warcup, J. H.

nella seem to have little effect.   Some other
orchids, however, are much less exacting in their
symbiotic partners.   On the other hand, whether symbiotic germination tests, alone, are an efficient way
of gathering information on specificity of the
orchid-fungus association in nature seems an open
question.   Study of germination of orchid seed in
the field would seem necessary.

ACKNOWLEDGMENTS

I wish to thank Dr. P. H. B. Talbot for advice
on taxonomy of Basidiomycetes; Professor J. Arditti,
University of California, Irvine, California;
Mr. R. Nash, Blackwood, South Australia and Mr. J.
W. Wrigley, Curator, Canberra Botanic Gardens, A.C.T.
for gifts of orchid seed.

REFERENCES

ARDITTI, J. (1967). Factors affecting the germination
   of orchid seeds.   *Bot. Rev.*, 33, 1-97.

BERNARD, N. (1909). L'evolution dans la symbiose.
   Annls. Sci. nat., 9, 1-96.

BURGEFF, H. (1936). *Samenkeimung der Orchideen*.
   Jena, G. Fischer.

BURGEFF, H. (1959). Mycorrhiza of orchids.   In *The
   Orchids*.   ed. C. L. Withner.   The Ronald Press,
   New York.

CURTIS, J. T. (1937). Non-specificity of orchid
   mycorrhizal fungi.   *Proc. Soc. exp. Biol. Med.*,
   36, 43-44.

CURTIS, J. T. (1939). The relation of specificity of

orchid mycorrhizal fungi to the problem of symbiosis. *Am. J. Bot.*, <u>26</u>, 390-399.

DERX, H. G. (1937). Tentative de synthèse d'une symbiose. *Annls. Sci. nat.*, <u>19</u>, 155-166.

DOWNIE, D. G. (1959). The mycorrhiza of *Orchis purpurella*. *Trans. Proc.bot. Soc. Edinb.*, <u>38</u>, 16-29.

HADLEY, G. (1969). Cellulose as a carbon source for orchid mycorrhiza. *New Phytol.*, <u>68</u>, 933-939.

HADLEY, G. (1970). Non-specificity of symbiotic infection in orchid mycorrhiza. *New Phytol.*, <u>69</u>, 1015-1023.

HARLEY, J. L. (1969). *The Biology of Mycorrhiza*. Leonard Hill, London.

HARRISON, C. R. & ARDITTI, J. (1970). Growing orchids from seeds. *Orchid Digest,* <u>34</u>, 199-204.

HARVAIS, G. & HADLEY, G. (1967a). The relation between host and endophyte in orchid mycorrhiza. *New Phytol.*, <u>66</u>, 205-215.

HARVAIS, G. & HADLEY, G. (1967b). The development of *Orchis purpurella* in asymbiotic and inoculated cultures. *New Phytol.*, <u>66</u>, 217-230.

KNUDSON, L. (1922). Nonsymbiotic germination of orchid seeds. *Bot. Gaz.*, <u>73</u>, 1-25.

KNUDSON, L. (1925). Physiological study of symbiotic germination of orchid seeds. *Bot. Gaz.*, <u>79</u>, 345-379.

ROGERS, D. P. (1944). The genera *Trechispora* and *Galzinia* (Thelephoraceae). *Mycologia,* <u>36</u>, 70-103.

SPRAU, F. (1937). Beiträge zur Mykorrhizenfrage.
Die Fructification eine aus *Orchis mascula* isolier-
ten Wurzelpiles *Corticium masculi* (nov.spec.).
Jber. wiss. Bot., 85, 151-168.

TALBOT, P. H. B. (1973). In *The Fungi. An Advanced
Treatise.* Vol. IVB. ed. Ainsworth, G. C.,
Sparrow, F. K. & Sussman, A. S. Academic Press,
New York & London.

WARCUP, J. H. (1957). Studies on the occurrence and
activity of fungi in a wheat-field soil.
Trans. Br. mycol. Soc., 40, 237-262.

WARCUP, J. H. (1971). Specificity of mycorrhizal
association in some Australian terrestrial orchids.
New Phytol., 70, 41-46.

WARCUP, J. H. (1973). Symbiotic germination of some
Australian terrestrial orchids. *New Phytol.,*
72, 387-392.

WARCUP, J. H. & TALBOT, P. H. B. (1966). Perfect
states of some Rhizoctonias. *Trans. Br. mycol.
Soc.,* 49, 427-435.

WHITNEY, H. S. & PARMETER, J. R. (1964). The perfect
stage of *Rhizoctonia hiemalis.* *Mycologia,* 54,
114-118.

SOME MYCOLOGICAL ASPECTS OF THE BIOLOGY OF MYCORRHIZA IN THE ERICACEAE

D. J. READ AND D. P. STRIBLEY

Department of Botany, The University, Sheffield, U.K.

INTRODUCTION

Members of the Ericaceae are normally infected by a mycorrhizal fungus. Two types of mycorrhizal association are found, known as the 'ericoid' and 'arbutoid' types. The ericoid type is the most important and is found in such genera as *Erica, Vaccinium, Rhododendron* and *Calluna* where infection typically consists of intracellular hyphal complexes in the outer cortical cells of the lateral 'hair' roots. The arbutoid mycorrhizas are of the ect-endo type and are found in the genera *Arctous, Arbutus,* and *Arctostaphylos.*

Routine methods have recently been developed for the isolation and culture of the endophyte of ericoid mycorrhizas (Pearson & Read, 1973). The view expressed repeatedly by Rayner (1913, 1922) that species of *Phoma* were the mycorrhizal fungi of Ericaceae and that the infection was systemic has been discounted (Harley, 1969; Pearson & Read 1973). One true mycorrhizal isolate has now been induced to fruit in pure culture and has been named *Pezizella ericae* sp. nov. (Read, 1974).

Since the long controversy surrounding the identity of the endophyte has been resolved interest now centres upon the biological importance of ericoid mycorrhizas.   Studies both of the extent of infection in individual plants and of the physiological role of mycorrhiza are required.

This paper presents an analysis of the distribution and anatomy of ericoid mycorrhizal infection in *Calluna vulgaris* and contrasts the pattern of mycorrhizal infection with that found in plants with vesicular-arbuscular infection.   The second paper (Stribley & Read, this volume) presents some results of investigations into the physiological relationship between host and endophyte.

MATERIALS AND METHODS

Turves containing *Calluna vulgaris* seedlings of ages ranging from approximately 2 months to one year were collected from heathland sites at Beeley Moor in Derbyshire (Nat. Grid Ref. SK292664 and SK293668). In the laboratory the seedlings were carefully teased out of the soil in order to retain the fine hair roots.   Camera lucida drawings were made of 6 typical seedlings in which the entire root system had been successfully extracted.   The major dimensions of the root systems and of the cortical cells were measured, before the entire root systems were excised and stained in cotton blue and lactophenol.   The staining enabled the distribution of mycorrhizal infection to be mapped.   Representative parts of the root systems were then mounted in methyl cellulose and serial transverse sections were cut in a cryostat.

By comparing the results of the mapping and sectioning techniques it was possible to categorise the various levels of intensity of mycorrhizal infection in each seedling.   The proportion of the root in which 98 to 100% of the cells were mycorrhizal was first estimated.   This portion was termed the fully mycorrhizal category (Category 1).   The proportions falling into 5 lower categories of infection was then estimated viz 75 to 98% infection (Category 2), 50 to

75% (Category 3), 25 to 50% (Category 4), 0 to 25% (Category 5) and no infection (Category 6).

The figures obtained from the observations were used in conjunction with the microscopical analysis to estimate the total volume of the root system occupied by fungal tissue and the total number of fungal entry points per unit root length and per plant.

RESULTS

Figures for total root lengths and for the proportions of each root system falling into the various categories are presented in Table 1.   The mean figure of the extent of mycorrhizal infection in the seedlings is 65%, and of this clearly the largest proportion falls into the categories representing above 75% intensity.   The categories with lower levels of infection represent only a small proportion (14%) of the total root length.   As revealed in Fig. 1, the main sites of heavy infection are the lateral 'hair' roots where fungal penetration begins about 0.5 mm behind the root cap and extends backward into older parts of the system.   In these 'hair' roots the great majority of the outer cortical cells become infected.   Figs. 2a and b show a typical hair root of Category 1 infection intensity in surface view and in transverse section.   Transverse sections reveal the great density of fungal material in the cortical cells and the high proportion of the total volume of root occupied by fungus in Category 1 infection zones.   Such sections (Fig. 2b) reveal that up to 80% of the total root volume is occupied by fungal material.   If the volume occupied by fungus in 100% infected root is 80% then by making proportionate reductions in the other categories and taking into account the proportion of the total root length which each category occupies it is possible

Read, D. J. & Stribley, D. P.

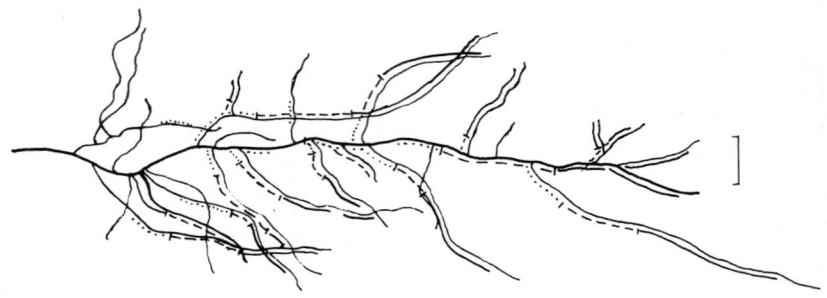

Fig. 1. Camera lucida drawing of the root system of seedling 1. Above each infected part of the root system the category of infection is indicated thus:- _____ Cat.1; ------ Cat. 2; ..... Cats. 3 to 5. For explanation of infection categories see text. Scale line is 1cm.

to calculate the volume of the total root system occupied by the endophyte. The results of such calculations (Table 2) indicate that in the six seedlings for which detailed measurements have been made 42.4% of the total root volume is occupied by fungal material.

Careful examination of individual infected cortical cells normally reveals at least one fungal entry point per cell, and a larger number is normal (Fig. 3a and b). As many as twelve entry points have been observed per cell. By measuring the length and diameter of roots and individual cortical cells in the various categories it is possible to estimate the number of entry points in any length of root.

Table 1. The total root lengths of six representative seedlings of *Calluna vulgaris*, and details of the extent of their mycorrhizal infection.

Seedling No.	Total root length cm	Proportion of root that is mycorrhizal %	% length of root in each infection category*					
			1	2	3	4	5	6
1	88.4	70	34	18	11	5	2	30
2	140.1	68	26	24	9	5	4	32
3	153.6	64	29	20	9	3	3	36
4	99.7	71	37	16	12	4	2	29
5	185.0	63	24	24	9	4	2	37
6	161.1	59	22	20	6	6	5	41
Mean	137.9	65	28	20	9	4	4	35

* For explanation of categories 1-6, see text.

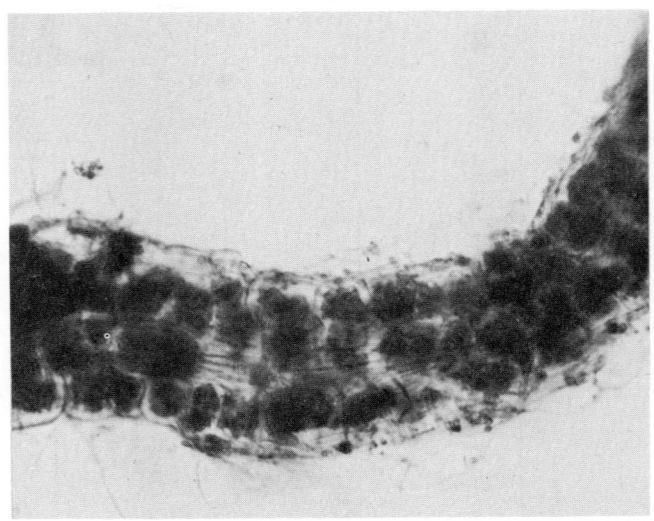

Fig. 2.   a) Surface view of Category one
root of *Calluna vulgaris* seedling

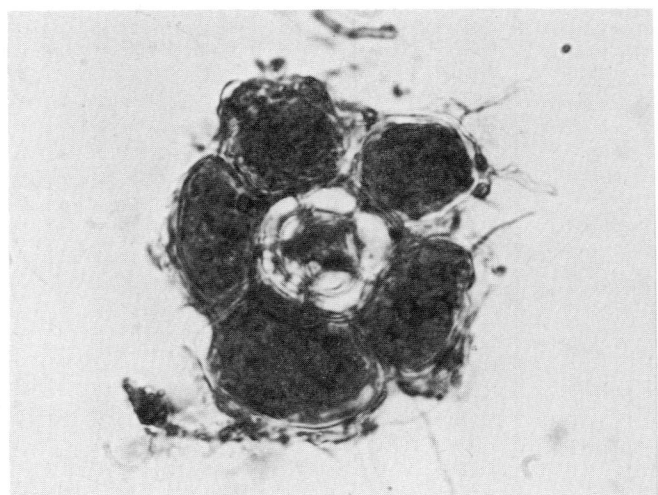

Fig. 2.   b) Transverse section of a
category one root showing the intensive
infection in all the cortical cells.

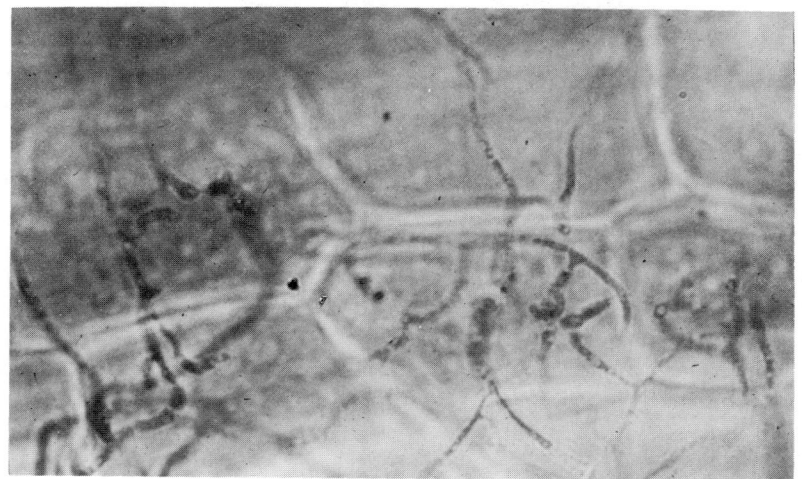

Fig. 3. a) Surface view of root showing numerous appressoria associated with entry points in cells of the Category one zone.

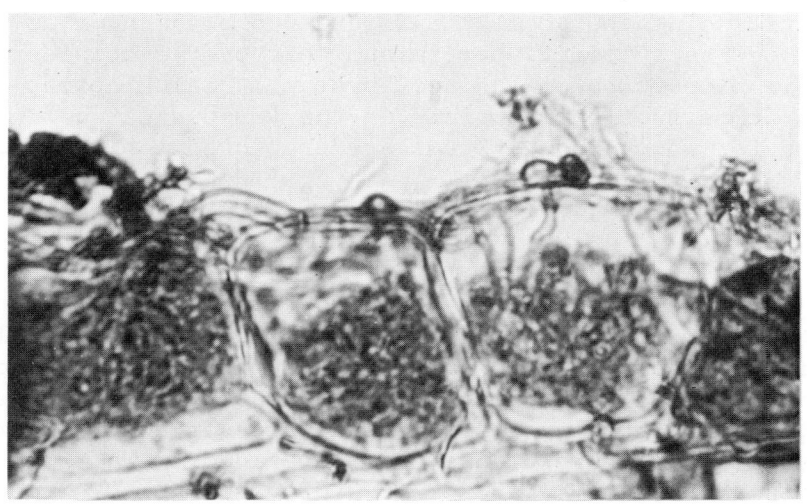

Fig. 3. b) Lateral view of cortical cells showing four entry points in one field of focus.

Read, D. J. & Stribley, D. P.

The diameter of typical hair roots of the seed-lings was in the range of 40-60 μm, and the circum-ference was in the range 124-188 μm. The size and shape of individual cells is variable. Cells of square surface dimensions of 20-30 x 20-30 μm are common behind the root tip. In this region of very heavy infection there may be from 5 to 10 cells around the periphery of the root. Further back along the root cortical cells often become rectang-ular with a surface dimension of 25-45 x 15-20 μm. Assuming that there are five cells around the circum-ference, each of 25 x 25 μm outer dimension, then in 1 cm of root there are $5 \times \dfrac{10,000}{25} = 2000$ cells.

If there is one entry point per cell, a figure of 2000 entry points per cm of root is obtained for Category 1 parts of the root. This figure is a very conservative one based on the assumption that there are only 5 cells around the circumference of the root and only one entry point per cell. Since there may be up to 10 cells around the circumference and 12 entry points per cell the figure can be as high as 48,000 entry points per cm and this over a consider-able proportion of the root system. Calculations of numbers of entry points in the various categories and for whole root systems are presented in Table 3.

DISCUSSION

It is important to consider the extent to which the mycorrhizal status of the seedlings is comparable with that of mature plants of *Calluna vulgaris* and of other genera with ericoid mycorrhizas. Unfortunate-ly it is virtually impossible to recover entire root systems of mature plants from the field. The fine roots of Ericaceae are particularly susceptible to breakage and any attempt to assess distribution of infection of mature plants will inevitably lead to gross errors. It is for this reason that field

Table 2. The calculated mean amounts of fungal tissue in each category of fungal infection and in the whole root system.

	Infection Category				
	1	2	3	4	5
Proportion of root falling into the category, mean of 6 seedlings, %.	28	20	9	4	4
Mean level of infection assumed for category, %.	100	87.5	62.5	37.5	12.5
Calculated volume occupied by fungus, %.	80	70	50	30	10
Calculated volume as proportion of whole root system, %.	22.4	14.0	4.5	1.2	0.3
Mean total volume occupied by fungus (6 seedlings), %.			42.4%		

Table 3. Calculated number of entry points in the various categories and in the entire root system.

	Infection Category				
	1	2	3	4	5
Number of entry points per cm	2000	1750	1250	750	250
Root length per category, cm	28	20	9	4	4
Number of entry points per category	56×10^3	35×10^3	11×10^3	3×10^3	1×10^3
Mean total number of entry points for seedlings assuming					
a) 1 entry point per cell			106×10^3		
b) 6 entry points			636×10^3		
c) 12 entry points			127×10^4		

seedlings were used in this study.   Observations
using older pot-grown plants from which most of the
roots can be recovered suggest that the proportion of
fine roots infected by endophyte is similar - about
60%.   More of the infection is in the categories of
lower infection intensity however.

The basic pattern revealed in *Calluna* is repeated in other ericaceous genera which have been examined.   In *Vaccinium* and *Rhododendron,* however, the
'hair' roots are often of more robust dimensions with
as many as 4 layers of cortical cells around the
stele and a circumference made up of 20 or more cells.

When ericoid infections are compared with that
found in other types of mycorrhiza, the major difference clearly lies in the amount of fungal tissue
involved.   In vesicular-arbuscular mycorrhizas
reports of numbers of entry points vary from 2 to 20
per mm in strawberry (Mosse, 1959) to 6 per cm in
onion (Sanders & Tinker, 1973).   Observations in
this laboratory indicate that the number in *Festuca
ovina* may be up to 15 per cm.   This compares with
well over 1,000 per cm over much of the *Calluna* root.

Within the root, despite the general lack of
inter-cellular ramification in ericaceous plants the
fungus forms a significant proportion of the total
tissue mass in most of the infected areas.   This
situation is in marked contrast with that in vesicular-arbuscular mycorrhizas in which, despite considerable internal ramifications, the ratio of fungal to
higher plant tissue is always small.   In transverse
sections of roots infected with *Endogone* it is often
difficult to detect fungal material.

In ecto-mycorrhizal species it has been estimated that 39% of the dry weight of the mycorrhizal
root may be attributable to fungal tissue   (Harley,
1969).   It is clearly not possible to make direct
measurements of the weight of endo-mycorrhizal fungi
but some comparison may be drawn from observations of
relative tissue proportions.   In heavily mycorrhizal
regions of *Calluna* roots it may be assumed that since

Read, D. J. & Stribley, D. P.

80% of the total volume of the root is occupied by fungal tissue a very high proportion of the weight of that tissue is attributable to the fungus. The proportion may be at least 80% since the fungal tissue is densely cytoplasmic relative to the neighboring vacuolate uninfected cells of the root.

It is difficult to envisage any living host tissue being unaffected physiologically by the penetration of such large masses of alien tissue. Indeed the invasion is so great that the survival and apparently healthy growth of the host is in many ways surprising. The major emphasis of research in this field should now be to investigate the basic physiology of the host-fungus relationship, with a view to elucidating the extent to which the association is truly mutualistic.

REFERENCES

HARLEY, J. L. (1969). *The biology of mycorrhiza.* Leonard Hill, London.

MOSSE, B. (1959). The regular germination of resting spores and some observations on the growth requirements of *Endogone* species causing vesicular-arbuscular mycorrhizas. *Trans. Br. mycol. Soc.,* 42, 274-286.

PEARSON, V. & READ, D. J. (1973). The biology of mycorrhiza in the Ericaceae. I. The isolation of the endophyte and synthesis of mycorrhiza in aseptic culture. *New Phytol.,* 72, 371-379.

RAYNER, M. C. (1913). The ecology of *Calluna vulgaris* L., *New Phytol.,* 12, 59-77.

RAYNER, M. C. (1922). Mycorrhiza in the Ericaceae. *Trans. Br. mycol. Soc.,* 8, 61-66.

READ, D. J. (1974). *Pezizella ericae* sp. nov. the
  perfect state of a typical mycorrhizal endophyte
  of Ericaceae.   *Trans. Br. mycol. Soc.,* 63,
  381-383.

SANDERS, F. E. & TINKER, P. B. (1973). Phosphate
  flow into mycorrhizal roots.   *Pestic. Sci.,*
  4, 385-395.

COMPARATIVE ASPECTS OF THE CARBON NUTRITION OF MYCORRHIZAS

D. H. LEWIS

*Department of Botany,
The University of Sheffield, Sheffield, U.K.*

INTRODUCTION

Since the carbohydrate physiology of mycorrhizas has been reviewed several times recently (Meyer, 1966; Harley, 1969, Harley & Lewis, 1969; Smith, Muscatine & Lewis, 1969; Lewis, 1970, 1973a, 1974; Hacskaylo, 1973; S. E. Smith, 1974), no attempt is made here to cover the same ground as these papers. Instead, the article attempts to recognize natural kinds of mycorrhizas about which generalizations can be made, considers similarities and differences between the carbon nutrition of those kinds which can be recognized and highlights some outstanding problems with particular reference to endomycorrhizas. Where appropriate, comparisons are made with pathogenic associations and free-living fungi.

NATURAL KINDS OF MYCORRHIZAS

In the introduction to the second edition of his 'Biology of Mycorrhiza', Harley (1969) noted that "errors arise ... from the tendency to generalize about mycorrhizal phenomena without first attempting to appreciate their diversity or to classify them

into natural kinds on the basis of similarity".   In
1973, I presented a case for abolishing the usual
framework of classifying mycorrhizas into ecto-,
endo- and ectendo-trophic kinds.   The substitution
of four main categories, which I termed sheathing,
vesicular-arbuscular, ericaceous and orchidaceous,
was suggested.   A fifth, miscellaneous, category was
also included for those associations of fungi and
absorbing organs about which insufficient is known
for generalizations to be made.   From the comments
on the scheme by S. E. Smith (1974) and correspondence and discussion with several mycorrhizologists,
it appears there are *at least* three serious faults
with this classification.   It does not recognize the
dichotomy of ericaceous mycorrhizas into ericoid and
arbutoid groups (Harley, 1969) which are worthy of
the same rank as the other main groups;   the linear
arrangement fails to indicate links between the de-
limited groups and there are valid nomenclatural
criticisms.   Firstly, my suggestion of 'sheathing'
as a substitute for 'ectotrophic' and its variants is
inadequate since a fungal mantle is also a characteristic of arbutoid, and even a few ericoid, mycorrhizas.   Alternatives include 'ectocellular' (Wilde &
Lafond, 1967) and 'ectomycorrhiza' (Peyronel *et al.*,
1969). Although both continue the false implication
that the opposites, 'endocellular' and 'endomycorrhiza', also apply to a natural group of mycorrhizas,
ectomycorrhiza, as adopted in the U.S.A. (Hacskaylo,
1971;   Marks & Koslowski, 1973), will be used here.
Secondly, Daft and Nicolson (1974) prefer 'arbuscular'
to 'vesicular-arbuscular' since vesicles are sometimes absent from *Endogone*-type mycorrhizas.   Fig. 1,
which includes nomenclatural variations used in the
past, shows the inter-relationships between the five
major groups of mycorrhizas.   In addition to the
linking features illustrated here, groups A and B are
linked via *Endogone lactiflua* which forms a sheath on
some conifers (Fassi, 1965;   Fassi *et al.*, 1969) and
E. eucalypti which forms a sheath on some *Eucalyptus*

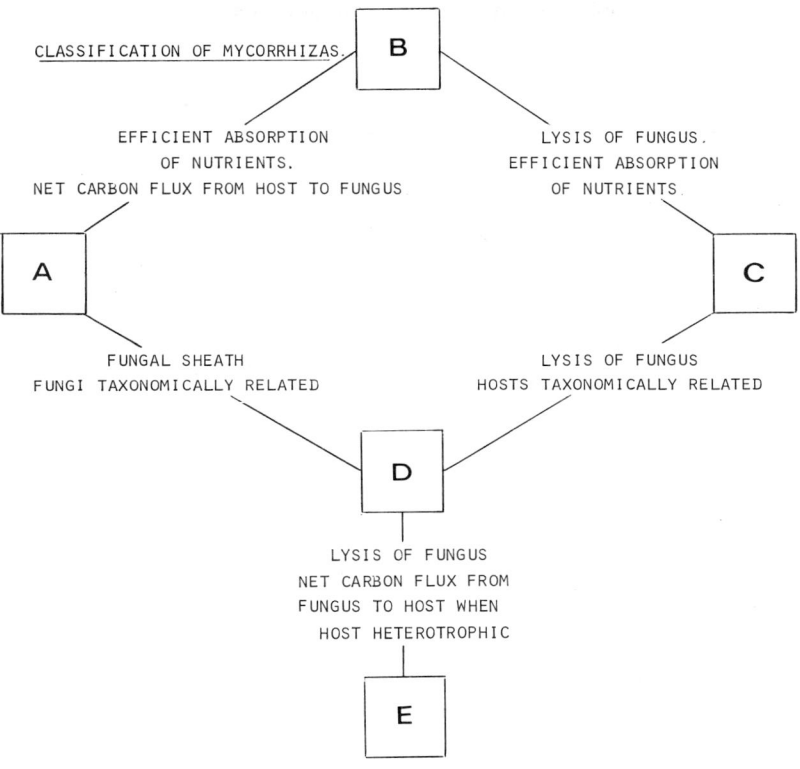

CLASSIFICATION OF MYCORRHIZAS.

B

EFFICIENT ABSORPTION
OF NUTRIENTS.
NET CARBON FLUX FROM HOST TO FUNGUS

LYSIS OF FUNGUS.
EFFICIENT ABSORPTION
OF NUTRIENTS.

A

C

FUNGAL SHEATH
FUNGI TAXONOMICALLY RELATED

LYSIS OF FUNGUS
HOSTS TAXONOMICALLY RELATED

D

LYSIS OF FUNGUS
NET CARBON FLUX FROM
FUNGUS TO HOST WHEN
HOST HETEROTROPHIC

E

NOMENCLATURE OF MYCORRHIZAS.

A: ECTO-; ECTOTROPHIC; ECTOCELLULAR; SHEATHING; HARTIGIAN.
B: ENDO-; ENDOTROPHIC; PHYCOMYCETOUS; VESICULAR-ARBUSCULAR;
   ARBUSCULAR.
C: ENDO-; ENDOTROPHIC; ERICACEOUS; ERICOID.
D: ECTENDO-; ECTENDOTROPHIC; ERICACEOUS; ARBUTOID.
E: ENDO-: ENDOTROPHIC, ORCHIDACEOUS.

Fig. 1. Classification and nomenclature
of mycorrhizas. The diagram of the classification indicates some traits which link
the various groups. The terms underlined
in the nomenclature are those which will be
used in this paper.

spp. (Warcup, 1975).   Groups A and E may also sometimes have common fungi since some orchids are
connected by rhizomorphs and hyphae to group A mycorrhizas (Campbell, 1963; Went, 1973).   Ericoid
mycorrhizas, e.g. those of *Vaccinium,* may also have
a loose sheath (Read, pers. comm.), further strength-
ening their alliance with the arbutoid type.

CARBON NUTRITION OF MYCORRHIZAL ASSOCIATIONS

*Physical sources of carbon for mycorrhizal
fungi and the direction of net flux of
carbohydrate in mycorrhizas*

Mycorrhizal fungi have three alternative but not
mutually exclusive sources of carbon; dead organic
matter in the soil, the roots they infect forming
mycorrhizas or a second autotrophic higher plant
which they infect in addition to their mycorrhizal
associate.   These alternatives will now be considered for each group of mycorrhizas in turn.
1.   <u>Group A</u>   Although a few fungi which form ectomycorrhizas can exist as saprophytes (see Meyer,
1966; Harley, 1969), most are ecologically obligate
biotrophs (Lewis, 1973a, 1974).   They depend on the
roots with which they form mycorrhizas as their
source of carbon.   Even for the species which
apparently can exist as saprophytes, it is possible
that the free-living strains are not mycorrhizal
(Lundeberg, 1970).   Direct evidence for supply of
carbon from host to fungus was given by Melin and
Nilsson (1957) and Lewis and Harley (1965c) proposed
that movement of soluble carbohydrates was one-way,
host to fungus.   Since digestion of hyphae is not a
marked feature in ectomycorrhizas, re-utilization by
the host of products of degradation of fungal storage
and structural polymers is also minimal.   Net flux

of carbohydrate (not to be confused with flux of carbon - see below) is therefore from host to fungus. Harley (1971, 1975) has commented on quantitative aspects of this movement in relation to the production of fruit bodies and the respiration rate in soils.

2.   Group B   No species of Endogonaceae which form arbuscular mycorrhizas has been axenically cultured. Like most of the ectomycorrhizal fungi, they are also ecologically obligate biotrophs although only a few studies have experimentally demonstrated the movement of photosynthetic products to these fungi *in vivo* (Ho & Trappe, 1973; Bevege *et al.*, 1975; Cox *et al.*, 1975).

The extent to which the lysis of hyphae releases utilizable carbohydrate to the host remains to be investigated.   The role of host and fungal (i.e. autolytic) enzymes in this lysis also has not been studied but Pegg and Vessey (1973) who demonstrated the presence of a chitinase in uninfected tomato plants alluded to its possible involvement in digestion of mycorrhizal fungi.   Despite the occurrence of lysis in arbuscular infections, the presence of abundant extra-matrical hyphae and spores indicates that net flux of carbohydrate is, like in ectomycorrhizas, also from host to fungus.   However, the saprotrophic potential of the extra-matrical hyphae in the soil requires investigation.

3.   Group C   The isolation and fruiting in culture of an ericoid mycorrhizal endophyte of *Vaccinium* and other Ericaceae, *Pezizella ericae,* and its use in experiments designed to elucidate nutritional interrelations in this ecologically most important symbiotic interaction by Read and co-workers are major achievements (Pearson & Read, 1973a, 1973b; Read, 1974; Read & Stribley, 1973, 1975; Stribley & Read, 1974a, 1974b, 1975).   The ecological status of *P. ericae* has yet to be fully investigated.   It appears to be widely distributed even in soils which do not support ericaceous genera (Pearson & Read,

1973a) but whether as active mycelium or spores is
not known.   Its status as an obligate or facultative
symbiont so remains unresolved.   Its relatively slow
growth rate in culture indicates a low competitive
saprophytic ability although it and similar isolates
can utilize at least some carbohydrate polymers,
e.g. pectin and carboxymethyl-cellulose, but not
native cellulose (Nieudorp, 1969; Pearson, 1971).
Stribley and Read (1974b) have also shown that it can
utilize the complex organic nitrogenous components of
a peaty soil.

The endophyte contains a mannose polymer in the
cell wall.   The proportion of this polymer in infected roots suggests that it may not be re-utilized by
the host following lysis of hyphae (Stribley & Read,
1974a).   As in arbuscular mycorrhizas, the mechanism
of lysis is unknown.   Host photosynthetic products
not only pass to the internal endophyte (Stribley &
Read, 1974a) but also to external hyphae (Pearson &
Read, 1973a).   However, in the latter paper, no
controls to discount leakage of metabolites from
roots to agar medium and subsequent uptake by hyphae
were included.   Stribley and Read conclude that the
direction of net flux of carbohydrate remains to be
elucidated.

Information is needed on the ability of the
fungus to synthesize carbohydrate from amino-
compounds derived from the complex nitrogenous compounds it can utilize.   Only if there is a considerable movement from fungus to host of carbon derived
from sources external to the root can my suggestion
(Lewis, 1973a) that ericoid hosts behave in a manner
comparable to angiospermous hemiparasites be upheld.
4.   Group D   It is unfortunate that there have
been so few experiments on the carbohydrate nutrition
of the arbutoid mycorrhizas since they occupy such an
important, cross-roads, position in the interrelations between the various mycorrhizal types
(Fig. 1).   The experiments of Bjorkman (1960) on
Monotropa hypopitys show not only that a single

fungus can form more than one kind of mycorrhiza but also that it can act as a bridge for the passage of nutrients to this non-photosynthetic plant. Campbell (1971) and Went (1973) indicate that tripartite arrangements may be a common feature of this kind of mycorrhiza.

Using light, and transmission and scanning electron microscopy, Lutz and Sjolund (1973) not only also confirmed the continuity of rhizomorphs and the fungal sheath in mycorrhizas of *Monotropa* but also demonstrated proliferations of cell walls of the host in regions of haustorial development. They likened these to the structure of transfer cells, commonly found in regions of active transport of metabolites between adjacent cells in terrestrial plants (Pate & Gunning, 1972). They emphasized the need for extension and more detailed analysis of experiments of the kind initiated by Björkman (1960). Among uninvestigated nutritional problems are, once again, the mechanisms of lysis and the extent of bilateral movement of nutrients between arbutoid 'host' and fungus. Only when the latter is answered can the association be classified as one of mutualism or of parasitism of the fungus by the 'host'. The nature of the physiological interactions with the auto-trophic arbutoid hosts, such as *Arbutus* and *Arctostaphylos,* is totally without experimental investigation. Whereas in heterotrophic plants with arbutoid mycorrhizas, net flux of carbohydrate must be from fungus to 'host', it remains to be seen whether or not *Arbutus* etc. are partially dependent on their fungi for carbon.

5. Group E At least to the plantlet stage in species that become photosynthetic and throughout the life of those that remain heterotrophic, there is a net movement of carbon from fungus to 'host'. I include 'host' in apostrophe here and in the previous paragraph since, from the point of view of movement of carbohydrate, it is the fungus which is the host (A. M. Smith, 1952; D. Smith *et al.*, 1969; Lewis,

1973a). The ultimate source of carbon for the permanent or transient heterotrophic phase of orchid development is either the soil or a second higher plant with which the orchid endophyte has a second symbiotic relationship. This, depending on species, may be either parasitic or mutualistic. As detailed by Hadley & Purves (1974) and Purves & Hadley (1975), there is a limited movement of photosynthetic products from orchids to fungi. (See also the unpublished experiment of S. E. Smith quoted by Harley, 1969, p.224). This is to be expected since, in balanced symbioses between orchid and endophyte (i.e. where stimulation of orchid growth occurs), there will be living hyphae of the fungus as well as digested pelotons in protocorms and roots. As these living hyphae will be in contact with host sugars, these will be absorbed and utilized. This biotrophic absorption is followed by lysis of hyphae and return of nutrients to the orchid, i.e. a transient biotrophic transfer from orchid to fungus is followed by necrotrophic transfer from fungus to orchid. Utilization of fungal trehalose by orchid tissue has been demonstrated by S. E. Smith (1967, 1972) and S. E. & F. A. Smith (1973). Although Burges (1939) showed that cell sap from infected tubers could induce lysis of cultured hyphae of *Rhizoctonia solani,* the role of orchid and fungal cytolytic enzymes has yet to be elucidated. It may be significant that the storage glucomannans of orchid tubers are often acetylated and that virtually all acetyl groups are linked to mannose residues (Buchala *et al.,* 1974). It may be that the enzymes which re-utilize these are of broad specificity and can degrade the acetylated glucosamine polymers (chitin) of fungal cell walls. If this is so, orchids may be regarded as pre-adapted to the lysis of fungal hyphae, a situation which may have influenced the evolution of the mycorrhizal habit in the Orchidaceae.

*Chemical nature of major carbon sources
for mycorrhizal fungi*

1.   Group A   The major carbon source of the ecologically obligate biotrophs of group A is probably
sucrose since this is the main translocated carbohydrate in higher plants.   According to Nelson
(1964) and Bevege, *et al.*, (1975), patterns of translocation are altered by such infections, a point
disputed by Ahrens and Reid (1973).   Nevertheless,
the utilization of sucrose following its hydrolysis
to hexoses is consistent with the experiments of
Harley and Jennings (1958) and Lewis and Harley
(1965a-c) with mycorrhizal roots of beech.   It is
therefore surprising that sucrose is sometimes a poor
source for the growth of some ectomycorrhizal fungi
in culture (Hacskaylo, 1973).   Increase in invertase, associated with an accumulation of hexoses at
infection sites, is a common feature of infections by
biotrophic fungi (Long *et al.*, 1975), but the contribution of fungal and host invertases is not known.
Since invertase levels can be altered by hormone
status (Kaufman *et al.*, 1968; Gaylor & Glasziou,
1969) and since mycorrhizal fungi produce hormones
(Moser, 1959;   Ulrich 1960;   Miller, 1967, 1971;
Laloue & Hall, 1973;   Slankis, 1973), it is possible
that fungal utilization of sucrose could be mediated
via stimulated synthesis and activity of a host
invertase.   In this context it is of interest that
Rhizobia in culture cannot utilize sucrose and
Robertson and Taylor (1973) suggest that, *in vivo*,
utilization is mediated via a host invertase.
2.   Group B   By analogy with ectomycorrhizas, it
might be assumed that the major endogenous carbon
source for arbuscular mycorrhizal fungi is also
sucrose.   However, in the absence of marked effects
on patterns of translocation (Bevege *et al.*, 1975)
and the current inability to check the *in vitro* utilization of sucrose by the fungus, this view may be
erroneous.   Nevertheless, as indirect evidence of

utilization of sucrose, the effects of infection on
levels of free hexoses and of invertase in roots may
act as pointers.   Other potential carbon sources
include *myo*-inositol and glycerol.   Mosse and
Phillips (1971) noted that, in dual cultures of *Trifolium parviflorum* and *Endogone mosseae,* the endophyte formed a profuse external mycelium when *myo*-
inositol was included in the external medium.   From
their studies of some ecologically obligate haustorial mycoparasites which, like the Endogonaceae, are
members of the Mucorales, Barnett and co-workers have
found that glycerol is the most effective carbon
source for *in vivo* growth (Barnett & Binder, 1973).
Although attempts to grow *Endogone in vitro* using
these carbon sources have not yet succeeded (Mosse,
pers. comm.), it may be that a balanced combination
of other necessary growth factors has not yet been
achieved.   Of special interest here could be sulphur
aminoacids which are required by obligate biotrophs
as diverse as haustorial mycoparasites and rust fungi
(Scott, 1972; Binder & Barnett, 1974).   The possible
utilization of glycerol and *myo*-inositol by mycorrhizal members of the Endogonaceae focusses attention
on their lipid metabolism.   Further clues to their
growth requirements may be indicated by a better
knowledge of the lipid composition of the oil droplets of hyphae and vesicles and of the abundant
membranes not only in the fungal plasmalemma around
the finely divided branches of the arbuscules themselves but also in the host plasmalemma which surrounds them (Cox & Sanders, 1974; Cox *et al.,* 1975,
Kaspari, 1975).   It may be significant that *myo*-
inositol is required for root organ cultures of
another mutualistic system rich in extra membranes,
the legume nodule (Raggio *et al.,* 1959).

As with infections of higher plants by coenocytic fungi in general, relatively little is known
about the carbohydrate composition of arbuscular
infections.   Both Hayman (1974) and Bevege *et al.*
(1975) comment on the absence of trehalose and

mannitol in such mycorrhizas.   The absence of polyols may be expected since they have not been recorded
from either free-living or parasitic coenocytic fungi
(Lewis & Smith, 1967;   D. Smith *et al.*, 1969).   The
failure to detect trehalose is perhaps more surprising.   Among coenocytic fungi, this disaccharide is a
major component of species from three groups of
slime-moulds (Clegg & Filosa, 1961;   Iwanoff, 1925;
Keen & Williams, 1968;   Williams *et al.*, 1968);
from the Peronosporales (Long & Cooke, 1974;   D. J.
Maclean, pers. comm.) and from several species of
the Mucorales to which the Endogonaceae belong
(Rudolph & Ochsen, 1969;   Assche *et al.*, 1972;
A. J. E. Lyon, pers. comm.).   The maltose, stated to
be present and metabolically active in *Zygorhynchus
moelleri* by Moses (1958), was probably a misidentif-
ication for trehalose.   Among factors which may
contribute to the failure to detect trehalose in
arbuscular mycorrhizas is the relatively low ratio of
fungal biomass to that of root, which would result in
a masking of fungal products by those of the root and
problems of distinguishing trehalose from inositols
on paper chromatograms.   Clear separations, however,
can be affected by gas chromatography (Holligan &
Drew, 1971).   This analytical technique, together
with other confirmatory tests such as acid and enzy-
mic hydrolysis (Lewis & Harley, 1965a, Stribley &
Read, 1974a) should be applied to extracts from
spores and extra-matrical hyphae.

3.   <u>Groups C and D</u>   Since the physical sources of
carbohydrates for mycorrhizal fungi of the Ericales
have not yet been certainly established, fully mean-
ingful discussion of their chemical sources will not
be possible until more is known of their ecology and
saprotrophic potential.   Nevertheless, some features
of their biotrophic nutrition have been established.

As far as the conifer-fungus-*Monotropa* association is concerned, it may be concluded from the
field experiments of Bjorkman (1960) that simple
sugars pass from the conifer to the fungus in a bio-

trophic manner as for other ectomycorrhizas.
Although C^{14}-glucose was injected into the trees, it
is probable that this was rapidly converted to
sucrose before being translocated to the roots.   In
the laboratory experiments of Stribley and Read
(1974a), mycorrhizal roots of *Vaccinium* contained a
significantly higher concentration of sucrose than
uninfected roots.   This effect is very similar to
that brought about by many biotrophic pathogens (see
Long *et al.*, 1975 for references) and suggests that
the endophyte induces altered patterns of translocation.   Following feeding of $^{14}CO_2$ to shoots of
uninfected and mycorrhizal plants, sucrose was most
heavily labelled in both.   Hexoses, which were
present in greater (but not significantly greater)
absolute amounts in mycorrhizal roots, were more
heavily labelled with ^{14}C in these than in uninfected
roots.   In the mycorrhizal roots, the fungal metabolites, mannitol, trehalose and insoluble mannans,
incorporated ^{14}C.   This pattern is very similar to
ectomycorrhizas (Lewis & Harley 1965 a-c; Bevege
et al., 1975) and to biotrophic pathogenic infection
of leaves (D. C. Smith *et al.*, 1969; Holligan *et
al.*, 1974).   The labelling pattern in sucrose and
hexoses in the roots again suggests a stimulated
activity of invertase.   As for biotrophic infections
in general, an accumulation of sucrose and stimulated
activity of invertase are not incompatible since the
sites of these apparently conflicting events are
metabolically compartmented at the infection sites
(Long *et al.*, 1975).
4.   Group E   The carbon nutrition of the mycorrhizal fungi of orchids contrasts markedly with that of
other mycorrhizal groups.   The fungi, commonly
virulently pathogenic to other organisms, can utilize
a wide range of simple and complex carbon sources
including, for some species, non-carbohydrates such
as lignin.   The fungal products, metabolized from
these varied substrates which may be living or dead,
are then passed to the orchid.   The proportion of

this transfer that is biotrophic as distinct from net
gain to orchid via fungal lysis remains to be established (Harley, 1969; Hadley & Williamson, 1971).
Of greater significance to the theme of this paper is
the nature of the reverse movement of carbohydrate
from orchid to fungus.   As noted above, living endophytic hyphae will absorb those sugars they encount-
er.   As detailed by Purves and Hadley (1975), the
experiments of Hadley and Lewis indicate that, as
expected, sucrose is translocated from shoot to root
in orchid plantlets.   Indirect evidence shows that
infection apparently results in the development of an
active invertase.   The possibility that some ectomycorrhizal fungi rely on a stimulated host invertase
for utilization of sucrose has been mentioned above.
In the case of orchids, the reverse may be true,
i.e. the orchid may rely on fungal enzymes to mobil-
ize its reserves.   For both stems and roots, several
studies have shown that growth rate and activity of
acid invertase are positively correlated (Hatch &
Glasziou, 1963; Glasziou & Bull, 1965; Lyne & ap
Rees, 1973.   Uninfected protocorms show a slow
growth rate and apparently have low activity of
invertase.   Growth rates can be stimulated either by
infection or by treatment with hormones (Hadley &
Harvais, 1968; Hadley, 1970; Hadley & Williamson,
1971).   Infection appears to result in an increase
in invertase activity and some combinations of hormones are, in other species as noted above, known to
stimulate the development of invertase.   Hadley has
likened the orchid protocorm with its massive storage
of starch to the seed of other plants.   Germination
of seeds results in a massive increase in the activity of a wide range of hydrolytic enzymes, the synthesis of which is often hormonally controlled
(Varner & Johri, 1968).   The effects of infection on
levels in protocorms of hormones and degradative
enzymes, invertase and amylase in particular, merit
further study.

Lewis, D. H.

Maintenance of Biotrophy

     With the exception of the Endogonaceae of arb-
uscular mycorrhizas which have never been cultured
and the fungi of arbutoid which have not been investigated physiologically, some ectomycorrhizal and
ericoid fungi and probably all orchidaceous mycorrhizal fungi are capable of *in vitro* utilization of
at least some of the complex polymers which constitute the cell walls of higher plants. Yet, *in vivo,*
they do not degrade the walls of their mycorrhizal
hosts and their necrotrophic potential is not
expressed. This must mean that, in mutualistic
unions, either cytolytic enzymes are not produced or
their activity is nullified. No evidence is yet
available that specific inhibitors of their activity
occur in mycorrhizas but such substances have been
extensively studied in pathogenic interactions (see
Wood, 1967 p.457-461). However, there is considerable circumstancial evidence for control of their
synthesis by catabolite repression. With reference
to ectomycorrhizal fungi, Melin (1948, 1953) and
Norkrans (1950) suggested that they produce cellulase
only when soluble sugars in the root are depleted.
This is a common mechanism for the control of synthesis of cytolytic enzymes of necrotrophic pathogens
(Horton & Keen, 1966a,b; Keen & Horton, 1966; Beihn
& Dimond, 1971; Patil & Dimond, 1968; Spalding
et al., 1973; Goodenough & Maw, 1974). I have
suggested (Lewis, 1973a, 1974) that this mechanism is
important in the maintenance of biotrophy in general.
Since so much work on arbuscular mycorrhizas is
conducted with onions, the work of Horton and Keen is
especially relevant.
     To investigate this suggestion, it would be of
interest not only to investigate the effects of
exogenous sugars on the *in vitro* production of cytolytic enzymes by the putative facultative ectomycorrhizal symbionts discussed above, but also to
investigate their production and effects *in vivo* in

relation to mycorrhizal development.   There have
been several studies of the effect of light intensity
on the development of ectomycorrhizal and arbuscular
infection and on the level of free sugars in the
roots (see Harley, 1969; Harley & Lewis, 1969;
Hacskaylo, 1973; Hayman, 1974).   Hacskaylo has
also made most interesting observations on the effect of photoperiod on ectomycorrhizal development.
However, there appear to have been no studies on the
effects *on pre-formed mycorrhizas* of treatments
likely to reduce levels of sugars in the roots,
e.g. shading or partial shoot excision.   If these
treatments do result in a lowered soluble carbohydrate status in the roots and if the mycorrhizal
fungi have the potential to produce cytolytic
enzymes, it follows that a biotrophic mutualistic
situation could be transformed into a necrotrophic
parasitic one as occurs in legume nodules under such
conditions (Thornton, 1930).   Such investigations
are ecologically relevant since the conversion of
effectively mutualistic mycorrhizal fungi of seedlings to pathogens following shading or subsequent
to grazing by herbivores could contribute to seedling
mortality.

How orchids control the production of cytolytic
enzymes by their endophytes is not known but clearly
must be an important feature in the synthesis of
symbioses, beneficial to the orchids.   Although
catabolite repression may be significant in established protocorms, roots and tubers, it is difficult
to see how this control mechanism can be operative
during the initial infection of seeds unless the
small number of cells of which they are composed
contain high concentrations of soluble sugars or
other metabolites effective in catabolite repression.

Two-way transport of carbon

By use of ^{14}C, bilateral transport of carbon
between fungi and roots has been demonstrated by

Reid & Woods (1969) and Reid (1971) for ectomycorrhizas of pine, by Stribley and Read (1974a) for
ericoid mycorrhizas and by Hadley and Purves (1974)
and Purves and Hadley (1975) for orchidaceous
mycorrhizas.   The study by Björkman (1960) of the
arbutoid *Monotropa* situation also showed that the
same fungus could simultaneously receive carbon from
one higher plant and donate it to another.   Reid's
experiments were also conducted with a tripartite
association but one in which donor and recipient
higher plants were the same species, *Pinus taeda*.
Lewis (1973b) speculated that arbuscular mycorrhizas
may form underground bridges capable of transport of
nutrients between dissimilar species, a factor which
could contribute to the phytosociological fidelity
of ecological associations of some higher plant
species.   Associations of this nature may also be
responsible for the existence of non-green monocotyledons and the non-green gametophytes of some bry-
ophytes and pteridophytes some of which have coenocytic and others septate mycorrhizal endophytes
(Harley, 1969).

Hacskaylo (1973) cited two-way transport of [14]C
as evidence against the one-way transport of carbohydrate in ectomycorrhizas which Lewis and Harley
(1965c) proposed to occur.   Lewis (1973b) pointed
out that carbon which moved from host to fungus as
carbohydrate could return as amino-compounds so that
bilateral transport of *carbon* should not be equated
with that of *carbohydrate*.   Some experimental evidence supporting this is now available.   Harley
(1964) and Carrodus (1966, 1967) showed that, in
beech mycorrhizas, absorbed ammonia is rapidly converted, principally in the sheath, to the amino-
compounds, glutamic and aspartic acids and, especially, glutamine.   In a series of experiments with
excised mycorrhizal roots of beech, Reid and Lewis
(unpublished) labelled the glutamine pool of the
fungus with ^{14}C using the method of Harley (1964) and
either simultaneously or sequentially supplied a

range of ^{12}C-amino-acids.   Then, in the manner of
the 'inhibition technique' of Drew and Smith (1967)
which has proved so valuable in identifying metabolites, mobile between the symbionts in lichens (Smith
et al., 1969), they analysed the media for ^{14}C-amino-
compounds.   Although only insignificant amounts were
released into water, considerable ^{14}C-glutamine was
released into media containing the amino-compounds
tested (alanine, glutamic and aspartic acids, and
glutamine).   This evidence is interpreted as
indicating that glutamine is a principal mobile nitrogen compound between fungus and host and that the
higher plant is effectively heterotrophic for nitrog-
en in the same manner as nodulated legumes.   Extension of this approach may prove useful in understanding interchange of amino-compounds between fungi and
hosts not only for ectomycorrhizas but for the range
of endomycorrhizal groups also.

Postscript - Björkman's Hypothesis

    In 1942, Björkman proposed that there was a
positive correlation between extent of ectomycorrhizal infection and the level of soluble sugars in
roots.   This hypothesis has stimulated a good deal
of investigation and an equal amount of controversy
(see reviews cited in the Introduction).   From the
above, three points are of crucial importance in any
re-evaluation.
    Firstly, initial infection will only be effect-
ive in roots where there is an adequate supply of
soluble sugars so that there will, therefore, be an
initial *causal* correlation between sugar levels in
roots and mutualistic association.   Secondly,
following the formation of these, levels of the host
sugars, sucrose and especially hexoses, will rise in
response to infection.   The principal factors
responsible are changes in patterns of translocation
and enhanced levels of invertase.   These changes are
therefore *effects* of infection.   Thirdly, when

analyses of infected material are being considered, it must be remembered that the soluble fungal components are non-reducing carbohydrates.

Proper assessment of these three factors may bring about a reconciliation between proponents of cause or effect in the correlations observed.

REFERENCES

AHRENS, J. R. & REID, C. P. P. (1973). Distribution of ^{14}C-labeled metabolites in mycorrhizal and non-mycorrhizal lodgepole pine seedlings. *Can. J. Bot.*, 51, 1029-1035.

ASSCHE, J. A. VAN, CARLIER, A. R. & BEKEERSMAEKER, H. I., (1972). Trehalase activity in dormant and activated spores of *Phycomyces blakesleeanus*. *Planta.*, 107, 327.

BARNETT, H. L. & BINDER, F. L. (1973). The fungal host-parasite relationship. *A. Rev. Phytopath.*, 11, 273-292.

BEVEGE, D. I., BOWEN, G. D. & SKINNER, M. F. (1975). This symposium, p. 149.

BIEHN, W. L. & DIMOND, A. E. (1971). Effect of galactose on polygalacturonase production and pathogenesis by *Fusarium oxysporum* f.sp. *lycopersici*. *Phytopathology*, 61, 242-243.

BINDER, F. L. & BARNETT, H. L. (1974). Amino acid requirements for axenic growth of *Tieghemiomyces parasiticus*. *Mycologia*, 66, 265-271.

BJÖRKMAN, E. (1942). Über die Bedingungen der Mykorrhizabildung bei Kiefer und Fichte. *Symb. botan. Upsal.*, 6, 1-190.

BJÖRKMAN, E. (1960). *Monotropa hypopitys* L. - an epiparasite on tree roots. *Physiol. Pl.*, 13, 308-327.

BUCHALA, A. J., FRANZ, G. & MEIER, H. (1974). A glucomannan from the tubers of *Orchis morio*. *Phytochemistry.*, 13, 163-166.

BURGES, A. (1939). The defensive mechanism in orchid mycorrhiza. *New Phytol.*, 38, 273-283.

CAMPBELL, E. O. (1963). *Gastrodia minor* Petrie - an epiparasite on Manuka. *Trans. R. Soc. N.Z. Bot.*, 2, 73-81.

CAMPBELL, E. O. (1971). Notes on the fungal assoc-iation of two *Monotropa* species in Michigan. *Mich. Bot.*, 10, 63-67.

CARRODUS, B. B. (1966). Absorption of nitrogen by mycorrhizal roots of Beech. I. Factors affecting the assimilation of nitrogen. *New Phytol.*, 65, 358-371.

CARRODUS, B. B. (1967). Absorption of nitrogen by mycorrhizal roots of Beech. II. Ammonium and nitrate as sources of nitrogen. *New Phytol.*, 66, 1-4.

CLEGG, J. S. & FILOSA, M. F. (1961). Trehalose in the cellular slime mould, *Dictyostelium mucoroides*. *Nature, London*, 192, 1077-1078.

COX, G. & SANDERS, F. E. T. (1974). Ultrastructure of the host-fungus interface in a vesicular-arbuscular mycorrhiza. *New Phytol.*, 73, 901-912.

COX, G., SANDERS, F. E. T., TINKER, P. B. H. & WILDE, J. (1975). This symposium, p. 297.

Heywood, V. H.)., pp. 151-172.   Academic Press. London.

LEWIS, D. H. (1974). Micro-organisms and plants: the

140　　　　　　　　　　Lewis, D. H.

invertase activity to sugar content and growth rate
in storage tissue of plants grown in controlled
environments.　*Pl. Physiol., Lancaster.*, 38,
344-348.

146　　　　　　　　　　Lewis, D. H.

lupini.　*Planta,* 112, 1-6.

RUDOLPH, H. & OCHSEN, B. (1969). Trehalose-Umsatz
wärmeaktivierter Sporen von *Phycomyces blakesleeanus.*　VI. Beitrag zur Kausalanalyse der
wärmeaktivierung von Piltzsporen.　*Arch. Mikrobiol.*, 65, 163-171.

SCOTT, K. J. (1972). Obligate parasitism by phytopathogenic fungi.　*Biol. Rev.*, 47, 537-572.

SLANKIS, V. (1973). Hormonal relationships in mycor-
rhizal development.　*Ectomycorrhizae.*, (Eds.
Marks, G. C. & Kozlowski, T. T.) pp. 232-298.
Academic Press, New York.

SMITH, A. M. (1952). The so-called saprophytic
orchids.　*Naturalist,* 843, 159-163.

SMITH, D., MUSCATINE, L. & LEWIS, D. (1969). Carbohydrate movement from autotrophs to heterotrophs
in parasitic and mutualistic symbiosis.　*Biol.
Rev.*, 44, 17-90.

SMITH, S. E. (1967). Carbohydrate translocation in
orchid mycorrhizas.　*New Phytol.*, 66, 371-378.

SMITH, S. E. (1972). Asymbiotic germination of orchid
seeds on carbohydrates of fungal origin.　*New
Phytol.*, 72, 497-499.

SMITH, S. E. (1974). Mycorrhizal fungi.　*Critical
Rev. Microbiol.*, 3, 275-313.

SMITH, S. E. & SMITH, F. A. (1973). Uptake of glucose, trehalose and mannitol by leaf slices of the
orchid *Bletilla hyacinthina.*　*New Phytol.*, 72,
957-964.

SPALDING, D. H., WELLS, J. M. & ALLISON, D. W. (1973).

KASPARI, H. (1975). This symposium, p.325.

KAUFMAN, P. B., GHOSHEH, N. S., LaCROIX, J. D., SONI, S. L. & IKUMA, H. (1973). Regulation of invertase levels in *Avena* stem segments by gibberellic acid, sucrose, glucose and fructose. *Pl. Physiol., Lancaster.*, 52, 221-228.

KEEN, N. T. & HORTON, J. C. (1966). Induction and repression of endopolygalacturonase synthesis by *Pyrenochaeta terrestris. Can. J. Microbiol.*, 12, 443-453.

KEEN, N. T. & WILLIAMS, P. H. (1969). Translocation of sugars into infected cabbage tissues during clubroot development. *Pl. Physiol. Lancaster.*, 44, 748-754.

LALOUE, M. & HALL, R. H. (1973). Cytokinins in *Rhizopogon roseolus.* Secretion of N[9-(β-D Ribofuranosyl-9H)purin-6-yl carbamoyl]threonine in the culture medium. *Pl. Physiol. Lancaster.* 51, 559-562.

LEWIS, D. H. (1970). Physiological aspects of symbiosis between green plants and fungi. *Lichenologist.*, 4, 326-336.

LEWIS, D. H. (1973a). Concepts in fungal nutrition and the origin of biotrophy. *Biol. Rev.*, 48, 261-278.

LEWIS, D. H. (1973b). The relevance of symbiosis to taxonomy and ecology, with particular reference to mutualistic symbioses and the exploitation of marginal habitats. *Taxonomy and Ecology.* (Ed. Heywood, V. H.)., pp. 151-172. Academic Press. London.

LEWIS, D. H. (1974). Micro-organisms and plants: the

142   Lewis, D. H.

evolution of parasitism and mutualism.   *Symp. Soc. gen. Microbiol.*, 24, 367-392.

LEWIS, D. H. & HARLEY, J. L. (1965a). Carbohydrate physiology of mycorrhizal roots of beech. I. The identity of endogenous sugars and utilization of exogenous sugars. *New Phytol.*, 64, 224-237.

LEWIS, D. H. & HARLEY, J. L. (1965b). Carbohydrate physiology of mycorrhizal roots of beech II. Utilization of exogenous sugars by uninfected and mycorrhizal roots. *New Phytol.*, 64, 238-255.

LEWIS, D. H. & HARLEY, J. L. (1965c). Carbohydrate physiology of mycorrhizal roots of beech. III. Movement of sugars between host and fungus. *New Phytol.*, 64, 256-269.

LEWIS, D. H. & SMITH, D. C. (1967). Sugar alcohols (polyols) in fungi and green plants. I. Distribution, physiology, and metabolism. *New Phytol.*, 66, 143-184.

LONG, D. E. & COOKE, R. C. (1974). Carbohydrate composition and metabolism of *Senecio squalidus* L leaves infected with *Albugo tragopogonis* (Pers.) S. F. Gray. *New Phytol.*, 73, 889-899.

LONG, D. E., FUNG, A. K., McGEE, E. E. M., COOKE, R.C. & LEWIS, D. H. (1975). The activity of invertase and its relevance to the accumulation of storage polysaccharides in leaves infected by biotrophic fungi. *New Phytol.*, 74, 173-182.

LUNDEBERG, G. (1970). Utilization of various nitrogen sources, in particular bound soil nitrogen, by mycorrhizal fungi. *Studia Forestalia Suecica.*, 79, 1-75.

LUTZ, R. W. & SJOLUND, R. D. (1973). *Monotropa uni-*

flora: Ultrastructural details of its mycorrhizal habit. *Am. J. Bot.,* 60, 339-345.

LYNE, R. L. & ap REES, T. (1971) Invertase and sugar content during differentiation of roots of *Pisum sativum. Phytochemistry.,* 10, 2593-2599.

MARKS, G. C. & KOZLOWSKI, T. T. (Eds.) (1973). *Ectomycorrhizae.* Academic Press, New York.

MELIN, E. (1948). Recent advances in the study of tree mycorrhiza. *Trans. Br. mycol. Soc.,* 30, 92-99.

MELIN, E. (1953). Physiology of mycorrhizal relations in plants. *A. Rev. Pl. Physiol.* 4, 325-346.

MELIN, E. & NILSSON, H. (1957). Transport of ^{14}C-labelled photosynthate to the fungal associate of pine mycorrhiza. *Svensk bot. Tidskr.,* 51, 166-186.

MEYER, F. H. (1966). Mycorrhiza and other plant symbioses. *Symbiosis. Vol. 1* (Ed. Henry, S. M.) pp 171-255. Academic Press New York.

MILLER, C. O. (1967). Zeatin and zeatin riboside from a mycorrhizal fungus. *Science N.Y.,* 157, 1055-1057.

MILLER, C. O. (1971). Cytokinin production by mycorrhizal fungi. *Mycorrhizae, Proc. 1st N. Am. Conf. Mycorrhizae.* pp. 168-174. Washington D.C.

MOSER, M. (1959). Beiträge zur kenntnis der Wuchsstoffbeziehungen im Bereich ectotrophen Mykorrhizen. *Arch. Mikrobiol.,* 34, 251-269.

MOSES, V. (1958). [14]C] Glucose metabolism in fungal cells. *J. gen. Microbiol.,* 20, 184-196.

MOSSE, B. & PHILLIPS, J. M. (1971). The influence of phosphate and other nutrients on the development of vesicular-arbuscular mycorrhiza in culture. *J. gen. Microbiol.*, 69, 157-166.

NELSON, C. D. (1964). The production and translocation of photosynthate-C[14] in conifers. *The formation of wood in forest trees.* (Ed. Zimmermann, M. W.). pp. 243-257. Academic Press, New York.

NIEUWDORP, P. J. (1969). Some investigations on the mycorrhiza of *Calluna, Erica,* and *Vaccinium.* *Acta bot. neerl.*, 18, 180-196.

NORKRANS, B. (1950). Studies in growth and cellulolytic enzymes of *Tricholoma,* with special reference to mycorrhiza formation. *Symb. bot. upsal.*, 11, 1-126.

PATE, J. S. & GUNNING, B. E. S. (1972). Transfer cells. *A. Rev. Pl. Physiol.*, 23, 172-196.

PATIL, S. S. & DIMOND, A. E. (1968). Repression of polygalacturonase synthesis in *Fusarium oxysporum* f.sp. *lycopersici* by sugars and its effect on symptom reduction in infected tomato plants. *Phytopathology.*, 58, 676-682.

PEARSON, V. (1971). *The Biology of Mycorrhiza in the Ericaceae.* Ph.D. thesis. University of Sheffield.

PEARSON, V. & READ, D. J. (1973a). The biology of mycorrhiza in the Ericaceae. I. The isolation of the endophyte and synthesis of mycorrhizas in aseptic culture. *New Phytol.*, 72, 371-379.

PEARSON, V. & READ, D. J. (1973b). The biology of mycorrhiza in the Ericaceae, II. The transport of carbon and phosphorus by the endophyte and the mycorrhiza. *New Phytol.*, 72, 1325-1331.

PEGG, G. F. & VESSEY, J. C. (1973). Chitinase activity in *Lycopersicon esculentum* and its relationship to the *in vivo* lysis of *Verticillium albo-atrum* mycelium. *Physiol. Pl. Path.,* 3, 207-222.

PEYRONEL, B., FASSI, A., FONTANA, A. & TRAPPE, J. M. (1969). Terminology of mycorrhizae. *Mycologia.,* 61, 410-411.

PURVES, S. & HADLEY, G. (1975). This symposium, p.175.

RAGGIO, N., RAGGIO, M., & BURRIS, R. H. (1959). Enhancement by inositol of the nodulation of isolated bean roots. *Science.,* 129, 211.

READ, D. J. (1974). *Pezizella ericae* sp. nov., the perfect state of a typical mycorrhizal endophyte of Ericaceae. *Trans. Br. mycol. Soc.,* 63, 381-383.

READ, D. J. & STRIBLEY, D. P. (1973). Effect of mycorrhizal infection on nitrogen and phosphorus nutrition of ericaceous plants. *Nature, London,* 244, 81-82.

READ, D. J. & STRIBLEY, D. P. (1975). This symposium, p.105.

REID, C. P. P. (1971). Transport of ^{14}C-labelled substances in mycelial strands of *Thelephora terrestris*. *Mycorrhizae. Proc. 1st N. Am. Conf. Mycorrhizae,* pp. 222.227. Washington D.C.

REID, C. P. P. & WOODS, F. W. (1969). Translocation of ^{14}C-labelled compounds in mycorrhizae and its implications in interplant nutrient cycling. *Ecology,* 50, 179-187.

ROBERTSON, J. G. & TAYLOR, M. P. (1973). Acid and alkaline invertases in roots and nodules of *Lupinus augustifolius* infected with *Rhizobium lupini.*   *Planta,* <u>112</u>, 1-6.

RUDOLPH, H. & OCHSEN, B. (1969). Trehalose-Umsatz wärmeaktivierter Sporen von *Phycomyces blakesleeanus.* VI. Beitrag zur Kausalanalyse der wärmeaktivierung von Piltzsporen.   *Arch. Mikrobiol.,* <u>65</u>, 163-171.

SCOTT, K. J. (1972). Obligate parasitism by phytopathogenic fungi.   *Biol. Rev.,* <u>47</u>, 537-572.

SLANKIS, V. (1973). Hormonal relationships in mycorrhizal development.   *Ectomycorrhizae.,* (Eds. Marks, G. C. & Kozlowski, T. T.) pp. 232-298. Academic Press, New York.

SMITH, A. M. (1952). The so-called saprophytic orchids.   *Naturalist,* <u>843</u>, 159-163.

SMITH, D., MUSCATINE, L. & LEWIS, D. (1969). Carbohydrate movement from autotrophs to heterotrophs in parasitic and mutualistic symbiosis.   *Biol. Rev.,* <u>44</u>, 17-90.

SMITH, S. E. (1967). Carbohydrate translocation in orchid mycorrhizas.   *New Phytol.,* <u>66</u>, 371-378.

SMITH, S. E. (1972). Asymbiotic germination of orchid seeds on carbohydrates of fungal origin.   *New Phytol.,* <u>72</u>, 497-499.

SMITH, S. E. (1974). Mycorrhizal fungi.   *Critical Rev. Microbiol.,* <u>3</u>, 275-313.

SMITH, S. E. & SMITH, F. A. (1973). Uptake of glucose, trehalose and mannitol by leaf slices of the orchid *Bletilla hyacinthina.*   *New Phytol.,* <u>72</u>, 957-964.

SPALDING, D. H., WELLS, J. M. & ALLISON, D. W. (1973).

Catabolite repression of polygalacturonase, pectin lyase and cellulase synthesis in *Penicillium expansum*. *Phytopathology.*, 63, 840-843.

STRIBLEY, D. & READ, D. J. (1974a). The biology of mycorrhiza in the Ericaceae. III. Movement of carbon-14 from host to fungus. *New Phytol.*, 73, 731-741.

STRIBLEY, D. & READ, D. J. (1974b). The biology of mycorrhiza in the *Ericaceae*. IV. The effect of mycorrhizal infection on uptake of ^{15}N from labelled soil by *Vaccinium macrocarpon* Ait. *New Phytol.*, 73, 1149-1155.

STRIBLEY, D. & READ, D. J. (1975). This symposium, p. 195.

THORNTON, H. C. (1930). The influence of the host plant in inducing parasitism in lucerne and clover nodules. *Proc. R. Soc. B.*, 106, 110-122.

ULRICH, J. M. (1960). Auxin production by mycorrhizal fungi. *Physiol. Pl.*, 13, 429-443.

VARNER, J. E. & JOHRI, M. M. (1968). Hormonal control of enzyme synthesis. *Biochemistry and Physiology of Plant Growth Substances*. (Eds. Wightman, F. & Setterfield, G.) pp. 793-814. Runge Press Ltd. Ottawa, Canada.

WARCUP, J. (1975). This symposium, p. 53.

WENT, F. W. (1973). Rhizomorphs in soil not connected with fungal fruiting bodies. *Am. J. Bot.*, 60, 103-110.

WILDE, S. A. & LAFOND, A. (1967). Symbiotrophy of lignophytes and fungi: Its terminology and conceptual deficiencies. *Bot. Rev.*, 33, 99-104.

WILLIAMS, P. H., KEEN, N. T., STRANDBERG, J. O., &
  McNABOLA, S. S. (1968). Metabolite synthesis and
  degradation during clubroot development in cabbage
  hypocotyls.   *Phytopathology.*, <u>58</u>, 921-928.

WOOD, R. K. S. (1967). *Physiological plant pathology.*   Blackwell, Oxford.

COMPARATIVE CARBOHYDRATE PHYSIOLOGY OF ECTO- AND ENDOMYCORRHIZAS

D. I. BEVEGE[1], G. D. BOWEN AND M. F. SKINNER[2]

Division of Soils, C.S.I.R.O., Glen Osmond, South Australia.

INTRODUCTION

No detailed studies on the carbohydrate physiology of vesicular-arbuscular mycorrhizas have yet been made, although carbohydrate levels in mycorrhizal and non-mycorrhizal plants under a range of environmental and nutritional conditions have been reported (Bevege, 1971; Hayman, 1974) and Ho and Trappe (1973) recorded the transfer of ^{14}C compounds from the higher plant to mycelium of *Endogone*. Although rather more studies have been made on ectomycorrhizas, the most detailed are those of Lewis and Harley (1965) with excised *Fagus sylvestris* mycorrhizas fed with ^{14}C-labelled sugars. Melin and Nilsson (1957) demonstrated the transfer of ^{14}C assimilate from *Pinus sylvestris* to the associated mycorrhizal fungus under axenic conditions. Nelson (1964), Shiroya *et al.*, (1962), Lister *et al.*, (1968) and Schweers

---

[1] Present address: Department of Forestry, INDOOROOPILLY, Queensland.

[2] Department of Botany, University of Melbourne, PARKVILLE, Victoria.

and Meyer (1970) presented evidence that ectomycorrhizas of *P. resinosa, P. strobus* and *P. sylvestris*
could act as sinks for photosynthates in plants
exposed to $^{14}CO_2$ but the detailed chemical fate of
the carbohydrate in these systems was not examined.
In *Fagus* mycorrhizas, sucrose transferred to the
fungus is stored principally as trehalose and mannitol, thus maintaining a concentration gradient for
further movement of sucrose (Lewis & Harley, 1965).

In view of the fact that polyols have not yet
been demonstrated as occurring in Mucorales, it was
appropriate that we examine the fate of photosyn-
thesized $^{14}CO_2$ in endomycorrhizal plants, specifically enquiring into the chemical partitioning of
assimilate translocated to the fungus, and whether
vesicular-arbuscular mycorrhizas act as significant
assimilate sinks.   It was also appropriate to
extend the detail of the Lewis and Harley studies to
Pinus, using intact plants assimilating $^{14}CO_2$.
This paper therefore reports the fate of ^{14}C-labelled
photosynthate in endomycorrhizal hoop pine (*Araucaria
cunninghamii* Ait.) and subterranean clover (*Trifolium
subterraneum* L. var. Bacchus March), comparing their
carbohydrate physiology with that of ectomycorrhizal
radiata pine (*Pinus radiata* D. Don).   Information is
also presented on the pattern of assimilation in
Endogone.

METHODS

Plant Growth

Pilot studies with hoop pine were conducted on
18 months old pot plants raised in organic amended
nursery soil infected with a mixture of *Endogone
araucareae* (Bevege 1971), *E. mosseae* and *E. macrocarpa.*   Otherwise, 40 months old seedlings were
raised from surface-sterilised seed in steam-sterilised peat/sand mix with complete nutrient solution

and inoculated with spores of *E. araucareae*;   control
plants received spore washings to ensure a common
microflora.   The plants were grown in 100 g of medium in 20 x 3 cm cotton-wool plugged glass tubes.

Radiata pine raised from surface-sterilised seed
were grown for six months in gamma-irradiated (2.5
Mrad) Mt. Burr sand (Stephens *et al.*, 1941) inoculated with *Rhizopogon luteolus*.   The plants were
grown in porcelain enamelled pots capacity 4 kg, with
four seedlings/pot, and fertilised with 30 kg ha^{-1}
superphosphate.

Clover plants from surface-sterilised seed were
grown for 52 days in gamma-irradiated Wanneroo sand
with complete nutrient solution (10 ml/plant of half
strength complete nutrient solution, Hoagland &
Arnon, 1938).   Inoculation with *E. mosseae* was made
on day 8 and again on day 22;   control plants received spore washings only.   Containers were 2.5 cm
diameter polythene tubes of 250 g medium capacity
with one plant/tube.

Hoop and radiata pines were raised under glasshouse conditions and the clover in a controlled
environment, 18 hr day (20 400 lumen m^{-2})/6 hr night
at 20°C/15°C respectively.   Labelling of clover was
conducted under the same conditions and that of hoop
and radiata pines was under 16 hr day (20 400 lumen
m^{-2})/8 hr night at a constant 20° and 22°C ± 1°C
respectively.

$^{14}CO_2$ *Labelling*

Shoots were isolated from roots by a mineral
oil/wax barrier (melting point 42°C) poured on to
the surface of the soil and sealing around the stem.
When labelling intermittently over several days, a
breather tube to ensure adequate root aeration was
inserted in the soil prior to preparing the wax
barrier;   this tube was temporarily sealed during
the labelling period each day.   At labelling,
shoots were enclosed in a polythene bag sealed

Bevege, D. I. *et al.*

around the container and the appropriate volume of $^{14}CO_2$ injected into the enclosed atmosphere with a gas syringe and the hole sealed with cellulose tape. Labelled $^{14}CO_2$ was generated from $Na_2^{14}CO_3$ with lactic acid (McDougall & Rovira, 1965) and unless otherwise stated 20 $\mu Ci^{14}CO_2$ per plant was introduced to give a final CO_2 concentration of 1% in a volume of 1 500 ml approximately. Plants were usually exposed to $^{14}CO_2$ for 2-4.5 hr as a single pulse label followed by harvest 18-20 hr later which included a dark period of 6-8 hr. In experiments with labelling over 6 days (see text) daily 2 hr pulses with 20 $\mu Ci^{14}CO_2$ were applied.

Analysis of ^{14}C-Labelled Photosynthate

Spatial distribution of ^{14}C activity in roots was examined by macroautoradiography (Kodirex X-ray film). Roots were carefully washed and *Endogone* hyphae was either completely harvested (clover) or sampled (hoop pine). Selected root material was lightly blotted dry, weighed, plunged into boiling 80% ethanol and extracted four times at 80°C over 20 hr. The ethanol soluble extract was cleared and deionised following the procedures of Somogyi (1945) and Lewis and Harley (1965). Extracts were concentrated to less than 10 ml under reduced pressure at 50°C and brought to 50 ml with distilled water; 2.5 ml carbonate-free $Ba(OH)_2$ was added followed by 2.5 ml 5% $ZnSO_4$ 3 mins later. The resulting precipitate (the metal-precipitated fraction) was filtered off and the now-clear and decolourized filtrate made up to standard volume. Half the filtrate was deionised by adding a 2:1 mixture of 1R-4B(OH): 1R-120(H) Amberlite exchange resin (the neutral soluble fraction). The ethanol insoluble material was hydrolysed with 1.5N H_2SO_4 under reflux for 2 hr, the hydrolysate was then neutralised with $BaCO_3$ and the filtrate deionised as before. Residue was digested with 5N LiOH for 4 hr.

Some further biochemical characterisation was attempted of compounds in the various fractions prepared.   The ethanol soluble fraction is recorded as "neutral", "ionised" and "metal-precipitated". The first two fractions are considered to consist mostly of carbohydrates, and of organic and amino acids respectively.   The composition of the "metal-precipitated" fraction is more complex and while we consider it to comprise lipids, lipid-protein and some soluble protein (Davies et al., 1964, and helpful discussion with Professor K. J. Scott), further definitive study is obviously needed.   Kjeldahl analyses indicated the crude ethanol extracts contained 20-25 % of total root nitrogen, or 0.18% of total root weight and we assume the source of this ethanol soluble nitrogen is predominantly protein and phosphatides as well as amino acids.   The neutral sulphuric acid hydrolysate is recorded as polysaccharide, while the residue is considered to contain cell wall material and protein.

All fractions from the above analysis were retained and ^{14}C activity measured by liquid scintillation counting.   Precipitates were counted following suspension in thixotropic gel.   Scintillation fluid was based on POPOP in toluene and 2 methoxyethanol.

Sugars in the neutral ethanol fraction and in the neutral hydrolysate were separated and putatively identified by one-way paper chromatography.   Double runs were made with each of two solvent systems: ethylacetate-acetic acid-water, and n propanol-ethylacetate-water (Lewis & Harley, 1965).   Chromatograms were scanned with a radiochromatogram strip scanner (Nuclear Chicago Actigraph III) and relative proportions of various sugars calculated from the respective areas under the trace.   Putative identities of sugars were further checked by co-chromatography with known compounds, using two-way thin-layer chromatography on activated boric acid buffered silica gel plates using n butanol-acetic acid-

diethyl ether-water and isopropanol-ethyl acetate-
water (Bevege, 1971).   Radioactive spots on the
plates were located by autoradiography on X-ray film.

SPATIAL DISTRIBUTION OF ^{14}C IN
HOOP PINE ROOTS

The root systems of hoop pine had three types of
rootlets   (a) unsuberized actively growing uninfected
laterals,   (b) terminal unsuberized infected short
roots and   (c) suberized infected short roots, either
terminal or sub-terminal to root type (b).   Auto-
radiograms (Figure 1) indicated that ^{14}C was concentrated in the unsuberized roots;   young mycorrhizas

---

Figure 1.   Spatial Distribution of ^{14}C in hoop pine
roots.
A.   Plants 18 months old, mixed inoculum from soil.
Two hour exposure to 20 μCi^{14}CO$_2$;   harvested 18 hr
later.
Note the order of concentration of label in root
types is (a) greater than or equal to (b) greater
than (c).
B1. Plants 40 months old, inoculated *Endogone araucareae*.   Two hour exposure to 120 μCi^{14}CO$_2$, harvested 24 hr later;   6 day exposure to X-ray film.
B2. Plants 40 months old, inoculated *Endogone araucareae*.   Two hour exposure daily for 6 days, harvested 24 hr later;   2 day exposure to X-ray film.
Note concentration of label in type (b) roots is
greater than (c) and labelling of extra-matrical
hyphae and spores of the endophyte.
Key to lettering:
a.   unsuberized actively growing uninfected laterals
b.   terminal unsuberized infected short roots
c.   suberized infected short roots
h.   extra-matrical hyphae
s.   zygospores

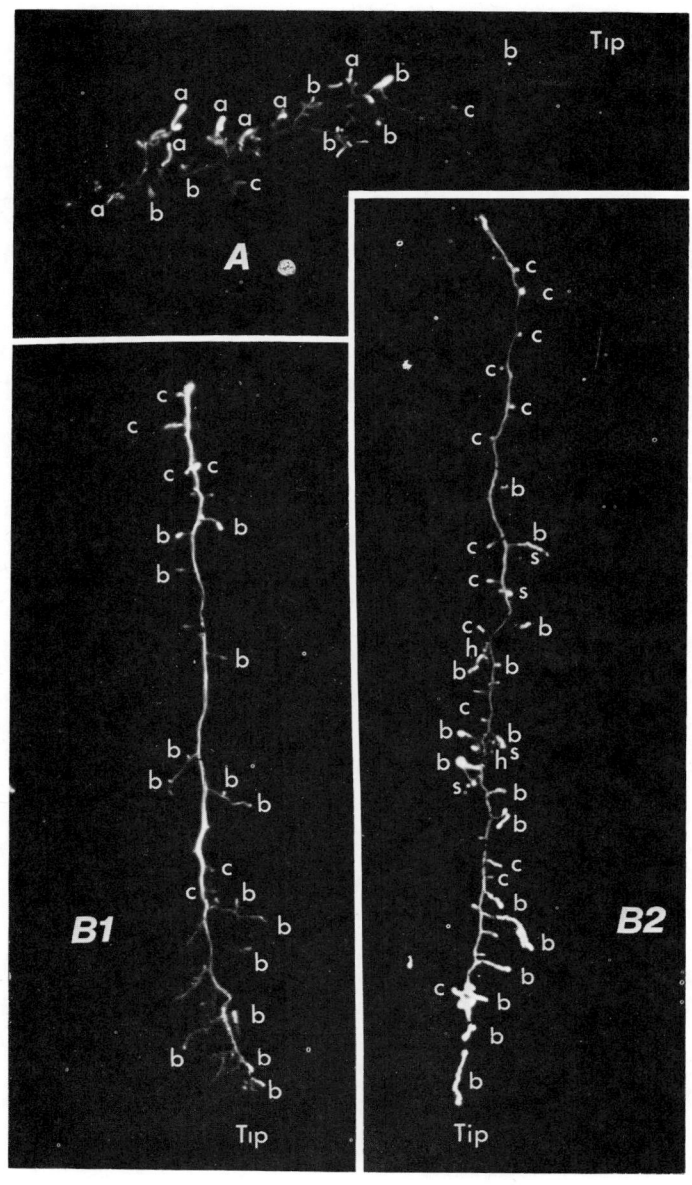

(type (b)) do not appear to be more active in this
regard than actively growing uninfected roots (type
(a)) and older suberized mycorrhizas (type (c)) are
less active than young mycorrhizas.   Type (b) roots
are in the active development stage with arbuscules
predominating over vesicles (see plate in Bevege,
1968) which in turn tend to predominate in the suberized roots; this pattern of preferential movement of
photosynthate to actively metabolising roots is consistent with observations made on wheat roots by
Rovira and Bowen (1973).

THE CARBOHYDRATE COMPONENTS OF MYCORRHIZAS

Hoop Pine

    Using co-chromatography, Bevege (1971) located
putative mannose, arabinose, glucose, galactose,
fructose, sucrose and raffinose in the neutral ethanol extract of hoop pine mycorrhizas.   As these were
also present in foliage, he concluded there was no
sugar in roots which might be distinctive for the
fungal endophyte.   Neither mannitol nor trehalose
was found.   Hayman (1974) also failed to detect
these fungal carbohydrates in endomycorrhizas of
onion.   Sulphuric acid hydrolysates of the ethanol
insoluble fraction of the hoop pine roots studied by
Bevege (1971) contained glucose, xylose and galactose
while perchloric acid hydrolysis yielded only glucose.
He concluded that the main polysaccharides were glu-
cosan and xylan with some galactan present, the
latter two probably arising from cell walls (Davies
et al., 1964).   Histochemical techniques in both
light and electron microscopy demonstrated starch in
the higher plant and glycogen in *Endogone* hyphae.
    In the present study, the neutral ethanol sol-
uble extract of mycorrhizas and uninfected roots contained similar levels of the same sugars in approximately the same proportions (Table 1).

By far the greatest activity was found in sucrose
followed by glucose;   other sugars were relatively
minor components.   Arabinose was not detected but
was one of the sugar moieties found in the acid
hydrolysate along with glucose, xylose and galactose.
No polyols, inositols or trehalose were detected.
Insufficient material was available to permit determination of carbohydrates in *Endogone* hyphae associated with roots.

Table 1.   Percent Composition of Root
Sugars of Hoop Pine[1]

| Sugar | Mycorrhizas | | Uninfected Roots | |
	1 day[2] label	6 day[3] label	1 day[2] label	6 day[3] label
Sucrose[4]	66.5	69.0	64.5	61.8
Glucose	16.0	7.2	10.5	16.8
Galactose	9.5	10.2	17.5	6.8
Mannose	5.5	5.8	7.5	7.0
Raffinose	2.5	4.5	0	4.2
Xylose	0	2.5	0	1.8
Fructose	0	0.7	0	1.5
Activity Counts/min/mg	449	4078	431	4398

1. inoculated with *Endogone araucareae*;   2 replica-
   tions/treatment, sample size 11–27 mg roots for
   analysis.

2. 24 hr after a 2 hr pulse of 120 $\mu ci^{14}CO_2$/plant.

3. labelled daily for 6 days with 20 $\mu Ci^{14}CO_2$ plant;
   harvested 24 hr after final pulse.

4. putative identifications based on co-chromatography with known carbohydrates.

Radiata Pine

In parallel investigations to those described above, the sugar composition of radiata pine mycorrhizas (endophyte *R. luteolus*) and uninfected roots on the same plants was examined. The host sugars of uninfected roots were sucrose, glucose and fructose, with sucrose predominating (Table 2). This parallels the observations made with other species of *Pinus* (Shiroya *et al.*, 1962, Schweers & Meyer 1970). In contrast to the uninfected roots, the fungal carbohydrates mannitol and trehalose dominated the mycorrhizas, making up 68% of the ^{14}C assimilate; there were also traces of xylose. This observation parallels the experience of Lewis and Harley (1965) who fed labelled sucrose to excised *Fagus* mycorrhizas with an unidentified endophyte. The polysaccharides contained the moieties glucose, arabinose, xylose and ribose; Lewis and Harley (1965) similarly detected these sugar components in the hydrolysate of *Fagus* mycorrhizas. In the present study we could detect no differences between the polysaccharides of mycorrhizas and uninfected roots. However, if one accepts that the primary polysaccharide of the fungus is glycogen (as suggested by Lewis & Harley) then an increase in the proportion of glucose is all that may be detected following hydrolysis.

The contrast between the carbohydrates of ecto- and endomycorrhizas is marked. The former contained four host-derived sugars plus two fungal carbohydrates, the latter eight sugars in all, none of which could be regarded as exclusively a fungal metabolite. While it is likely that sucrose is the main carbohydrate transferred to the fungus and passive movement of sucrose along a concentration gradient occurs in both systems, it seems this is achieved in quite different ways. The ectomycorrhizas keep sucrose concentration low by converting it to trehalose and mannitol, thus becoming a highly effective sink, but other systems operate with vesicular-arbuscular mycorrhizas.

Table 2.   Percent Composition of Root
Carbohydrates of Radiata Pine[1]

| Carbohydrate | Mycorrhizas | | Uninfected Roots[2] | |
	1 day[3] label	6 day[4] label	1 day[3] label	6 day[4] label
Sucrose[5]	6.5	7.2	56.0	ND[6]
Glucose	18.5	21.5	34.5	ND
Fructose	5.8	3.9	9.5	ND
Xylose	1.5	0	0	ND
Trehalose	52.2	45.7	0	ND
Mannitol	15.5	21.7	0	ND
Activity Counts/min/mg	79754	15492	3543	1408

1. inoculated with *Rhizopogon luteolus*; 4 replica-
   tions/treatment, sample size 6 mycorrhizas or
   rootlets for analysis.

2. roots from same plants as mycorrhizas.

3. 24 hr after a 2 hr pulse of 120 μCi $^{14}CO_2$/plant.

4. labelled daily for 6 days with 20 μCi $^{14}CO_2$/plant;
   harvested 24 hr after final pulse.

5. putative identifications based on co-chromatography with known carbohydrates.

6. not determined; samples lost during chromatography.

^{14}C INCORPORATION INTO MYCORRHIZAS
AND UNINFECTED ROOTS

The incorporation of ^{14}C into roots of three
specific systems was investigated (a) radiata pine
infected with *R. luteolus*, comprising mycorrhizas and
uninfected roots on the same plants, (b) hoop pine
infected with *E. araucareae* compared to uninfected

Bevege, D. I. *et al.*

plants of equivalent development and (c) clover infected with *E. mosseae* compared with uninfected plants.

Radiata Pine

Plants were pulse labelled with 20 $\mu Ci^{14}CO_2$ per plant daily for 6 days or with 120 $\mu Ci^{14}CO_2$ per plant on the sixth day and harvested on the seventh day. As Rangnekar and Forward (1972) have pointed out, the pattern of distribution of the ^{14}C-label becomes established within 3 days in *P. resinosa,* so that the difference between 24 hr and 7 day distribution patterns enables a comparison of the rate of metabolic incorporation between mycorrhizal and uninfected plants.

After 24 hr, mycorrhizas had 15 times the activity of uninfected roots, demonstrating considerable capacity to attract photosynthate; 74% of this was in the carbohydrate fraction compared to 51% in uninfected roots (Table 3). The difference represents fungal carbohydrate, as 68% of mycorrhiza carbohydrate was in trehalose and mannitol i.e. 50% of the total label. Only slight proportional differences were found between the levels of other components of the system although uninfected roots tended to have more assimilate converted to structural material, polysaccharide and acid at this time. This suggests translocation to roots in an ectomycorrhizal plant will be greater than in a non-mycorrhizal plant of the same size. This point has yet to be demonstrated, and it is unfortunate that the data of Nelson (1964), while undoubtedly indicating increased translocation to roots of mycorrhizal plants, is based on plants of different sizes and root/shoot ratios.

After 7 days, mycorrhizas still had 8 times the activity of other roots but the difference in the proportions in the respective carbohydrate pools had narrowed; fungal carbohydrate still comprised 67% of total soluble in mycorrhizas with mannitol

Table 3.   Fractionation of Assimilate of Radiata Pine[1]
            (counts/min/mg fresh wt.)

Fraction	Mycorrhizas		Uninfected Roots[2]	
	1 day[3] label	6 day[4] label	1 day[3] label	6 day[4] label
EtoH Soluble				
Metal-Precipitated	3904(3.6)	3618(11.6)	540(7.7)	562(15.6)
Ionised	757(0.7)	410(1.3)	632(9.0)	587(16.2)
Neutral	79754(74.3)	15492(49.6)	3544(50.6)	1408(39.0)
H₂SO₄ hydrolysable				
Neutral	17935(16.7)	8100(25.9)	1447(20.7)	644(17.8)
Ionised	472(0.4)	1029(3.3)	0(0)	0(0)
Residue	4450(4.1)	2592(8.3)	841(12.0)	408(11.3)
Total	107271(100)	31240(100)	7004(100)	3610(100)
Fresh Wt. (mg) of 6 rootlets	1.91		1.04	

1. inoculated with *Rhizopogon luteolus*; 4 replications/treatment, sample
    size 6 mycorrhizas or rootlets for analysis.   Percentage of total
    ^{14}C in brackets.

2. roots from same plants as mycorrhizas.

3. 24 hr after a 2 hr pulse of 120 μCi $^{14}CO_2$/plant.

4. labelled daily for 6 days with 20 μCi $^{14}CO_2$/plant; harvested 24 hr after
    final pulse.

increasing relative to trehalose.   Uninfected roots
maintained a high proportion of their assimilate in
non-carbohydrate pools but mycorrhizas retained relatively more in their polysaccharide pool.   The
reduction in activity over seven days was probably
due to respiration loss; Ursino and Paul (1973)
measured a 40% respiration loss of ^{14}C-assimilate
within 48 hr, in *P. strobus*.   In the present study
losses from mycorrhizas were 70% (derived principally
from the carbohydrate and glycogen fractions) and 50%
from uninfected roots, indicating a higher metabolic
rate in the former than the latter; Schweers and
Meyer (1970) also found respiration increased in
mycorrhizas of *P. sylvestris* compared to uninfected
roots.

Hoop Pine

Plants were pulse labelled as for radiata pine.
After 24 hr, activity in mycorrhizas was no greater
than that of uninfected roots and both categories had
over 80% of the ^{14}C activity in the soluble carbohydrate fraction, i.e. about 53% of the total label
in sucrose.   After 7 days, overall activity had
increased 11-fold in both kinds of roots indicating
that respiratory loss of assimilate was small over
the period (Table 4) especially when compared with
ectomycorrhizas of radiata pine.   This result reflects the relatively slow metabolic rate of hoop pine
compared to radiata pine.   Increased activity after
7 days in hoop pine was measured in all pools, but
was mainly in the soluble carbohydrate fraction
(74-77%) with some proportional increase in the polysaccharide and cell wall/protein pools; there was
little change in the metal-precipitated fraction and
a slight drop in the proportion of acids.
The striking feature is the close similarity
between the relative sizes of the various pools in
mycorrhizas and uninfected roots, from which we
conclude that there is little physiological differ-

Table 4. Fractionation of Assimilate of Hoop Pine and Endogone[1] (Counts/min/mg fresh wt.)

	Mycorrhizas[2]		Uninfected Roots[2]		Endogone Hyphae[4]	
	1 day label	6 day[3] label	1 day label	6 day[3] label	1 day label	6 day[3] label
EtoH soluble						
Metal-Pre-cipitated	31 (6.1)	429 (7.0)	25 (4.9)	334 (5.5)	969 (48.7)	2799 (39.8)
Ionised	28 (5.4)	80 (1.5)	33 (6.4)	140 (2.4)	634 (31.2)	862 (12.2)
Neutral	449 (82.2)	4078 (74.1)	431 (83.8)	4398 (77.7)	290 (14.6)	530 (7.5)
H_2SO_4 hydrolysable						
Neutral	13 (2.6)	627 (10.0)	14 (2.7)	441 (7.5)	0 (0)	835 (11.9)
Ionised	5 (0.9)	20 (0.2)	2 (0.4)	19 (0.3)	0 (0)	35 (0.5)
Residue	14 (2.7)	448 (7.1)	9 (1.8)	394 (6.6)	107 (5.4)	1980 (28.1)
Total	540 (100)	5683 (100)	513 (100)	5725 (100)	2000 (100)	7041 (100)
Fresh Wt. (mg) of sample	27.2	18.5	16.9	10.8	NOT DETERMINED	

1. inoculated with *Endogone araucareae*; 2 replications/treatment. Percentage composition in brackets.
2. 24 hr after a 2 hr pulse of 120 µCi $^{14}CO_2$/plant.
3. labelled daily for 6 days with 20 µCi $^{14}CO_2$/plant; harvested 24 hr after final pulse.
4. hyphal counts are for the total sample.

Bevege, D. I. *et al.*

entiation between the two kinds of root. Further-
more, as uninfected and mycorrhizal plants were the
same size, there is no evidence for mycorrhizas
influencing the rate or pattern of translocation.
This conclusion is supported by the data of Table 1
which shows that both kinds of roots have an identical suite of sugars, there being slight differences
only in the relative proportions of glucose and
sucrose. The very striking differences in labelling
patterns between the plant roots and the extramatrical hyphae shown in Table 4 are discussed below.

Clover

Plants were inoculated at a specific point on
the root system and at the time of harvest systemic
infection was restricted to this inoculated zone.
In the discussion below, the mycorrhizal roots are
those from this zone while uninfected roots comprise
those from the corresponding zone of uninoculated
plants. Total fresh weight and ^{14}C activity were
determined for roots from the inoculation zone (the
subtending roots), the remaining root system, shoots
and *Endogone* hyphae external to the infected subtending roots (extramatrical hyphae). There was extensive proliferation of roots and hyphae over the
experimental period. The weight of subtending roots
was 14% of the total root weight and that of extra-
matrical hyphae 6% of subtending root weight.

Plants were exposed to $^{14}CO_2$ for 4.5 hr followed
by 18.5 hr assimilation time. Activity was assessed
in the ethanol soluble and acid hydrolysable fractions only, as the hoop pine studies indicated the
label in the residue fraction was quite small (less
than 3%) after 24 hr. Only 6% and 12% respectively
of measured activity was located in the soluble
carbohydrate pools of mycorrhizas and uninfected
roots, whereas the incorporation into the metal-
precipitated and polysaccharide fractions together
accounted for 90% and 76% respectively.

Table 5.   Fractionation of Assimilate of Clover[1]

Fraction	Mycorrhizas			Uninfected Roots			Endogone Hyphae		
	count/min/mg	total[2] count	per cent	count/min/mg	total count	per cent	count/min/mg	total count	per cent
EtoH soluble									
Metal-precipitated	743	440	(71.9)	984	116	(52.4)	2101	84	(56.2)
Ionised	7	3	(0.5)	35	4	(2.0)	220	9	(5.9)
Neutral	64	35	(5.7)	237	27	(12.5)	296	12	(7.9)
H_2SO_4 hydrolysable									
Neutral	204	121[3]	(19.8)	423	43	(19.6)	704	28	(18.8)
Ionised	23	13	(2.1)	173	30	(13.5)	419	17	(11.2)
Total	1042	612	(100)	1853	220	(100)	3741	150	(100)

1. inoculated with *Endogone mosseae*; labelled 18 hr prior to harvest with 20 µCi $^{14}CO_2$/plant.   5 replications/treatment. Percentage composition in brackets.

2. Total activity in fractions measured (counts/min x10^{-3}) for the total weight of roots or hyphae in the inoculated zone and the corresponding zone in controls.

3. differences mycorrhizas vs uninfected significant at 10 percent level.

This pattern differs markedly from that of hoop pine
where the carbohydrate component made up 80% of activity.   Differences are probably due to variation in
host physiology;   clover is a more vigorously growing
species than hoop pine and would therefore be expected to incorporate assimilate more rapidly.   The
polysaccharide of mycorrhizas and uninfected roots
did not exhibit very great differences on a proportional basis although in terms of total activity roots
from mycorrhizal plants had nearly 3 times the ^{14}C
level of roots from uninfected plants;   this however
was a consequence of the greater weight of roots as
mycorrhizal plants were much larger than uninfected
ones (Table 6).

Endogone

   Endogone araucareae hyphae were also fractionated after 24 hr and 7 days (Table 4).   While the
total counts for each sample cannot be compared
directly because hyphal weight was not recorded, the
proportions of ^{14}C-assimilate in the various pools
are of interest.   After 24 hr the metal precipitated
and ionised pools contained 80% of the assimilate
between them and the soluble carbohydrate pool only
15%;   there was no transfer at this stage into glycogen.   After 7 days, these fractions still accounted for 52% of the assimilate, the decrease being
accounted for to some extent by a relative increase
in the cell wall plus protein fraction.   Glycogen
increased and the proportion of soluble carbohydrates
decreased.   Comparing the soluble carbohydrates of
mycorrhizas and hyphae, it is evident that the high
level of sugars in the mycorrhizas was not matched
by a corresponding level in the external hyphae.
After 7 days, host polysaccharide (predominately
starch) and that of hyphae (i.e. glycogen) both
comprised about 10% of their respective labels.   The
big difference in the relative proportions of carbohydrate and non-carbohydrate between host and

Table 6.   Effect of Mycorrhizas on Growth and Assimilation of $^{14}CO_2$ by Clover[1].

Plant Part	Mycorrhizas	Uninfected Roots	Significance[2]
(a) *Fresh Wt.* (g)			
Tops	0.21 ± 0.05	0.14 ± 0.02	NS
Roots	4.47 ± 0.69	1.77 ± 0.20	*
Total Plant	4.68 ± 0.71	1.91 ± 0.20	*
Hyphae	0.04 ± 0.01	——	
(b) *Total Activity*[3] (counts/min X10^{-3})			
Tops	9756 ± 996	7110 ± 756	
Roots	3294 ± 468	2010 ± 198	(*)
Total Plant	13050 ± 1320	9120 ± 960	(*)
Hyphae	150 ± 42	——	

1. Plants 52 days old inoculated with *Endogone mosseae* on days 8 and 22;   labelled 18 hr prior to harvest with 20 μCi $^{14}CO_2$/plant. 5 replications/treatment.   Means ± standard error.

2. NS not significant;   (*) difference significant at 10 percent level;   * difference significant at 5 percent level.

3. Activity in ethanol-soluble and H_2SO_4-hydrolysable fractions;   no counts were made on residues.   See text.

external hyphae is evidence for the relatively small
amount of hyphae present within the root.

Hyphae of *E. mosseae,* after an assimilation
period of 18.5 hr, incorporated assimilate into
metal-precipitated and polysaccharide pools (56% and
19% respectively).   The 8% in the soluble carbohydrate pool corresponded to the 6% in the equivalent
pool in the mycorrhizas (Table 5) and there was also
a close correspondence between the relative sizes of
the polysaccharide pools (18% and 20% respectively).
The clover plant and fungus therefore incorporated
assimilate in the same manner.

Although both species of *Endogone* diverted a
high proportion of assimilate in 24 hr to metal-
precipitable material, *E. araucareae* had a relatively
high label in the ionisable fraction but not in polysaccharides, but the reverse occurred in *E. mosseae.*
Whether these differences between species are condi-
tioned by host physiology or are due to the age and
state of the mycelium is unknown; as both fungi
form mycorrhizas with clover (Bevege & Bowen, 1975),
a study of their relative carbohydrate physiology
when growing in conjunction with a common host is
worthy of further study.

CONCLUSION

The concept of ectomycorrhizas acting as a
physiological sink for assimilate is well established
(Harley & Lewis, 1969).   Lewis and Harley (1965)
elucidated the mechanism cf the transfer from host to
fungus in *Fagus sylvatica* and our studies confirm
this for the *Pinus radiata* - *Rhizopogon luteolus*
ectomycorrhiza (see Table 2) namely the one-way
transfer of carbohydrate from host to endophyte where
it is converted to specific fungal carbohydrates,
trehalose and mannitol, and ultimately the storage
polysaccharide glycogen, none of which can be util-
ised by the host plant.   In this way a concentration

gradient of sugar, predominantly sucrose, is maintained from host to endophyte.   Smith *et al.*, (1969)
pointed out the generality of this in many host-
parasite relations of plants and Jankiewicz *et al.*,
(1969) discussed similar phenomena involving insect
parasites.

The storage of carbohydrates in vesicular-
arbuscular mycorrhizas appears to be small: trehalose and mannitol or other polyols were not found.
Some storage occurred as glycogen.   The high proportion in the metal-precipitated fraction, which
probably contains lipid, lipo-protein and soluble
protein, needs further study, and is consistent with
the suggestion of Cox *et al.*, (1975) that lipid may
serve an important storage function in these mycorrhizas.   In *E. araucareae,* the large label in the
ionised fraction containing organic and amino acids
(at 1 day) and the great increase in label in the
cell wall plus protein fraction over 6 days suggests
a significant part of the assimilate "sink" is
related to growth of hyphae rather than storage.

The diversion of assimilate to a "growth" sink
will be determined by conditions for mycelial
growth.   Such a sink would be far less than the
"storage" sink (and growth sink) in ectomycorrhizas.
However, diversion of assimilate by this means can
be quite significant.   In clover, where infection
was deliberately restricted to a localised part of
the root system (see above), activity per unit
weight of hyphae was nearly 4 times, and total activity 24%, that of subtending roots.   A removal of
this magnitude from the entire root system of a sys-
temically infected plant plus some respiratory loss
by hyphae would represent a considerable drain on
host carbohydrate supply;   in the present instance,
the localised removal by hyphae represented only 1%
of total root weight and 4% of total root activity
(Table 6).   In hoop pine the relative amount of
hyphae was very much less.   Obviously wide differences in hyphal production occur (see also Bevege &

Bowen, 1975) and this aspect is deserving of much
closer study.

The extramatrical *Endogone* hyphae in the clover
experiment was only 1% of the total weight of the
plant and probably reflects a relatively small diversion of assimilate to the hyphae (see Table 6, b).
This small diversion to the fungus increased phosphate absorption and resulted in a fresh weight
increase of 2.77 g (150% increase in plant growth)
in the test soil used - a handsome return on the
investment! The greater translocation of ^{14}C
assimilate to the roots of mycorrhizal plants was
due to greater amounts of root in the mycorrhizal
system, not to large fungal sinks.

On the evidence of this study we propose
sucrose and glucose to be the principal sugars transferred and that low levels of sucrose in the fungal
cytoplasm, maintaining a concentration gradient for
further passive movement or facilitated diffusion,
are achieved by its conversion into glycogen and
eventual incorporation into lipid, protein and amino
and organic acids involved in the active growth of
the fungus (excess acids would be stored in vacuoles).

Both *Endogone* species had high incorporation
into the metal-precipitated fraction in common; we
do not know if this fraction also serves a storage
function (similar to glycogen) and this needs
further study. After 24 hr, *E. mosseae* had little
label in organic acids and some 20% in polysaccharides whereas *E. araucareae* was quite the reverse
with 30% in the acid fraction. Whether these
differences are a host induced response rather than
a specific fungal characteristic remains to be demonstrated, as organic acids might be expected to play
a balancing role in cation uptake by mycelium.

Further work along these lines is warranted as
is a definition of the carbohydrate physiology of
typical structural features of endomycorrhizas,
namely vesicles, arbuscules and chlamydospores. The
role of vesicles has yet to be satisfactorily

explained; are they temporary storage organs?   If
so, are storage compounds in vesicles different from
those in hyphae?   There is some evidence (Bevege,
1971) that vesicle frequency maximises under conditions of supraoptimal nitrogen supply to the host,
but whether they store this excess nitrogen, and if
so in what form, is unknown.   In our experience
vesicles can accumulate a limited quantity of phosphate, but have never been observed to fulfil any
infective role and tend to be formed late in the
development of the infection (Bevege & Bowen, 1975;
Bowen *et al.*, 1975).   What storage compounds
besides glycogen characterise spores as distinct
from the somatic hyphae?   The presence of occluded
pores in many *Endogone* spores would infer that
mature spores are effectively isolated from further
material transfer from the body of hyphae.

ACKNOWLEDGMENTS

These studies were supported in part by a
research grant from the Nuffield Foundation to
D.I.B., and the Radiata Pine Fund.   The technical
assistance of Miss Barbara Arnott throughout is
gratefully acknowledged.

REFERENCES

BEVEGE, D. I. (1968). A rapid technique for clearing
    tannins and staining intact roots for detection of
    mycorrhizas caused by *Endogone* spp., and some
    records of infection in Australasian plants.
    Trans. Br. mycol. Soc., <u>51</u>, 808-810.

BEVEGE, D. I. (1971). Vesicular-arbuscular mycor-
    rhizas of *Araucaria* : aspects of their ecology and
    physiology and role in nitrogen fixation.
    Ph.D. Thesis.   University of New England.
    Armidale Australia, 351 pp.

BEVEGE, D. I. & BOWEN, G. D. (1975). This symposium p. 77.

BOWEN, G. D., BEVEGE, D. I. & MOSSE, BARBARA. This symposium, p. 241.

COX, G. C., SANDERS, F. E. T., TINKER, P. B. H. & WILD, J. A. (1974). This symposium, p. 297.

DAVIES, D. D., GIOVANELLI, J. & AP REES, T. (1964). *Plant Biochemistry*. Botanical Monographs Vol. 3., ed. James, W. O., Blackwell Scientific Publications, Oxford. 454 pp.

HARLEY, J. L. & LEWIS, D. H. (1969). The physiology of ectotrophic mycorrhizas. *Adv. microb. Physiol.*, 3, 53-81.

HAYMAN, D. S. (1974). Plant growth responses to vesicular-arbuscular mycorrhiza. VI. Effect of light and temperature. *New Phytol.*, 73, 71-80.

HO, I. & TRAPPE, J. M. (1973). Translocation of ^{14}C from *Festuca* plants to their endomycorrhizal fungi. *Nature, London*, 244, 30-31.

HOAGLAND, D. R. & ARNON, D. I. (1938). The water-culture method for growing plants without soil. Circ. Calif. agric. exp. Stat., No. 347.

JANKIEWICZ, L. S., PLICH, H. & ANTOSZEWSKI, R. (1969). Preliminary studies on the translocation of ^{14}C-labelled assimilates and $^{32}PO_3^{3-}$ towards the gall evoked by *Cynips (Diplolepis) Quercus-folii* L. on oak leaves. *Marcellia*, 36, 163-172.

LEWIS, D. H. & HARLEY, J. L. (1965). Carbohydrate physiology of mycorrhizal roots of beech. *New Phytol.*, 64, 224-269.

LISTER, G. R., SLANKIS, V., KROTKOV, G. &
NELSON, C. D. (1968). The growth and physiology of
Pinus strobus L. seedlings as affected by various
nutritional levels of nitrogen and phosphorus.
Ann. Bot. (Lond.), 32, 33-43.

McDOUGALL, BARBARA & ROVIRA, A. D. (1965). Carbon-14
labelled photosynthate in wheat root exudates.
Nature, London, 207, 1104-1105.

MELIN, E. & NILSSON, H. (1957). Transport of
C^{14}-labelled photosynthate to the fungal associate
of pine mycorrhiza. *Svensk bot. Tidskr.*, 51,
166-186.

NELSON, C. D. (1964). The production and translocation of photosynthate - ^{14}C in conifers.
In *The Formation of Wood in Forest Trees*.
ed. Zimmerman, M. H., Academic Press, New York,
243-257.

RANGNEKER, P. V. & FORWARD, D. F. (1972). Foliar
nutrition and growth of red pine : distribution of
photoassimilated carbon in seedlings during bud
expansion. *Can. J. Bot.*, 50, 2053-2061.

ROVIRA, A. D. & BOWEN, G. D. (1973). The influence
of root temperature on ^{14}C assimilate profiles in
wheat roots. *Planta* 114, 101-107.

SCHWEERS, W. & MEYER, F. H. (1970). Einfluss der
mycorrhiza auf den transport von assimilaten in
die wurzel. *Ber. Dt. bot. Ges.*, 83, 109-119.

SHIROYA, T., LISTER, G. R., SLANKIS, V., KROTKOV, G.
& NELSON, C. D. (1962). Translocation of the
products of photosynthesis to roots of pine seedlings. *Can. J. Bot.*, 40, 1125-1135.

SMITH, D., MUSCATINE, L. & LEWIS, D. (1969).
Carbohydrate movement from autotrophs to hetero-
trophs in parasitic and mutualistic symbiosis.
Biol. Rev., <u>44</u>, 17-90.

SOMOGYI, M. (1945). Determination of blood sugar.
J. biol. Chem., <u>160</u>, 69-73.

STEPHENS, C. G., CROCKER, R. L., BUTLER, B. &
SMITH, R. (1941). A soil and land use survey of
the Hundreds of Riddoch, Hindmarsh, Grey, Young,
and Nangwarry, County Grey, South Australia.
CSIRO Bull. 142.

URSINO, D. J. & PAUL, J. (1973). The long-term fate
and distribution of ^{14}C photoassimilated by young
white pines in late summer. *Can. J. Bot.*,
<u>51</u>, 683-687.

MOVEMENT OF CARBON COMPOUNDS BETWEEN THE PARTNERS IN ORCHID MYCORRHIZA

SANDRA PURVES AND G. HADLEY

Botany Department, University of Aberdeen, Aberdeen, U.K.

INTRODUCTION

In sheathing (ectotrophic) and vesicular-arbus- cular (endotrophic) mycorrhizas, both of which are regarded as mutualistic symbioses, the host provides the endophyte with carbon compounds and in return benefits from an enhanced uptake of phosphorus and nitrogen. Orchid mycorrhizas show a uniform pattern of infection and are clearly endotrophic, but are unusual in some respects. It is not yet known with certainty whether host photosynthate can be transferred to the endophytic fungus, or whether the fungus plays any part in uptake of ions by the root. Indeed most experimental work has been done with het- erotrophic protocorms, and not with root-bearing adult plants.

Orchid seeds fail to germinate and grow satis- factorily in the absence of suitable infecting fungi, the principal function of which is to transfer carbon compounds from the substrate into the germin- ating seed. Although the nutritional requirements of asymbiotic orchids have been studied extensively (see Arditti, 1967; Hadley & Harvais, 1968;

Hadley, 1970a; Harvais, 1972), little comparative
investigation of development with and without symbiotic fungi has been carried out.   In general, the
rate of development of asymbiotic protocorms is much
less than that of protocorms infected by suitable
endophytes (Hadley, 1970a; Hadley & Williamson,
1972) and the claims by Harvais (1972) to the contrary are not substantiated by comparative tests in
the experiments he describes.

Orchid endophytes mostly belong to the form-
genus *Rhizoctonia* and can make use of complex carbohydrates, especially cellulose (Hadley, 1969).
Some which make better growth on undefined nitrogen
sources such as yeast extract are known to have a
requirement for thiamine, p-aminobenzoic acid and
certain amino acids (Stephen & Fung, 1971a, 1971b;
Ong, 1973; Hijner & Arditti, 1973).

CARBON NUTRITION OF MYCORRHIZAL SEEDLINGS

Smith (1966) used cellulose as a carbon source
and found that infected seedlings of *Dactylorhiza
praetermissa* and *D. purpurella* grew well when cellulose was present in the medium, while seedlings in the
absence of cellulose made no growth after seven
weeks.

Harvais and Hadley (1967b) investigated this
problem and Hadley (1969) compared drip-feeding of
glucose with the use of cellulose in the medium.
The long-term supply of small aliquots of glucose,
or alternatively the provision of cellulose, enhanced
the growth of populations of infected protocorms.
These results indicate, indirectly, that carbon compounds derived from complex carbohydrates in the
surrounding medium are being transferred by the
fungus into orchid tissue, thereby promoting growth.
Similarly, Purves (unpublished) demonstrated that
seeds of *Goodyera repens* could neither germinate nor
grow on a Pfeffer 1% cellulose medium, while infected protocorms on this medium grew better than non-

infected ones on Pfeffer medium with glucose and
potato extract, the best medium for asymbiotic
growth.

TRANSLOCATION

Smith (1966) showed that a known endophyte of
Dactylorhiza purpurella, viz. *Rhizoctonia repens,*
could translocate ^{14}C and ^{32}P in an established
mycelium, and she demonstrated that both tracers
could be detected in mycorrhizal seedlings when the
fungus alone was fed with ^{14}C D-glucose and ^{32}P
orthophosphate.   In similar work the present authors
used ^{14}C with *Goodyera repens* (Fig. 1).   But in many
of our (unpublished) experiments translocation into
replicate protocorms varied, and there was wide
variation between replicate petri dishes (Table 1)
which may reflect differences in the translocating

Table 1. ^{14}C translocated into protocorms
of *Goodyera repens* in 48 hours, expressed in
dpm per mg fresh weight of tissue.

		Dish				
		1	2	3	4	5
Protocorm	1	523	452	1407	408	636
	2	979	999	801	622	530
	3	755	1606	1473	559	472
	4	200	557	1805	525	636
Mean		614	904	1122	529	569

Analysis of variance shows that variation
between dishes is greater than that within
dishes at the 95% level of confidence.

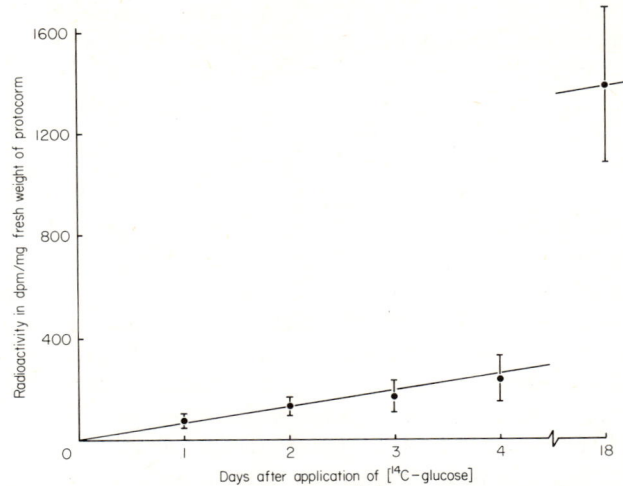

Figure 1. Accumulation of ^{14}C in infected proto-corms of *Goodyera repens.* Points are means for samples (each of two protocorms) from five separate plates.

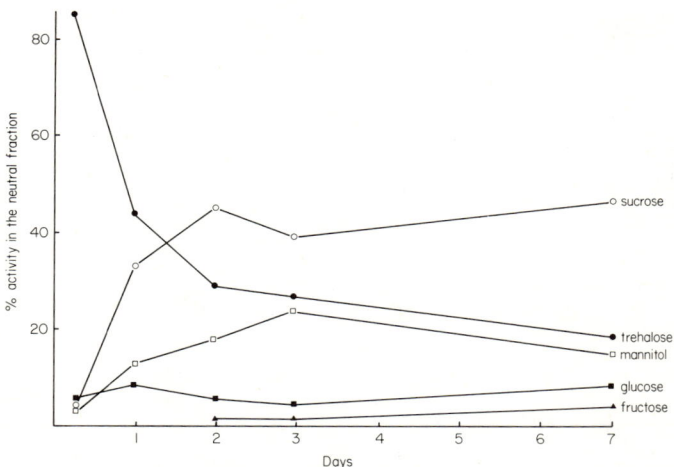

Figure 2. Proportion of ^{14}C in components of the neutral fraction of protocorms of *Dactylorhiza purpurella* infected by *Rhizoctonia* sp. (strain Rs 10). (Redrawn from Smith (1967)).

Table 2. Percentage germination and size of protocorms of
Goodyera repens after 8 weeks in darkness at 21°C.
Measurements (mean of 25) in μm.

	Pfeffer	Glucose (1.0%)	Fructose (1.0%)	Sucrose (1.0%)	Trehalose (1.0%)	Mannitol (1.0%)
Percentage germination	0	84	71	77	75	0
Length	233±25*	423±66	437±79	417±64	425±49	No germination
Breadth	121±12	229±30	242±36	230±24	231±24	

* Control; ungerminated seed

ability of individual cultures of the fungus.

In addition, translocation experiments have been confined to heterotrophic protocorms. It is not known whether carbon compounds are translocated through fungal mycelium into photosynthetic proto-corms, or adult plants, or under what conditions this might occur.

FORM IN WHICH CARBON COMPOUNDS MOVE

Smith (1967) established that trehalose and mannitol were peculiar to the endophyte (mannitol to certain endophytes only), and sucrose to the orchid. She followed the incorporation of ^{14}C, supplied to the fungus as ^{14}C glucose, into various sugars (Fig. 2). Over seven days labelling in trehalose declined, with a corresponding increase in the label-ling of sucrose, suggesting that trehalose may be a transfer compound. An indirect approach to this problem is to demonstrate the ability of non-infected orchid tissue to use trehalose as a carbon source. Subsequently (1973) Smith found that seeds of *Dacty-lorhiza purpurella* could germinate and grow in tre-halose almost as well as in glucose. Mannitol was inhibitory. Ernst (1967) reported growth of a *Phalaenopsis* cultivar on mannitol, but did not use carbohydrate-free controls. Our experiments have shown that *Goodyera repens* makes comparable growth on glucose, fructose, sucrose and trehalose. It would not germinate or grow on mannitol (Table 2).

For these results to constitute positive evid-ence that trehalose is a transfer compound, it would have to be shown that it is absorbed without external hydrolysis, as suggested by Ernst *et al.*, (1971). As Smith (1973) points out, similar results would be obtained if external hydrolysis were followed by rapid absorption of glucose. Smith and Smith (1973) demonstrated both uptake and metabolism of glucose and trehalose by leaves of *Bletilla hyancinthina*.

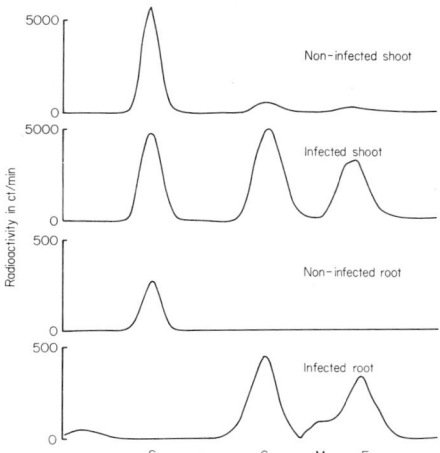

Figure 3. ^{14}C labelling of sugars in shoots and
roots of plantlets of *Dactylohriza purpurella* two
days after exposure to $^{14}CO_2$. Scan of paper chromatograms showing activity in sucrose (S), glucose (G),
mannitol (M) and fructose (F).

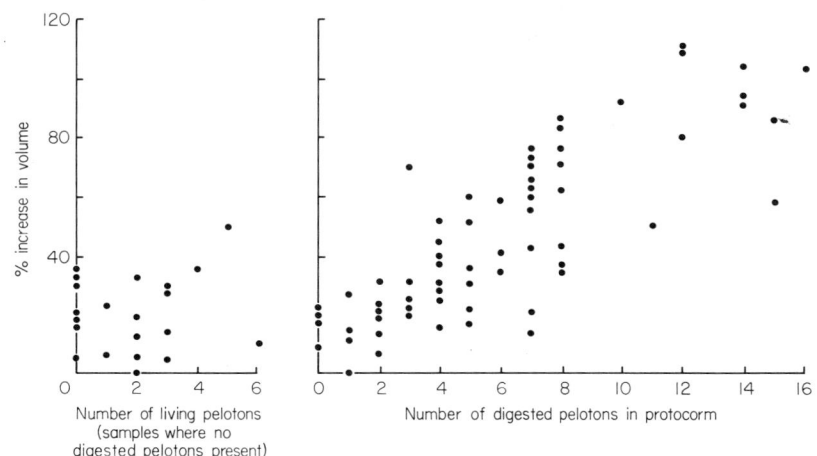

Figure 4. Increase in size of infected protocorms
in relation to (right) number of digested pelotons
and (left, where no digested pelotons present)
number of undigested ones.

Mannitol was absorbed but not metabolized and is therefore unlikely to be a transfer compound.

A more direct approach to the problem might be to use a system analogous to that first developed for lichen symbiosis, the 'inhibition technique' (see Smith *et al.*, 1969). This technique has been applied by Yuen (see Lewis, 1970) and Holligan and Lewis (personal communication) to parasitic associations of rusts and higher plants.

EFFECTS OF INFECTION ON HOST METABOLISM

Burgeff (1959) described the disappearance of starch from infected regions and suggested that the fungus transformed starch into storage compounds in hyphae. Hadley and Williamson (1971), in an analysis of events following infection of protocorms of *Dactylorhiza purpurella,* reported that starch hydrolysis was stimulated soon after infection. This suggested that the infecting fungus was causing an active mobilization of reserves which correlated with the stimulation of growth of the protocorm. In non-infected (asymbiotic) protocorms starch breakdown did not occur. Unpublished observations (Hadley) on many temperate and tropical species suggest that starch can accumulate in asymbiotic conditions on media containing glucose or other sugars. Quantitative information for *Goodyera repens* (Table 3) proves the very close correlation between infection, starch hydrolysis and the stimulation of growth.

Hadley and Lewis fed green plantlets of *Dactylorhiza purpurella* with $^{14}CO_2$, and found (Fig. 3) that in shoots of uninfected plantlets there was heavy labelling of sucrose with little incorporation into hexoses. Sucrose, but not hexoses, was present in the roots. Infected plantlets, in contrast, tended to show more heavy labelling of hexoses than sucrose in the shoot and their roots also contained hexoses, but no sucrose. The metabolic changes

Table 3.   Size* and starch content of
infected and non-infected protocorms of
Goodyera repens on deionised water (DIW),
Pfeffer 0.1% dextrose (PO.ID) and Pfeffer
1% dextrose (PID).   Size measurements
(volume) in mm^3 x 10^3.   Starch measured
on Scale O (none present in cells) to 10
(cells of entire protocorm filled with
starch grains).   Figures are means of 25.

Weeks					Medium			
		DIW		PO.ID		PID	
		size	starch	size	starch	size	starch
5	Non-infected	–	–	6.2	4	6.2	9
	Infected	5.9	< 1	7.6	0	12.4	0
9	Non-infected	–	–	7.1	8	12.3	10
	Infected	6.5	0	6.3	0	182.0	0
12	Non-infected	6.4	1	7.8	9	12.7	10
	Infected	6.7	0	7.4	0	169.4	0

*Volume measurements, calculated from $\dfrac{\pi B^2 (3L - B)}{12}$

where L = length; B = breadth

effected by the fungus may therefore operate through
the inversion of sucrose.   Presumably this is mediated by increased invertase activity in infected
tissue, and similar findings have been reported for
biotrophic situations such as rust infections of
higher plants (Kuprevich & Shcherbakova, 1964;

Purves, Sandra & Hadley, G.

Lewis, private communication).

TRANSFER OF METABOLITES FROM FUNGUS TO HOST

Concepts of biotrophic and necrotrophic nutrition

There are three possible ways in which nutrients can be transferred from fungus to orchid: (i) digestion of fungal hyphae containing nutrients derived from the external medium; (ii) transfer of substances from fungus to orchid across a living interface; (iii) absorption of substances made available in the external medium by the action of the fungus. Experiments with *Dactylorhiza purpurella* (Hadley & Williamson, 1971) and *Goodyera repens* (Purves, unpublished) tend to discredit the third possibility. Infection in any population of protocorms occurs in only a proportion of individuals, and these alone show an increase in growth rate. Uninfected individuals in the presence of the fungus maintain the same growth rate as asymbiotic protocorms in sterile nutrient media.

Although the term 'digestion' is used here it is not known whether death of the hyphae is induced by the host or by autolytic breakdown (see Williamson, 1973). Lewis (1973) considers that nutrition of the host is 'necrotrophic parasitism *by the orchid* and cannot be regarded as a mutualistic symbiosis'. We think that the association is more complex than this, and that there may be both biotrophic (transfer across a living orchid-fungus interface) and necrotrophic (digestion of dead fungus) activity involved in the movement of nutrients from fungus to orchid, even if the movement of materials is undirectional.

Relationship between digestion and growth

Any evidence that the stimulation of growth precedes digestion of the fungus would favour the

biotrophic hypothesis.   Mollison (1943) claimed
that a growth stimulus occurred before digestion in
Goodyera repens, but Purves (unpublished) has failed
to confirm this; in fact digestion could occur in as
little as 48 hours after infection, at which stage
it was impossible to record any definite growth stimulus.   Nevertheless Hadley & Williamson (1971)
obtained slight evidence that growth of *Dactylorhiza
purpurella* was stimulated before digestion occurred
(Fig. 4).   They sampled protocorms at various times
shortly after infection, and by means of a squash
technique made a visual count of the numbers of digested and undigested pelotons.   Some protocorms
showed as much as 50 per cent increase in volume
where only undigested pelotons were present.   The
relationship between volume and number of digested
pelotons also suggested that an increase in size may
precede digestion.

Metabolic changes preceding digestion

We are currently characterising some sugars in
the *Goodyera repens* system by thin layer chromatography, following Smith's (1967) findings with
Dactylorhiza purpurella that trehalose is peculiar to
the fungus and sucrose to the orchid.   After feeding
^{14}C glucose to infected protocorms via the fungus we
hope to demonstrate labelling in sucrose in protocorms in which digestion has not commenced.

RECIPROCAL MOVEMENT FROM ORCHID TO FUNGUS

Burgeff (1959), discussing the growth of hyphae
from infected plantlets into the surrounding soil
stated that 'A reversal in the direction of flow of
building substances occurs ...'.   Smith (1967) presumed that this reciprocal flow (from orchid to
fungus) occurred because the fungus could grow out
from infected protocorms on to a carbohydrate-free

medium and could cross a diffusion barrier.   Some
unpublished experiments by Smith, Harley and Lewis
(see Harley, 1969; Smith *et al.,* 1969) in which
green plantlets of *Dactylorhiza purpurella* were fed
with $^{14}CO_2$ showed no incorporation into fungal metabolites when the plantlets were transferred to carbohydrate-free plates and the fungus allowed to grow
out.

Further experiments by Lewis and Hadley (unpublished), again using *Dactylorhiza purpurella,* showed
that fungal metabolites became labelled when $^{14}CO_2$
was fed to infected plantlets.   But non-infected
control plantlets leaked nutrients to the surrounding agar, which could explain the appearance of ^{14}C
in mycelium growing out from the infected ones.
Also, since whole plantlets had been exposed to
$^{14}CO_2$, labelling of the fungal metabolites could
originate from dark fixation of $^{14}CO_2$ by the fungus
present.

These experiments led directly to those of
Hadley and Purves (1974), in which non-infected
plantlets of *Goodyera repens* were found to leak negligible amounts of radioactivity into the medium.
The quantity of ^{14}C fixed varied but the pattern of
distribution of label was similar in non-infected
and infected plantlets (Table 4) and did not alter
over a four week period.   Transferring plantlets
from light to dark conditions made no difference.
Very little ^{14}C moved to the rhizomes, indicating
that the fungus in the rhizomes was not acting as a
'metabolic sink', a concept suggested by Lewis and
Harley (1965b) for beech mycorrhiza.   The endophyte
readily grew into the surrounding (carbohydrate-free)
Pfeffer agar and carried small amounts of radioactivity with it (Table 5).   When the green tops
alone of infected plantlets were exposed to $^{14}CO_2$,
very little radioactivity could be detected in the
rhizomes and none in the emerging fungus.   When
rhizomes alone were exposed, small quantities of
radioactivity were detected in the emerging mycelium.

Table 4. Distribution of ^{14}C in infected and non-infected plant-lets of *Goodyera repens* (pooled samples of five).

	Infected plantlets			Non-infected plantlets		
	Fresh wt (mg)	^{14}C (cpm/mg fresh wt)	% total ^{14}C in plantlets	Fresh wt (mg)	^{14}C (cpm/mg fresh wt)	% total ^{14}C in plantlets
Shoot apex and leaves	33	7445	71.2	18	6926	68.1
Stem	50	1712	24.7	61	923	30.8
Rhizome	87	165	4.1	35	57	1.1

Purves, Sandra & Hadley, G.

Table 5. Movement of ^{14}C from infected and non-infected plantlets of *Goodyera repens* plated on Pfeffer agar after exposure to $^{14}CO_2$ (Radioactivity expressed as cpm for a sample of five 6- mm agar discs).

Treatment	Distance from source (mm)	Days after plating				
		2	5	8	11	15
Non-infected plantlets	5	0	2	1	15	11
	25	0	0	0	0	0
	50	0	0	0	0	0
Infected plantlets	5	281*	74*	46*	55*	39*
	25	0	63*	11*	16*	11*
	50	0	0	0	14*	0.1*

* Fungus present in disc

Nevertheless the fungus could readily transport ^{14}C from the remains of killed protocorms (Table 6). These results suggested that, under the conditions of the experiments, the fungus could not use and transport photosynthate from the orchid. Since the fungus itself was found to carry out dark fixation of $^{14}CO_2$ the small amounts of radioactivity detected in the emerging mycelium probably originated from this source. The possibility that the fungus is obtaining carbohydrate from so-called non-nutrient agar cannot be completely ignored, since Purves (unpublished) has shown that distilled water agar supports the growth of *Rhizoctonia goodyera-repentis* for several transfers. Also, Hadley and Williamson (1971) showed that deionized water agar supported the healthy development of freshly infected protocorms of *Dactylorhiza purpurella*.

In agar-free cultures *Rhizoctonia goodyera-*

Table 6.   Translocation of ^{14}C by *Rhiz-octonia goodyera-repentis* from killed (macerated) plantlets of *Goodyera repens* across a diffusion barrier on to Pfeffer agar.
(Radioactivity expressed as cpm for a sample of five 6-mm discs)

Distance from source (mm)	Days after plating			
	5	8	11	15
5	0	1688*	650*	205*
25	0	0	143*	104*
50	0	0	0	101*

* Fungus present in disc

repentis readily grows from infected protocorms over glass wool moistened with distilled water, and experiments attempting to provide a 'metabolic sink' for host products are being carried out by allowing the mycelium emerging from infected protocorms to infect other protocorms, when ^{14}C may move via the fungal mycelium to the 'sink' protocorm.   So far there is no evidence for such movement.

EXCHANGE OF NUTRIENTS OTHER THAN CARBOHYDRATES

Another aspect of reciprocal transfer is that of specific nutrient requirements of the endophyte which may be satisfied by the host.   Hijner and Arditti (1973) reported that seeds of an *Epidendrum* cultivar released p-aminobenzoic acid and that a *Rhizoctonia* sp. (originally isolated from *Cymbidium*) could grow on Knudson's medium in which the seeds had germinated,

or to which thiamine (or its thiazole component) and
p-aminobenzoic acid had been added.   Referring to
the findings of Mariat (1952) that the pyrimidine
moiety of thiamine had a significant effect on the
growth of a *Cattleya* hybrid, Hijner and Arditti
suggested that exchange of vitamins or their components may be involved, but their hypothesis has not
been fully tested.

DISCUSSION

The length of the heterotrophic stage of orchid
seedlings is variable; in some tropical orchids
leaves may be formed quickly and infection in the
adult plant may be infrequent.   It seems unlikely
that the fungus plays an important part in the carbon
nutrition of the mature orchid in view of the limited
extent of infection (Hadley & Williamson, 1972) and
the photosynthetic vigour of many orchids.   At the
other extreme, the so-called saprophytic orchids must
be totally dependent upon infection throughout their
life.
The carbon nutrition of orchid mycorrhizas contrasts sharply with that of other mycorrhizal systems.   In sheathing mycorrhizas ^{14}C-sucrose applied
to the host was rapidly converted to fungal products
in the sheath (Lewis & Harley, 1965b), but there was
no reciprocal flow from fungus to host (Harley, 1969).
Unlike orchid endophytes, these fungi are unable to
utilize complex carbohydrate substrates efficiently
in culture (Lewis & Harley, 1965a) and it seems
reasonable that, in nature, they normally rely upon
host photosynthate.   Fungi of vesicular-arbuscular
mycorrhizas are probably unable to break down complex
carbohydrates, and Ho and Trappe (1973) have recently
obtained evidence of translocation of ^{14}C into
Endogone spores.   In ericaceous mycorrhiza the
fungus has been shown to accumulate ^{14}C from ^{14}CO$_2$
supplied to the host, but labelled glucose supplied

to the fungus was not transferred to mycorrhizal
seedlings and the endophyte cannot utilize cellulose
efficiently (Pearson & Read, 1973).

Work on vesicular-arbuscular and sheathing
mycorrhizas has indicated that infection is a positive advantage to the host when growing in a low
phosphate regime (Daft & Nicolson, 1966; Harley,
1969). In ericaceous mycorrhiza, transport of ^{32}P
via fungal mycelium into plantlets has been demonstrated by Pearson and Read (1973). In these mycorrhizas the autotroph provides the fungus with carbohydrate and itself benefits from increased mineral
absorption in low nutrient regimes. Our sole knowledge of mineral nutrition on orchid mycorrhiza comes
from the work of Smith (1966) who showed that ^{32}P is
transported into heterotrophic protocorms in culture.

So far, evidence acquired for the orchid system
points to a situation in which a higher plant has
evolved to exploit, and may even be a parasite on, a
fungus. Only direct evidence of reverse flow of
nutrients, possibly other than carbohydrates, can
establish whether the system is mutualistic.

REFERENCES

ARDITTI, J. (1967). Factors affecting the germination
    of orchid seeds. *Bot. Rev.*, <u>33</u>, 1-97.

BURGEFF, H. (1959). In *The Orchids*. Ed. Carl. L.
    Withner. The Ronald Press Company, New York.

DAFT, M. J. & NICOLSON, T. H. (1966). Effect of
    Endogone mycorrhiza on plant growth. *New Phytol.*,
    <u>65</u>, 343-350.

ERNST, R. (1967). Effects of carbohydrate selection
    on the growth rate of freshly germinated *Phalaenop-
    sis* and *Dendrobium* seed. *Am. Orch. Soc. Bull.*,
    <u>36</u>, 1068-1073.

ERNST, R., ARDITTI, J. & HEALEY, P. L. (1971).
Carbohydrate physiology of orchid seedlings   II.
Hydrolysis and effects of oligosaccharides.
Am. J. Bot., 58, 827-835.

HADLEY, G. (1969). Cellulose as a carbon source for
orchid mycorrhiza.   *New Phytol.*, 68, 933-939.

HADLEY, G. (1970a).   The interaction of kinetin,
auxin and other factors in the development of
north temperate orchids.   *New Phytol.*, 69,
549-555.

HADLEY, G. & HARVAIS, G. (1968). The effect of
certain growth substances on asymbiotic germination
and development of *Orchis purpurella*.   *New
Phytol.*, 67, 441-445.

HADLEY, G. & PURVES, S. (1974). Movement of [14]Carbon
from host to fungus in orchid mycorrhiza.   *New
Phytol.*, 73, 475-482.

HADLEY, G. & WILLIAMSON, B. (1971). Analysis of the
post-infection growth stimulus in orchid mycorrhiza.   *New Phytol.*, 70, 445-455.

HADLEY, G. & WILLIAMSON, B. (1972). Features of
mycorrhizal infection in some Malayan orchids.
New Phytol., 71, 1111-1118.

HARLEY, J. L. (1969). *The Biology of Mycorrhiza*.
Leonard Hill, London.

HARVAIS, G. (1972). The development and growth
requirements of *Dactylorhiza purpurella* in asymbiotic cultures.   *Can. J. Bot.*, 50, 1223-1229.

HARVAIS, G. & HADLEY, G. (1967). The development of
Orchis purpurella in asymbiotic and inoculated
cultures.   *New Phytol.*, 66, 217-230.

HIJNER, J. A. & ARDITTI, J. (1973). Orchid mycor-
    rhiza: vitamin production and requirements by the
    symbionts. *Am. J. Bot.,* 60, 829-835.

HO. I. & TRAPPE, J. M. (1973). Translocation of ^{14}C
    from *Festuca* plants to their endomycorrhizal fungi.
    Nature, New Biology, 244, 30-31.

KUPREVICH, V. & SHCHERBAKOVA, T. (1964). The physio-
    logy of a diseased plant: invertase activity.
    Tenth International Botanical Congress. Abstracts
    of papers, p.409, Edinburgh.

LEWIS, D. H. (1970). Physiological aspects of sym-
    biotic associations between fungi and other
    plants. *Lichenologist,* 4, 326-336.

LEWIS, D. H. (1973). Concepts in fungal nutrition
    and the origin of biotrophy. *Biol. Rev.,* 48,
    261-278.

LEWIS, D. H. & HARLEY, J. L. (1965a). Carbohydrate
    physiology of mycorrhizal roots of beech.
    I. Identity of endogenous sugars and utilisation
    of exogenous sugars. *New Phytol.,* 64, 224-237.

LEWIS, D. H. & HARLEY, J. L. (1965b). Carbohydrate
    physiology of mycorrhizal roots of beech. III.
    Movement of sugars between host and fungus.
    New Phytol., 64, 256-269.

MARIAT, F. (1952). Recherches sur la physiologie des
    embryons d'orchidée. *Rev. Gen. Bot.,* 59,
    324-377.

MOLLISON, J. E. (1943). *Goodyera repens* and its
    endophyte. *Trans. Proc. bot. Soc. Edinb.,*
    33, 391-403.

Purves, Sandra & Hadley, G.

ONG, SOON-HO (1973). Nutritional requirements of certain orchid endophytes. *M.Sc. Thesis, Aberdeen University.*

PEARSON, V. & READ, D. J. (1973). The biology of mycorrhiza in the *Ericaceae.* II. The transport of carbon and phosphorus by the endophyte and the mycorrhiza. *New Phytol.,* 72, 1325-1331.

SMITH, S. E. (1966). Physiology and ecology of orchid mycorrhizal fungi with reference to seedling nutrition. *New Phytol.,* 65, 488-499.

SMITH, S. E. (1967). Carbohydrate translocation in orchid mycorrhizas. *New Phytol.,* 66, 371-378.

SMITH, S. E. (1973). Asymbiotic germination of orchid seeds on carbohydrates of fungal origin. *New Phytol.,* 72, 497-499.

SMITH, D. MUSCATINE, L & LEWIS, D. H. (1969). Carbohydrate movement from autotrophs to heterotrophs in parasitic and mutualistic symbiosis. *Biol. Rev.,* 44, 17-90.

SMITH, S. E. & SMITH, F. A. (1973). Uptake of glucose, trehalose and mannitol by leaf slices of the orchid *Bletilla hyacinthina.* *New Phytol.,* 72, 957-964.

STEPHEN, R. C. & FUNG, K. K. (1971a). Nitrogen requirements of the fungal endophytes of *Arundina chinensis.* *Can. J. Bot.,* 49, 407-410.

STEPHEN, R. C. & FUNG, K. K. (1971b). Vitamin requirements of the fungal endophytes of *Arundina chinensis.* *Can. J. Bot.,* 49, 411-415.

WILLIAMSON, B. (1973). Acid phosphatase and esterase activity in orchid mycorrhiza. *Planta,* 112, 149-158.

SOME NUTRITIONAL ASPECTS OF THE BIOLOGY OF ERICACEOUS MYCORRHIZAS

D. P. STRIBLEY AND D. J. READ

Department of Botany, University of Sheffield, Sheffield, U.K.

INTRODUCTION

Having discussed the basic biological features of ericaceous mycorrhizas (Read & Stribley, this volume) it is important to consider the significance of the symbiosis in the life of the host plant. The most obvious approach is to examine the effect of mycorrhizal infection upon plant growth and mineral nutrition. Unfortunately, uninfected control plants can only be grown on soils from which the indigenous endophyte has been eliminated by some form of sterilization. Most sterilization procedures however markedly change the chemical properties of the soil. Heat in particular may cause the release of toxic substances which can lead to many problems in experiments with ericaceous plants (see Harley, 1969, for a fuller discussion). For this reason soils sterilised by γ-irradiation have been employed in recent studies (Read & Stribley, 1973). Soil treated in this way was less toxic to ericaceous plants and it was shown that mycorrhizal infection significantly increased the dry weight and nitrogen content of cranberry plants compared to uninfected controls when grown on moorland soil completely sterilized by 3.0 Mrads of γ-irradiation. The percentage phosphorus content of infected plants was only slightly increas-

ed.   Similar results were obtained with heather,
Calluna vulgaris L. Hull, and it was concluded that
mycorrhiza had a beneficial effect on nitrogen nutrition and growth of the infected plant.   However,
these experiments are open to several criticisms.
Complete sterilization by γ-irradiation led to considerable increases of soil fertility and at the same
time the hyphae of the endophyte were free from competition from other microbes.

We have now examined the effects of a less
severe partial sterilization by γ-irradiation on
nutrient release in heathland soils and further
studied some aspects of the growth and mineral nutrition of mycorrhizal plants on different irradiated
soils.

To help us understand the nutritional relationships between host and endophyte, the growth studies
have been complemented with laboratory experiments
on movement of isotopes of nitrogen and carbon
between the symbionts.

RESULTS

Growth Studies

Growth on partially sterilized soils

Soils were selected from different parts of
Calluna-dominated podsolic profiles developed over
millstone grit in the Sheffield region.   Three soils
were used: a) the A_O horizon of a peaty podsol from
Hallam Moor (HM) (Grid ref. SK271864);   b) the
leached B horizon of a podsol from Fox House (FH)
(Grid ref. SK267868);   c) the C horizon from a podsol
at Rivelin (R) (Grid ref. SK284809).   These soils
were chosen because analysis had shown them to vary
widely in nutrient content.   The peaty soil (HM) had
the highest level of organic carbon, nitrogen (total
and exchangeable) and phosphorus ($NaHCO_3$-extractable)

followed by FH and R in descending order.   It was
found that when aseptically-germinated cranberry seedlings were grown in moorland soils which had been
partially sterilized by 0.8 Mrads of γ-radiation,
their roots remained non-mycorrhizal for at least six
months.   Partially sterilized (0.8 Mrads) soils show
increases in fertility (see below) accompanied by a
progressive release of inorganic nutrients particularly exchangeable ammonium, and these effects are more

Table 1.   Changes in exchangeable ammonium
in three moorland soils treated with
γ-radiation (0.8 Mrads).

| Soil* | Treatment | Exchangeable NH_4^+-N μg/g dry wt of soil | | |
		Before irrad-iation	Immed-iately after irrad-iation	After 15 weeks' incubat-ion+
Hallam Moor (4.82)	Irradiated	14.1	16.5	66.7
	Control	14.1	–	21.4
Fox House (0.81)	Irradiated	8.2	12.2	26.5
	Control	8.2	–	5.0
Rivelin (0.16)	Irradiated	9.9	7.8	12.0
	Control	9.9	–	4.8

+ Soils were kept in open pots in a glasshouse
and sub-irrigated twice weekly.

* Figures in brackets are the percentage
content (dry weight basis) of organic carbon
in the untreated soils.

pronounced in soils of high organic content (Table 1).   Microbial activity seems to be involved in nutrient release since the nutrient level of soil treated with a sterilizing dose of radiation (3.0 Mrads), although higher than that of untreated soil, remains constant over a long period of time if the soil is kept sterile (Stribley, unpublished results).

Dry weights and nitrogen contents of plants grown on these soils are being analysed by the computer program of Hunt and Parsons (1974) which computes various growth-functions from curves fitted to logarithms of raw data.

Some of the results obtained so far are given in Table 2.   After fifteen weeks there are clear differ-

Table 2.   Dry weights (fitted values) of mycorrhizal and non-mycorrhizal plants of *V. macrocarpon* after fifteen weeks' growth on irradiated and untreated moorland soils. Values are the natural logarithms of dry weights in milligrammes.

Soil	Treat-ment*	Whole-plant dry weight ($\log_e$ mg)	95% confidence limits	
			Lower	Upper
Fox House	F	4.0188	3.8101	4.2275
	M	5.3396	5.0978	5.5805
	NM	4.3278	4.0761	4.5795
Hallam Moor	F	3.9166	3.7477	4.0854
	M	7.0381	6.8118	7.2644
	NM	7.0209	6.8957	7.1461
Rivelin	F	4.1821	4.0339	4.3304
	M	5.1512	4.9845	5.3179
	NM	3.9113	3.6680	4.1547

* F, mycorrhizal plants on untreated soil;
  M, mycorrhizal plants on irradiated soil;
  NM, non-mycorrhizal plants on irradiated soil.

ences in *fitted* values for dry weights between mycorrhizal and uninfected plants grown on the soils poor
in organic matter (FH and R) but not on the peaty
soil (HM) where growth is enormously stimulated as a
result of γ-irradiation.   All plants on irradiated
soils are growing well and none appear to be suffer-
ing from phosphorus deficiency.   In every case
mycorrhizal plants are growing better on irradiated
soil than on field soil.

     Other computed curves have shown that, on the
poor soils (FH and R) only, mycorrhizal plants tend
to contain a higher percentage of nitrogen than do
non-mycorrhizal plants and that their roots are more
efficient at absorbing nitrogen (from the specific
absorption rate: μg N absorbed per mg root weight
per week).

Growth on sterile, ^{15}N-labelled soil

     The above experiments were repeated using ^{15}N-
labelled soil to test the hypothesis that mycorrhizal
plants had utilized a fraction of soil nitrogen
unavailable to uninfected plants.   The results are
given in Table 3 and confirm the earlier report that
infection improved growth and increased the percentage nitrogen content (on a dry weight basis) of
Vaccinium macrocarpon.   Plants grown in the presence
of either a species of *Trichoderma* or of *Aspergillus*
(common genera of saprophytic fungi) did not show
enhanced growth, indicating that growth stimulation
was a specific property of the endophyte.

     The distribution of the label in various nitrog-
en fractions of the soil was determined.   The ^{15}N
enrichment of the exchangeable ammonium was much
greater than that of the organic fractions.   No
nitrate was detected in the soil used.   Mycorrhizal
plants grown on this soil were found to have a lower
^{15}N enrichment than that of uninfected plants, indicating they had absorbed unlabelled nitrogen from a
source unavailable to the latter (Stribley & Read,

Table 3. Nitrogen content, yield and ^{15}N-excess of shoots of mycorrhizal and non-mycorrhizal plants of *Vaccinium macrocarpon* after six months' growth on ^{15}N-labelled soil.

Growth stage	N Content % oven dry weight	Yield mg oven dry wt	Total N mg/ plant	^{15}N-Excess Atom %
1. Sterile seedlings	0.94	4.23	0.04	0
2. Three months after inoculation				
Mycorrhizal	1.10	14.80	0.16	15.90
Non-mycorrhizal	1.00	12.45	0.11	17.36
Inoculated with:				
Controls A) *Trichoderma* sp.	0.90	12.20	0.11	N.D.
B) *Aspergillus* sp.	0.88	8.00	0.07	N.D.
C) Dead endophyte	N.D.	11.80	N.D.	N.D.
3. Six months after inoculation				
Mycorrhizal	1.20	30.32[+]	0.36[+]	15.38[+]
Non-mycorrhizal	0.98	20.97[+]	0.21[+]	20.03[+]
Inoculated with:				
Controls A) *Trichoderma* sp.	0.94	18.80	0.18	N.D.
B) *Aspergillus* sp.	0.82	16.20	0.13	N.D.
C) Dead endophyte	N.D.	19.10	N.D.	N.D.

N.D. = Not determined. Each figure represents a mean of fourteen plants, except at stage one, where thirty plants were analysed. + indicates figures significantly different within the growth stage at $P < 0.001$.

1947b).   Whether this potential is expressed in
natural ecological conditions will be investigated in
further experiments.

Translocation Studies

Translocation of ^{15}N

In view of the possibility that mycorrhizal
plants may have access to organic sources of nitrogen
in soils, an experiment was designed to examine the
capacity of the endophyte to absorb a labelled organic nitrogen source and translocate it into mycorrhizal host plants.

^{15}N-glutamine was fed to the endophyte on a
split plate (Fig. 1) and the isotope was monitored in

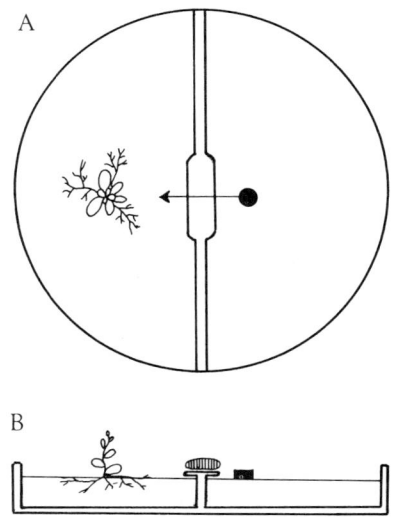

Fig. 1.   Surface view (A) and plan view (B)
of the split plate system used in the translocation studies.   Full circles represent
initial inoculum position, arrow represents
direction of fungal growth, and the cross
hatched area shows the position at which
lanolin was applied.

the host seedling 96 hours after its application to
the other side of the diffusion barrier.   Four bulk
samples of each of 30 experimental plants were assayed for ^{15}N content which was found to be 5.93, 6.34,
6.49 and 6.25 atom per cent.   In two control groups
where the isotope was added together with the inhibitor 2,4-dinitrophenol, the ^{15}N content of the shoots
was found not to be different, within experimental
error, from the natural abundance of the isotope
(0.3623 and 0.3661 atom per cent).

Translocation of carbon from fungus to host

Further translocation experiments employing the
split-plate method were designed to investigate the
movement of carbon compounds from fungus to host.
Under conditions which exclude diffusion and
capillary movement of solutes, we have found that
^{14}C, derived from either amino acids or glucose, is
translocated into tissues of the host plant when fed
to external hyphae of the endophyte (Table 4).
Translocation is greatly enhanced when the host
plant is deprived of carbon dioxide.   Very little
isotope moved into plants whose roots were closely
invested with but not infected by the endophyte.

Translocation of carbon from host to fungus

Pearson and Read (1973) demonstrated movement
of ^{14}C from the host plant to external mycelium of
ericaceous mycorrhiza but the possibility that the
fungus had absorbed labelled metabolites leaked from
the host roots cannot be discounted.   We (Stribley
& Read, 1974a) have re-investigated, under conditions
designed to minimize incorporation of ^{14}C-labelled
exudates, the possibility that the endophyte gains
carbon from its host.   Gas and paper chromatography
revealed that, as in beech mycorrhizas (Lewis &
Harley, 1965a), but in contrast to the situation in
vesicular-arbuscular mycorrhizas (Hayman, 1974),

Table 4.   Accumulation of ^{14}C in non-assimilating $(-CO_2)$ and assimilating $(+CO_2)$ shoots of mycorrhizal seedlings of *Vaccinium* 96 hours after feeding the endophyte with various ^{14}C-labelled compounds.

^{14}C-labelled compound	Specific Activity (CPM/mg) in shoot after 96 hours		
	Non-assimilating		Assimil-ating+
	Experiment	Control*	
Glucose	883	15	180
Glutamine	694	32	94
Glutamic acid	414	–	28
Leucine	287	–	40
Glycine	588	–	44
Protein Hydrolysate	912	–	117

Each figure represents the mean of fifteen seedlings

* Plants with roots invested with but not infected by the endophyte (non-assimilating plants only)

+ Corrected for atmospheric transfer of $^{14}CO_2$

infected roots contained the specifically fungal carbohydrates mannitol and trehalose.   These sugars incorporated much label when the host was fed $^{14}CO_2$ (Table 5).   The principal soluble carbohydrate in both infected and uninfected roots was sucrose;   it was also the most heavily labelled.   Much ^{14}C was also incorporated into an insoluble polymer of mannose which also appeared to be a fungal carbohydrate.

Table 5. Distribution of ^{14}C within soluble carbohydrates from mycorrhizal and non-mycorrhizal roots of *Vaccinium*, 72 hours after feeding $^{14}CO_2$ to the shoots (values as counts/min/mg dry weight extracted tissue*).

| | Sugar | | | | | |
	Glucose	Fructose	Sucrose	Mannitol	Trehalose	Myo-Inositol
Non-mycorrhizal	41(±12)	9(±3)	747(±141)	–	–	5
Mycorrhizal	80(±15)	26(±9)	370(±99)	108(±42)	37(±13)	3

* Each figure is the mean of seven observations. Figures in brackets represent the 95% confidence limits of the mean.

DISCUSSION

The beneficial effects of mycorrhiza on growth and nitrogen nutrition of the cultivated cranberry suggest that nutrients pass from the mycobiont into the host.   At the cellular level, this is indicated by the lysis which hyphae undergo in epidermal cells of the root (Nieuwdorp, 1969).   Our experiments with ^{15}N suggest that external hyphae of infected roots absorb soil nitrogen from inorganic, and perhaps organic, sources and conduct it into the host.   In view of the evidence that amino acids pass between partners in some symbiotic associations (Lewis, 1974) it may be worthwhile to consider the possible involvement of these compounds in the nitrogen physiology of ericaceous mycorrhizas.   Carbon derived from amino acids and nitrogen derived from the amide glutamine can pass from endophyte to host and suggests transfer of amino acids *per se*. Interpretation of such experiments however is rendered difficult because of the readiness with which fungi utilize amino acids as a carbon source.

The growth data so far obtained support the view that mycorrhizas may be advantageous to ericaceous plants growing in the nutrient-poor soils of their natural environment.   While such mycorrhizal associations clearly benefit the host plant, they may also benefit the mycobiont.   The experiments with $^{14}CO_2$ suggest that the carbohydrate physiology of mycorrhizas of ericaceous autotrophs may be similar to that of beech mycorrhizas where the fungus gains carbon from sucrose synthesised by the host (Lewis & Harley, 1965b).   Although carbon also passes from the endophyte to cranberry plants, movement in this direction is large only when the host is deprived of carbon dioxide i.e. when it is a powerful sink. However, present information does not allow us unequivocally to predict the *net* movement of carbon, especially in view of the lack of knowledge on the saprophytic activities of the endophyte (Stribley &

Read, 1974a).   On the basis of the work described in
the present paper we nevertheless tentatively propose
that the mycobiont of the ericoid mycorrhiza is
functionally like an extension of the (heterotrophic)
root system it infects.   It gains energy from the
host in the form of carbohydrates but benefits the
autotroph by absorbing from the soil nutrients such
as nitrogen and phosphorus (Pearson & Read, 1973b)
and releasing them, by mechanisms unknown, into cells
of the root.

REFERENCES

HARLEY, J. L. (1969). *The Biology of Mycorrhiza.*
  Leonard Hill, London.

HAYMAN, D. S. (1974). Plant growth responses to
  vesicular-arbuscular mycorrhiza.   VI. Effect of
  light and temperature.   *New Phytol.,* 73, 71-80.

HUNT, R. & PARSONS, I. T. (1974). A computer program
  for deriving growth-functions in plant growth-
  analysis.   *J. appl. Ecol.,* 11, 297-307.

LEWIS, D. H. (1974). Micro-organisms and plants:
  the evolution of parasitism and mutualism.
  Symp. Soc. gen. Microbiol., 24, 367-392.

LEWIS, D. H. & HARLEY, J. L. (1965a). Carbohydrate
  physiology of mycorrhizal roots of beech.
  I. The identity of endogenous sugars and utilization of exogenous sugars.   *New Phytol.,* 64,
  224-237.

LEWIS, D. H. & HARLEY, J. L. (1965b). Carbohydrate
  physiology of mycorrhizal roots of beech.
  III. Movement of sugars between host and fungus.
  New Phytol., 64, 256-269.

NIEUWDORP, P. J. (1969). Some investigations on the mycorrhiza of *Calluna, Erica* and *Vaccinium*. *Acta bot. neerl.*, <u>18</u>, 180-196.

PEARSON, V. & READ, D. J. (1973). The biology of mycorrhiza in the Ericaceae.   II. The transport of carbon and phosphorus by the endophyte and the mycorrhiza.   *New Phytol.*, <u>72</u>, 1325-1331.

READ, D. J. & STRIBLEY, D. P. (1973). Effect of mycorrhizal infection on nitrogen and phosphorus nutrition of ericaceous plants.   *Nature, London,* 244, 81-82.

READ, D. J. & STRIBLEY, D. P. (1975). This symposium.

STRIBLEY, D. P. & READ, D. J. (1974a). The biology of mycorrhiza in the Ericaceae.   III. Movement of carbon-14 from host to fungus.   *New Phytol.*, <u>73</u>, 731-741.

STRIBLEY, D. P. & READ, D. J. (1974b). The biology of mycorrhiza in the Ericaceae.   IV. The effect of mycorrhizal infection on uptake of ^{15}N from labelled soil by *Vaccinium macrocarpon* Ait.. *New Phytol.*, <u>73</u>, 1149-1155.

MEMBRANE STRUCTURE AND TRANSPORT PROBLEMS CONSIDERED IN RELATION TO PHOSPHORUS AND CARBOHYDRATE MOVEMENTS AND THE REGULATION OF ENDOTROPHIC MYCORRHIZAL ASSOCIATIONS

H. W. WOOLHOUSE

Department of Plant Sciences,
University of Leeds, U.K.

INTRODUCTION

There is a substantial measure of agreement between workers on mycorrhizas that the vesicular-arbuscular group of these symbionts are of particular significance with respect to phosphate uptake by the host.   The reason for this is that available phosphate is frequently present at low concentrations in the soil relative to the other major nutrients and diffuses only slowly (Lewis & Quirk, 1965; Vasey & Barber 1963).   Mycorrhizal fungi compensate for this deficiency by exploring a greater volume of soil than the roots can do and presenting a greater surface area for phosphate uptake.   There are a variety of experiments which support this simple view (Baylis, 1972; Hayman & Mosse, 1971; Sanders & Tinker, 1971; Gerdemann, 1965).   More elaborate hypotheses concerning such matters as the fungus possessing uptake mechanisms with a higher phosphate-binding affinity or tapping insoluble phosphate sources which the root is unable to metabolise have not been substantiated (Bieleski, 1973).

It is probable that there is a concomitant move-

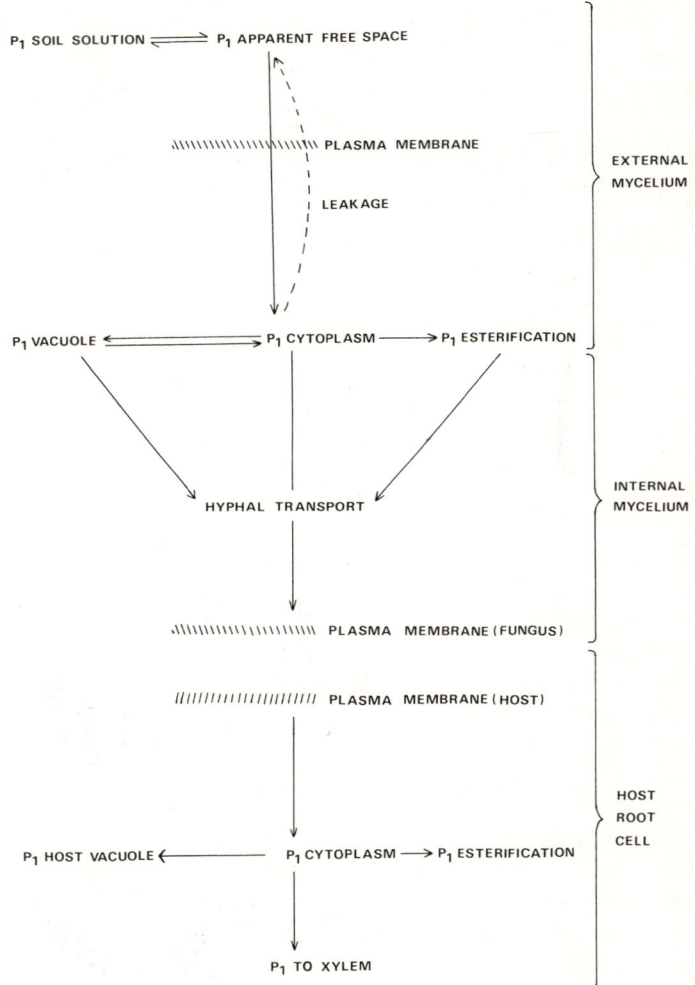

Figure 1. A diagrammatic representation of an end-otrophic mycorrhizal system. The juxtaposition of the plasma membranes of a host cell and a fungal arbuscule are shown. The prevailing directions of phosphate movement are shown, see also Fig. 8.

ment of sugars from host to fungus which provides a
significant part of the carbon-requirement of the
fungus (Bowen, 1973).   The transport pathways for
phosphate and carbohydrate in endotrophic mycorrhizal
associations are summarised in Figure 1.

This paper examines three basic problems posed
by this system:-

1. The mechanism of movement of phosphate from
soil solution into fungal cytoplasm and from fungal
cytoplasm to host cell cytoplasm.

2. The mechanism of movement of carbohydrates
from host cell cytoplasm to the cytoplasm of the
hyphae.

3. The mechanism by which the concentration of
available phosphate regulates the extent of fungal
penetration of the host, its growth within the host
and hence the subsequent movement of carbohydrates
from host to fungus.   A new hypothesis of regulation
is presented in this connection.

As a prelude to the examination of these questions it is necessary that we should consider briefly,
current views concerning the structure of the membranes at which these transport functions and acts of
host-mycorrhizal fungus recognition are carried out.

MEMBRANE STRUCTURE

Structural models of membranes are derived in
large measure from the original lipid bilayer model
of Danielli and Davson (1935).   The model has under-
gone a long sequence of modifications to accommodate
new findings.   The greatest constraint has undoubtedly been the need to take account of the heterogeneity of membrane surfaces inferred from experiments
on a variety of specific membrane transport carriers,
hormone receptors, immunoglobulins, and "patches"
possessing a greater permeability to water than the
rest of the membrane surface.

Recent studies suggest that many biological

membranes consist essentially of a non-polar hydrophobic region, a less well defined polar: non-polar
zone containing an array of charged groups, ordered
water molecules and bound ions and a polar surface
region containing carbohydrates, liposaccharide
groups and the polar regions of some protein molecules.   (Hechter, 1965; Finean, 1969; Pinto da
Silva & Branton, 1970; Spatz & Strittmatter, 1971;
Marchesi *et al.*, 1972; Radda & Vanderkooi, 1972).
    One model which accommodates many of these
recent findings is the fluid mosaic model of
Singer and Nicholson (1972).   This model satisfies
thermodynamic considerations, such as the need to
have the non-polar lipid chains away from water and
the polar and ionic groups in contact with water.
The proteins have hydrophilic and hydrophobic
regions and it is supposed that the former will
become aligned towards outer surfaces of the membrane
with the hydrophobic region facing inwards.   Some
of the large MW proteins have a sub-unit structure
and may extend through the whole thickness of the
membrane, the gap between the sub-units providing
fine pores for the passage of water and other small
molecules.   The smaller proteins are confined to
one surface of the membrane only (Guidotti, 1972).
It is probable that many of the proteins which
protrude on the external surfaces of the plasma
membranes of both plant and animal cells carry coval-
ently bound antigenic polysaccharides in which glucosamine, fucose, galactose, mannose and arabinose
residues are prominent components.   It is important
to stress that although this model has a certain
elegance and accommodates a number of observations
it still lacks critical experimental verification.

PHOSPHATE TRANSPORT

    The concentration of inorganic orthophosphate
(P_i) in most soil solutions is lower than that of all

other major nutrients, being in the range 0.01 - 2
μMolar.   It is worth noting *en passant* what a larger
proportion of published studies have been carried
out at P_i concentrations orders of magnitude greater
than this.
    Now we may take as a reasonable estimate for the
internal concentration of P_i in cytoplasm and vacuole
a value of say 5mM (Hall & Hodges, 1966);   this is an
estimate of the concentration of P_i in the vacuoles
of higher plant cells but I will suppose that the
concentration in the cytoplasm is of a similar magnitude.   If concentrations of P_i of this magnitude are
realistic for the fungal hyphae then it follows that
P_i must be moved into the hyphae against a concentration gradient of the order of 10^3 to 10^4.   If we
take the value for the electrical potential difference between hyphal cytoplasm and medium to be
similar to the values obtained for higher plant
cells i.e. of the order - 100 mV (Higinbotham *et al.*,
1967);   then for hyphae at equilibrium with the soil
solution with respect to phosphate we may apply the
Nernst equation:-

$$\Delta E = \frac{RT}{zF} \log \frac{[c_{Pi}]_o}{[c_{Pi}]_i}$$

Where ΔE is the electrical potential difference
between hyphal vacuole and soil, z the valency of the
ion (taken as $H_2PO_4^-$ in this instance), $(c_{Pi})_o$ the
concentration of P_i in the soil solution and $(c_{Pi})_i$
the concentration of the P_i internal to the plasmalemma of a soil hyphae of the mycorrhizal fungus.
Despite the approximate nature of these assumptions
for the magnitude of the gradients of P_i concentrations, and doubts concerning the applicability of the
Nernst equation in these circumstances, it is non-the
-less clear that phosphate is so far removed from a
state of electrochemical potential equilibrium that
the operation of a phosphate pump is imperative for

Woolhouse, H. W.

the maintenance of these gradients of phosphate concentration.

The existence of such a phosphate pump or carrier is inferred from a variety of studies with bacteria, fungi and higher plants. It is found for example that the uptake of phosphate by barley roots is affected by oxygen concentration (Fig. 2).

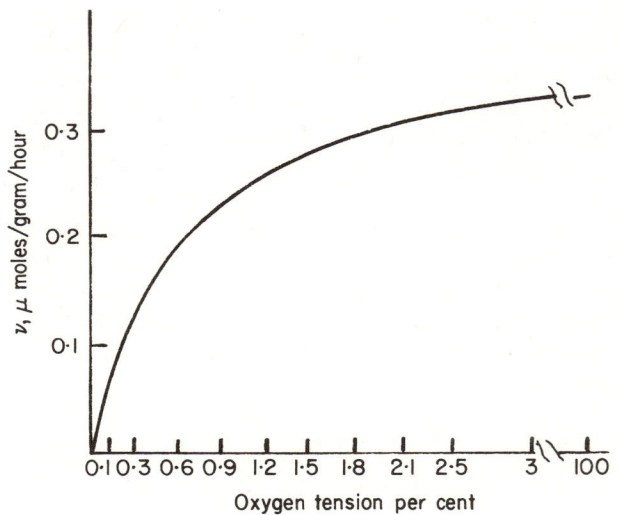

Figure 2. The effect of oxygen concentration in the solution on P_i uptake by excised barley roots. (After Hopkins, 1956).

Figure 2 shows that the rate of phosphate uptake is saturated at about 2% of oxygen and is half-maximal at 0.03% oxygen, suggesting that aerobic respiration mediated by the cytochrome system may be involved in the operation of the phosphate pump.

The existence of a phosphate pump is also inferred from the inhibition of P_i uptake in barley seedlings and mycorrhizas by such metabolic poisons as azide and 2:4 dinitrophenol (Russell & Martin,

1953; Harley & Brierly, 1954).   Thirdly we may
note the complex relationship between ion concentration and the rate of uptake which is found for phosphate and many other ions; the example shown in
Figure 3 is for potassium uptake, the case of phosphate is analogous.

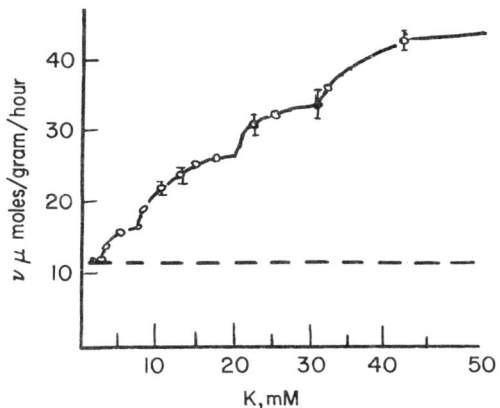

Figure 3.   The relationship between potassium uptake by barley roots and external
concentration.   Concentration of $CaSO_4$ in
the medium was 0.5mM.   (After Epstein &
Rains, 1965).

This type of curve has frequently been analysed in
terms of Michaelis - Menton kinetics in which the
analogy is drawn with the reaction between enzyme
and substrate, according to the relationship:-

$$E \quad + \quad S \rightleftharpoons ES \longrightarrow E \quad + \quad Product$$

and the reaction between the compound which is to be
transported and a "carrier" located within the particular cell membrane, according to the relationship:-

$$X_o \quad + \quad C \underset{\longleftarrow}{\overset{\longrightarrow}{\rule{0pt}{1.2em}}} CX \longrightarrow C \quad + \quad X_i$$

Where X_o is the compound outside the cell, X_i is the compound inside the cell, C is the carrier, and CX the carrier - complex.   In this instance the conversion from substrate to product is represented by the changed position of X from X_o to X_i.

Some of the most convincing evidence for the existence of these postulated carriers has developed from the pioneer work of Skou on the isolation of the Mg-dependent ATPases from a variety of animal membranes (Skou, 1965).   Subsequent work, particularly that involving the use of red-cell "ghosts", has led to the further characterisation of this carrier system.   In the wake of this work a number of other binding proteins have been isolated.

Of particular interest in the present context is the isolation of a specific phosphate-binding protein from the cell membrane of *E. Coli* (Medweczky & Rosenberg 1969, 1970).   Over 80% of this protein could be released from the cell membrane by means of the osmotic shock treatment of Neu and Chou (1967).   The purified protein (MW 42,000 ± 1000 Daltons) was shown to have a 1:1 stoichiometry of phosphate binding. Cells of *E. coli* which had been cultured routinely at 37°C when plunged into cold water at 3°C for five minutes lost the capacity for phosphate uptake.   If these cells were then treated with 250 µg of the purified phosphate-binding protein per 500 Klett units of cells, the uptake of phosphate by these cells could be in large measure restored (Fig. 4a).

Medweczky and Rosenberg went on to isolate mutants lacking this phosphate-binding protein and showed that addition of the P_i-binding protein to cold-shocked mutant cells restored their capacity for P_i uptake (Figure 4b).

Figure 4c shows the results of an experiment in which an antiserum (rabbit) was made to the binding protein and a range of concentrations of antiserum

were reacted with the binding protein and in each
case its ability to bind P_i was then checked.   It is
seen that control serum did not affect the binding of
P_i to the protein, whereas antiserum to the binding
protein, when reacted with binding protein, eliminated its ability to bind phosphate.   When the P_i-
binding protein was reacted with its specific antiserum, which knocks out its P_i-binding capacity, then
concommitantly its ability to restore the phosphate
uptake capacity of cold-shocked cells was lost
(Figure 4d).

I have dwelt on this pioneer study at some
length because it goes a long way towards proving the
existence of a binding protein which is quite clearly
implicated in the uptake of phosphate and it points
the way towards further work with higher plants and
mycorrhizas.   Another study which suggests the
existence of a P_i carrier in higher plants is that of
Weigl (1967, 1968) showing the influence of external
anion concentration on the rate of efflux of the same
anion from within the leaves of *Elodea* and from roots
of *Zea mays*.   These tissues were charged with $^{32}PO_4$
or ^{36}Cl for a selected period and the time course of
efflux of these ions was then followed in the presence of a range of external concentrations of the
same ion.   The efflux was specific to the external
concentration of the same anion, i.e. external chloride would stimulate chloride efflux but not phosphate
efflux and increasing external phosphate concentration would stimulate release of internal phosphate
but not of chloride (Figure 5).

Phosphate efflux is largely independent of external phosphate at concentrations up to 100 μM but
thereafter becomes concentration-dependent.   Weigle
argues that this is not an effect of phosphate on
membrane permeability and suggests that the results
are most simply explained in terms of a mobile
carrier.

I have examined this question of phosphate
transport at some length because it seems probable

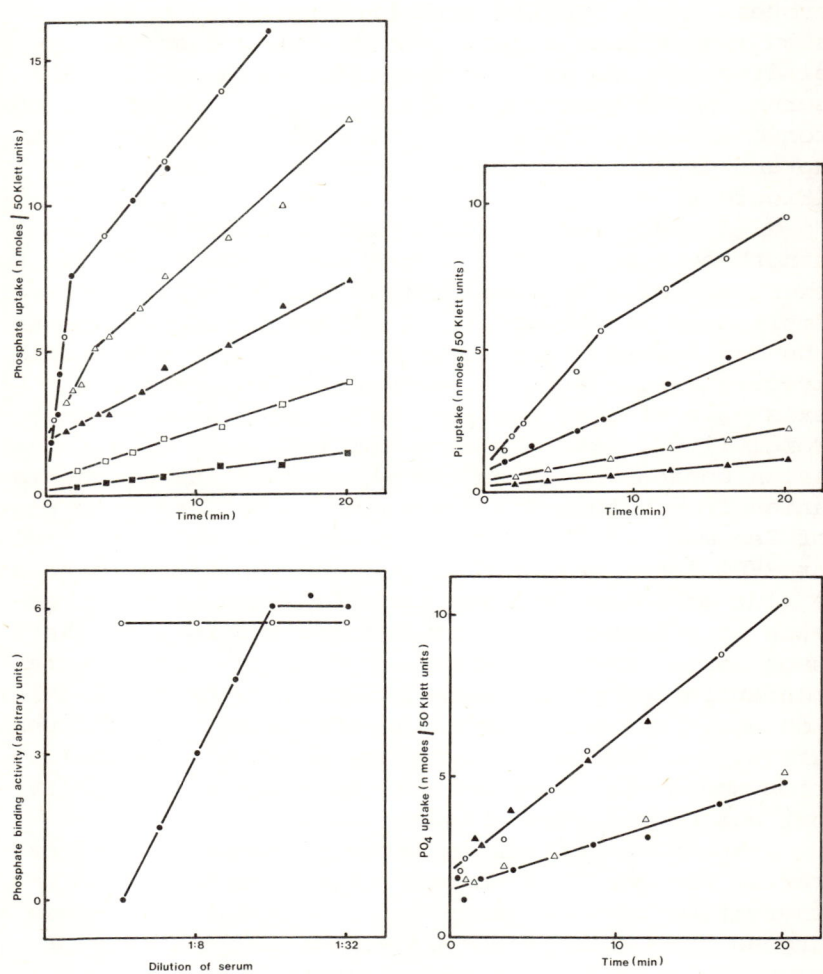

Fig. 4a. The stimulation by phosphate-binding protein of phosphate uptake in cold-shocked *E. coli*.
Cold-shocked cells were incubated at 500 Klett units
with 250 µg phosphate-binding protein for 2 min at
$0°$, and then diluted with 9 vol. of uptake medium.
Phosphate uptake was measured at 50 µM phosphate.
●——●, control cells alone; ○——○, control cells
with phosphate-binding protein.   Cold-shocked cells
along: ▲——▲, total uptake; ■——■ $^{32}P_i$ incorporated into acid-insoluble material.   Cold-shocked
cells with phosphate-binding protein: △——△, total
uptake; □——□ $^{32}P_i$ incorporated into acid-insoluble material.
Fig. 4b. The increase by phosphate-binding protein
of the rate of phosphate uptake in cold-shocked
cells.   Cold-shocked cells alone: ●——●, total
uptake ▲——▲, ○——○, total uptake; △——△, $^{32}P_i$
incorporated into acid-insoluble material.
Fig. 4c. The inhibition of phosphate-binding protein
by rabbit anti-phosphate-binding protein serum.
Phosphate-binding protein (0.2 ml, 100 mg/ml) was
incubated with 0.2 ml dialysed rabbit serum of the
appropriate dilution at $4°C$ for 8 h.   ○——○, control serum; ●——●, antiserum.
Fig. 4d. The inhibition of phosphate-binding protein
-stimulated P_i uptake in cold-shocked E. coli by
rabbit antiserum to phosphate-binding protein.   The
phosphate-binding protein was incubated with 0.2 ml
dialysed rabbit serum for 2 h at $0°C$.   The conditions for uptake are described in the legend for
Fig. 4a.   ●——●, cold-shocked cells alone; ○——○,
cold-shocked cells with phosphate-binding protein;
▲——▲ cold-shocked cells with phosphate-binding
protein and control serum; △——△ cold-shocked cells
with phosphate-binding protein and antiserum.
(After Medweczky & Rosenberg, 1970).

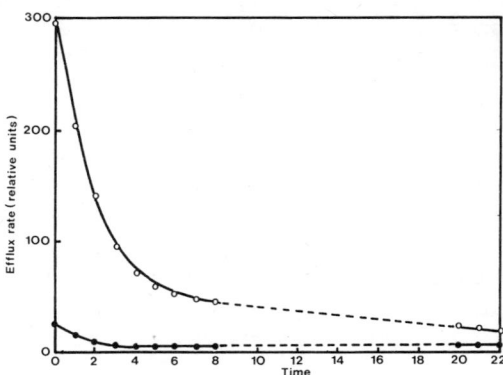

<u>Figure 5.</u>   The effect of external chloride
concentration on chloride efflux from roots
of maize.
o———o ^{36}Cl efflux into 5 x 10^{-2}M chloride.
●———● ^{36}Cl efflux into 5 x 10^{-5}M chloride.
(After Weigl, 1968).

that P_i uptake from a dilute soil solution into the
mycorrhizal fungus will require a mechanism involving
this general type of phosphate-binding protein.   I
would further suggest that the techniques are now
available for carrying out analogous experiments with
higher plants and fungi.   If one were able to
measure the number of P_i-binding protein molecules
per unit area of hyphal surface, as was done in the
case of *E. coli* (Medweczky & Rosenburg 1970), then
one would have a very precise and critical measure of
the increased uptake - potential afforded by the
external mycelium of an endotrophic mycorrhiza.

In the transport scheme as presented in Figure 1
we see that following transport of phosphate along
the fungal hyphae there arises the question of rever-
sing the transport of phosphate out from the mycorrhizal arbuscule into the apparent free space (AFS)
and into the host cell across its plasmalemma.
Discussion of this problem is deferred to a later
section since it is convenient to think about this

alongside the movement of carbon compounds in the opposite direction.

CARBON TRANSPORT FROM HOST TO FUNGUS

The relatively small standing biomass of external hyphae of vesicular-arbuscular mycorrhizal fungi in the soil suggests that, even if they had high respiration rates, they would not comprise very large sinks for organic carbon compounds from the host plant. This is borne out by the work of Bevege *et al.*, (1975) who demonstrate a relatively weak movement of carbon from host to fungus. They also find that there is no significant change in the spectrum of sugars between infected and uninfected plants, suggesting that transport gradients in these instances are probably not coupled to the formation of polyhydric alcohols or other compounds as in the case of ectotrophic mycorrhizas (Lewis & Smith, 1967). There is a lingering implication in the literature of endotrophic mycorrhizas that the main exchange of materials, particularly from fungus to host are dependent upon the deterioration and ultimate breakdown of the arbuscules. There is no doubt that plant cell membranes can under certain circumstances become leaky, for review see Simon (1974), so that some solutes can diffuse out and be lost to the cell. However diffusion of solutes in water is a slow process and it would be a mistake to assume that there is no need for the involvement of carrier mechanisms in the transport of sugar from host to fungus, simply because the fungus is a relatively weak carbohydrate sink. Carbohydrate transfer by diffusion would also break down if, as seems likely, there is a bulk flow of water associated with the transpiration stream taking place in the opposite direction. Similarly in the pathway of transport of phosphate, soil solution $\rightarrow$ fungal hyphae $\rightarrow$ AFS $\rightarrow$ host cell, diffusion is unlikely to account for the fluxes observed in

view of the probable concentration gradients involved.　One is led to conclude therefore that the
transport of carbon from host to fungus and of phosphate from fungus to host must involve both a sequence of carrier-mediated steps.　Carrier-mediated
transport may be of two types, a passive mediated
transport mechanism (Fig. 6) cr an active mediated
transport mechanism (Fig. 7).

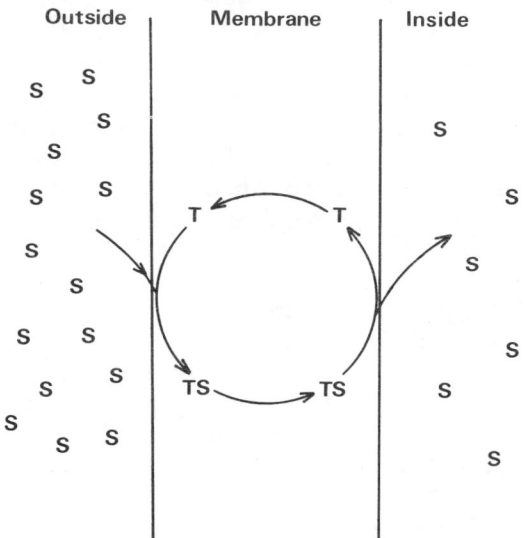

Figure 6.　Diagram of a passive mediated
carrier transport system
S　=　compound being transported
T　=　transport protein

Fig. 8 is an elaboration of Fig. 1 giving diagrammatic representation of a hypothesis for the
exchange of carbon and phosphorus between an endotrophic mycorrhiza and host cell in which the movement of materials is envisaged as involving 5 essential membrane transport steps two of which are
suggested to involve passive mediated transport
carriers and three involving active mediated trans-

port carriers.   Before presenting the argument for
this hypothesis it will be helpful to turn to the
third problem which was posed at the outset of this
paper, that is the question of the role of phosphate
supply in regulating the extent of mycorrhizal infection and of arbuscule development.   This involves a further consideration of the relationship of
the host and fungal surfaces which is also important
in relation to certain aspects of this model of
active and passive mediated transport systems.

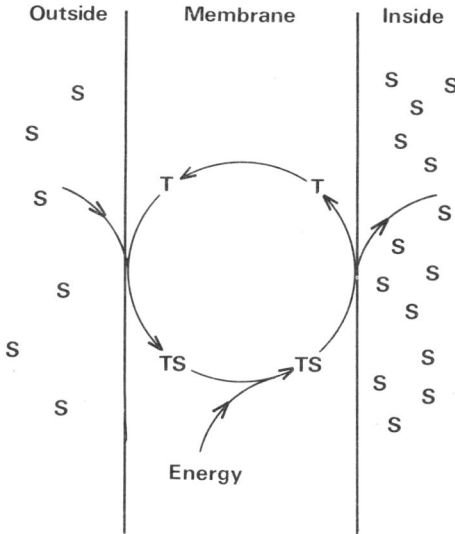

Figure 7.   Diagram of an active mediated
carrier transport system
S = compound being transported
T = transport protein

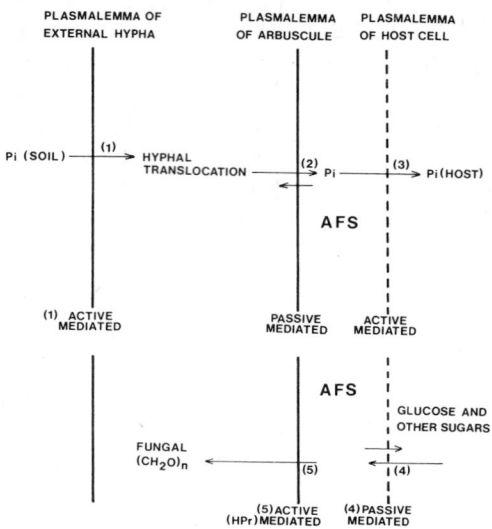

Figure 8.   The pathways of P_i and carbohydrate transport in an endotrophic mycorrhizal system.   Five pumps are postulated
in this system indicated by the numbers
1-5.   The suggested pumps are as follows:
1. An active mediated phosphate transport
system at the plasmalemma of the external
hyphae.
2. A passive (uncoupled) mediated phosphate transport system at the plasmalemma
of the arbuscule.
3. An active mediated phosphate transport
system at the plasmalemma of the host cell.
4. A passive mediated sugar transport system
at the plasmalemma of the host cell.
5. An energy-linked phosphorylating sugar
transport system at the plasmalemma of the
arbuscule.

PHOSPHATE SUPPLY AND THE REGULATION OF MYCORRHIZAL INFECTION AND ARBUSCULE DEVELOPMENT

There have been a number of studies showing that when the amount of available phosphate in a soil is low, heavy infections of vesicular-arbuscular mycorrhizas are developed (Sanders & Tinker, 1973).   If plants are grown in the presence of high concentrations of available phosphate, few if any arbuscules are formed.   The following testable hypothesis may account for this phenomenon.   The process of arbuscule formation may be envisaged as involving 3 steps:

(i)   The host plant emits a signal indicating that phosphate is in short supply.

(ii) The fungus receives the signal and responds by growing into the host cells and forming arbuscules.

(iii) The host plant modifies the fungus in such a way as to reverse the direction of phosphate transport so that phosphate passes out from the fungus and across into the host.

There are a number of observations which are of interest in relation to this hypothesis.

(i) Signal from the host induced by low phosphate supply:-

Several species of algae, fungi and higher plants when placed in a medium containing a low concentration of phosphate produce a 5-50 fold increase in the amount of non-specific acid phosphatases associated with the cell surfaces, for references see Woolhouse (1969) and Bieleski (1973).   In higher plants the phosphatase which is induced by a low phosphate supply includes new isozymes not present in the phosphate - sufficient cell.   The role of these inducible enzymes is not clear, they are associated with the golgi apparatus, plasmalemma and cell walls

and are not released into the soil.   This observation taken in conjunction with the low concentrations
of phosphate esters in soil makes it unlikely that
these enzymes play a significant role in the mobilis-
ation of insoluble phosphates.   Alternatively they
may be involved in regulating the development of the
internal mycelium and arbuscules of mycorrhizal
fungi.   Recent findings in the related field of
plant pathology suggest possible models for such a
regulatory mechanism;   they rest on the view that the
surface phosphatases function in a manner analagous
to lectins or lectin modifiers.   Lectins are plant
glycoproteins which are able to agglutinate erythro-
cytes by binding to sugar residues which form part of
specific glycoprotein receptors on the surface of the
plasma membrane.   Plant lectins will also inhibit
the growth of bacteria and fungi by binding to sugar
moieties on their surface glycoproteins.   Thus
lectins from the seeds of *Ricinus communis* will bind
to fucose and galactose which form minor components
of the glycoproteins of some fungi;   when this
happens growth of the fungi is inhibited (Sharon &
Lis, 1972).   Allen and Neuberger (1973) have purified and characterised a hydroxyproline-rich glycoprotein lectin from potatoes which binds to molecules
containing N-acetyl-glucosamine, a great many soil
fungi contain this component in their cell walls.
It has recently been shown that this lectin will
inhibit germ tube elongation in *Botrytis cinerea* and
will cause lysis of the hyphae tips of *Neurospora*
(Callow, unpublished).

Now there is a lot of evidence that most of the
hydroxyproline – rich glycoproteins in plants are
associated with the cell walls.   Somewhat regret-
tably, they have been given the name extensins and
speculatively implicated in the mechanism of cell
elongation (Lamport, 1965).   I suggest that the
hydroxy proline-rich glycoproteins of plant cell
walls represent in fact a spectrum of surface lectins

whose normal function is to protect the cells against invading fungi.   Under normal conditions they carry out this function but the phosphatases induced by phosphate deficiency disrupt this mechanism.   This brings us to stage 2 of the recognition system.

(ii) Fungal response to the low-phosphate signal from the host

The essence of stage 2 of this phosphate supply – fungal invasion model, is to suggest that the induced phosphatases modify the lectin glycoproteins of the host cell surface so that they no longer inhibit the growth of the mycorrhizal hyphae.

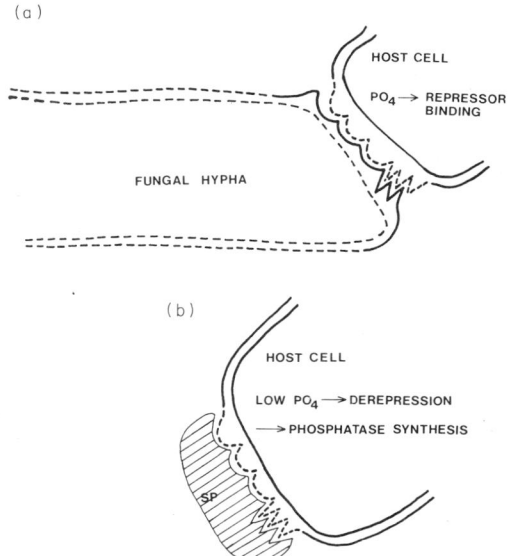

Fig. 9a and 9b.   Diagrammatic representations of the lectin-blocking hypothesis of phosphate regulation of mycorrhizal infection.

Figures 9a and b depict two ways in which this might

arise.   In the first model (Fig. 9a) the induced
phosphatases are shown as forming dimers with the
surface lectins so modifying their structure so that
they no longer bind to the N-acetyl glucosamine -
containing components of the fungal cell walls.   In
the second model (Fig. 9b) the phosphatases are
suggested to act either in the golgi bodies or at the
cell surface where they may degrade the phosphate
esters which serve as precursors of the carbohydrate
moiety of the lectins.   This would again lead to
structural modification of the lectins so that they
no longer inhibit growth of the fungi.   An additional mode of action of the lectins may be to prevent
dissolution of host cell wall since it has been shown
that they are able to inhibit the activity of fungal
polygalacturonases (Albersheim & Anderson, 1971).

*(iii) Host cell-mycelium interactions which modify
the direction of mediated transport*

There is a further effect of phosphate deficiency in higher plants which may influence the host-
fungus transport system;   this concerns the recent
finding that phosphate deficiency leads to a sharp
decline in cytokinin production by plant roots
(Waring, Personal Communication).   Mothes (1960)
reported that the treatment of tobacco leaves with
kinetin delayed senescence at the site of application
and enhanced the transport of materials applied to
other parts of the leaf lamina, into the zone of
application.   Subsequent studies have also implicated kinetin in the maintenance of membrane integrity
and the strength of the membrane transport function.
Feng (1973) has shown that kinetin treatment of epidermal cells of *Allium cepa* increased the permeability
to urea and thiourea, compounds which are generally
held to move through an aqueous "protein pore"
channel in the membrane.   Permeability to malonamide
and dimethyl urea, which show a much stronger oil to

water partition, was not affected by kinetin.
Circumstantial evidence from such phenomena as the
"green islands" around pathogen-infected leaves,
suggests that the increased cytokinin content of
these tissues, which is possibly produced by the fungus, may alter the normal direction of translocation
in favour of the pathogen.   It may be therefore that
in the case of carbohydrate transport, the reduced
cytokinin content of the host plant cells under low
phosphate supply and an enhanced cytokinin production
in the fungus, enchances the carrier activity of the
fungal plasmalemma relative to that of the host.
Thus if the carbohydrate carrier in the host cell
membrane was of the passive type, it would now be
working down a gradient of concentration from host
cytoplasm into the AFS from where carbohydrate was
rapidly taken up by the fungus.   Figure 9 depicts
the suggested alignment of host and fungal carriers
required to achieve this transfer.   Whether or not
this appears as a reasonable hypothesis for carbohydrate release from the host cell, it will no doubt
seem somewhat arbitrary to align a passive carrier in
the host plasmalemma with an active carrier in the
fungal plasmalemma.   I make this suggestion for two
reasons, firstly that an active pump in the fungus,
being vectorial, would protect the fungus against
resorption of its carbohydrates at least until the
time of breakdown of the arbuscule.   Secondly there
are specific types of active carbohydrate carriers
which have been described from bacterial and fungal
cells which could meet other constraints which may be
expected to arise in the fungal hyphae.   These we
may now consider.

THE MECHANISM AND REGULATION OF THE DIRECTION OF CARBOHYDRATE AND PHOSPHATE TRANSPORT BETWEEN HOST AND MYCORRHIZAL FUNGUS

Carbohydrate Transport

Bevege, *et al.* (1975) provide evidence of

accumulation of glycogen and lipid within the hyphae
of vesicular-arbuscular mycorrhizas.   They suggest
that the significance of the transformation of host
plant sugars into these compounds is to conserve the
gradient for "downhill" transport of carbohydrates
from the AFS into the arbuscules.   It is important
to recall however what such a compartmentation system
implies in terms of energy requirements.   Suppose we
take the case of glucose entering a fungal hypha;
in order to go from glucose to glycogen the following
sequence of reactions would be required:-

$$\text{Glucose + ATP} \xrightarrow{\text{(Hexokinase)}} \text{Glucose-6-PO}_4 \text{ + ADP}$$

$$\text{Glucose-6-PO}_4 \xrightleftharpoons{\text{(Phosphoglucomutase)}[1]} \text{Glucose-1-PO}_4$$

$$\text{Glucose-1-PO}_4 \text{ + ATP} \xrightarrow{\text{(ADPG pyrophosphorylase)}[2]} \text{ADP-glucose + pyrophosphate}$$

$$\text{Glycogen(n) + ADPG} \xrightarrow{\text{(Glycogen synthetase)}[3]} \text{Glycogen (n+1) + ADP}$$

[1] Phosphoglucomutase requires the presence of glucose
1:6 diphosphate to convert it into the active phosphorylated form.

[2] This reaction is generally held to be pulled in the
direction of ADPG formation with a standard free
energy of -7.0 Kcal, by the presence of a pyrophosphatase.

[3] Glycogen synthetase is an allosteric enzyme possess
-ing complex regulatory properties. In its phosphorylated, or D-form, the enzyme requires glucose-6-PO$_4$
as a positive modulator.

Consideration of this reaction sequence, involving the consumption of ATP for the production of
phosphorylated intermediates leads me to suggest that
one should look for an active sugar transport system
into the fungus of the type described by Kundig and
Roseman (1971) for bacteria or Van Steveninck (1968)
for yeast, in which the carbohydrate is phosphorylated as it is transported so that it is accumulated as
the phosphate ester.

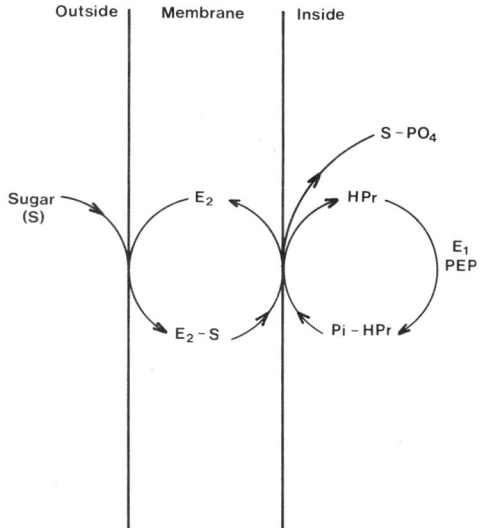

Figure 10.   A model of the phosphoenol
pyruvate-linked mechanism of active phos-
phorylative carbohydrate transport.

Fig. 10 represents the bacterial system which
involves 3 distinct enzymes.   The first enzyme,
located in the cytoplasm, known as the histidine protein (HPr) is phosphophorylated by a second cytoplasmic enzyme designated Enzyme I using phosphoenol-
pyruvate as the phosphate donor according to the
reaction:-

$$PEP + HPr \xrightarrow[Mg^{2+}]{\text{Enzyme 1}} Pyruvate + HPr \sim PO_4$$

The HPr is phosphorylated on nitrogen 1 of the imid-
azole ring of one of its histidine residues.   The
phosphorylated HPr then reacts with the sugar to be
transported, in a reaction catalysed by a third
protein, Enzyme 2.   This enzyme carries out a trans-
phosphorylation reaction of PO_4 from $HPr{\sim}PO_4$ to the
sugar, at the same time changing conformation within
the membrane so that the product, the phosphorylated
sugar, is delivered to the inside of the membrane.
There are a variety of E_2 enzymes specific for different sugars.   The model is supported by genetic
evidence.   Mutants affecting E and HPr lack the
capacity for any sugar transport, whereas mutants of
E_2 affect the transport of only one particular sugar.
      There are indications that yeasts contain both
passive and active sugar transport systems (Rothstein
& Van Steveninck, 1966; Van Steveninck & Rothstein,
1965; Van Steveninck & Booij, 1964).
      By using ^{14}C-labelled 2-deoxy-D-glucose, a sugar
which can be actively transported but not further
metabolised by yeast, it was shown that much of the
sugar appeared in the cells initially as 2-deoxy-D-
glucose 6-phosphate although some was rapidly recon-
verted to the free sugar after entry into the cells,
by the action of a phosphatase (Van Steveninck,
1968).   By means of inhibitor studies it was shown
that much of the activity of this phosphorylating
sugar transport system could be retained when glycol-
ysis was blocked.   The alternative source of energy
for this system was shown to be a pool of polyphosphate from which ATP was generated according to the
reaction:-

$$ADP + (HPO_3)_n \longrightarrow ATP + (HPO_3)_{n-1}$$

as described by Kornberg, Kornberg & Sims (1956).
If the preliminary indications that phosphate transport within the mycorrhizal hyphae takes place in the
form of polyphosphate granules (Cox *et al.*, 1975),
prove correct, then the potential interest of this

polyphosphate-coupled phosphorylating transport
system becomes apparent.   In any event these 2
models of phosphorylation - linked sugar transport
are closely analogous one to the other.   Either
mechanism would meet the requirements for active
sugar transport into the fungus as set out in Fig. 8.
in four respects.   Firstly by preventing the backward loss of sugar from the fungus, secondly by
delivering the sugar in a phosphorylated form readily
available for glycogen synthesis, thirdly in the case
of glucose, the phosphorylated product would serve as
an allosteric activator of glycogen synthetase and
finally, the active carriers would be specific for
the sugars which are observed to be transported.

Phosphate transport

There remains the question of accommodating the
transport of phosphate in this scheme.   I have
already considered steps 1 and 3 shown in Figure 8
suggesting the functioning of an active carrier at
the plasmalemma of the outer fungal hyphae and at the
host cell plasmalemma.   This leaves us with step 2
(Fig. 8), the movement of phosphate out from the
fungus to the AFS.   I have postulated here a passive
phosphate carrier to mediate the Pi secretion from
the arbuscules.   There are several points which
might be mentioned in connection with the feasibility
of operation of this system.   (a) If polyphosphate
does prove to be the form in which phosphate is
carried along the hyphae, then the pyrophosphatase
involved in driving the glycogen synthetase reaction
at the arbuscule end of the hyphal pipe, might also
degrade polyphosphate and so provide the locally high
P_i concentration needed to drive the passive P_i
carrier in the direction of the AFS,   (b) Being a
passive carrier, this would be independent of an
energy source, for which there would be severe competition from the sugar transporting and glycogen
synthesising reactions,   (c) the postulated high

Woolhouse, H. W.

level of cytokinin in the hyphae could serve to render the arbuscule plasma membrane impermeable to the reabsorption of phosphate once released into the AFS.   It is known for example that kinetin depresses the uptake of phosphate by beetroot discs (Palmer, 1967).   (d) Finally it could be that a further function of the lectin-like phosphatase at the surface of the host cell is to modify the plasma membrane of the arbuscule in the region of the phosphate carrier system so that it becomes uncoupled in which form it can only act as a passive carrier.

CONCLUSION

In this paper I have indulged in a large measure of speculation and have drawn on a wide range of biochemical evidence.   My apology for this extravagance is that I have tried to provide a framework, based on more recent biochemical evidence, which suggests some potentially fruitful avenues for a deepening of the investigation of regulatory aspects of the metabolism of these fascinating mycorrhizal associations.

REFERENCES

ALBERSHEIM, P. & ANDERSON, A. J. (1971). Proteins from plant cell walls inhibit polygalacturonase excreted by plant pathogens.   *Proc. natn. Acad. Sci. U.S.A.*, 68, 1815-1819.

ALLEN, A. K. & NEUBERGER, A. (1973). The purification and properties of lectin from potato tubers, a hydroxyproline-containing glycoprotein. *Biochem. J.*, 135, 307-314.

BAYLIS, G. T. S. (1972). Minimum leaves of available phosphorus for non-mycorrhizal plants (short communication).   *Pl. Soil.*, 36, 233-234.

BEVEGE, D. I., BOWEN, G. D. & SKINNER, M. F. (1975).
This symposium, P. 149.

BIELESKI, R. L. (1973).   Phosphate pools, phosphate
transport and phosphate availability.   *A. Rev. Pl
Physiol.*, 24, 225-252.

BOWEN, G. D. (1973). Mineral nutrition of ectotrophic
mycorrhizae.   In *Ectomycorrhizae*, eds. Marks, G. C.
& Kozlowski, T. T., pp. 151-205, Academic Press,
New York.

COX, G. C., SANDERS, F. E., TINKER, P. B. & WILD, J.
(1975). This symposium, p. 297.

DANIELLI, J. F. & DAVSON, H. (1935). A contribution
to the theory of permeability of thin films.
J. Cell. Physiol., 5, 495-508.

EPSTEIN, E. & RAINS, D. W. (1965). Carrier-mediated
cation transport in barley roots: kinetic evidence
for a spectrum of active sites.   *Proc. natn. Acad.
Sci. U.S.A.*, 53, 1320-1324.

FENG, K. A. (1973). Effects of kinetin on the permeability of *Allium cepa* cells.   *Pl. Physiol.*,
Lancaster, 51, 868-870.

FINEAN, J. B. (1969). Biophysical contributions to
membrane structure.   *Q. Rev. Biophys.*, 2, 1-23.

GERDEMANN, J. W. (1965). Vesicular-arbuscular mycorrhizae formed on maize and tuliptree by *Endogone
fasciculata*.   *Mycologia*, 57, 562-575.

GUIDOTTI, G. (1972). Membrane proteins. *A. Rev.
Biochem.*, 41, 731-752.

HALL, J. R. & HODGES, T. K. (1966). Phosphorus metabolism of germinating oat seeds.   *Pl. Physiol.*,
Lancaster, 41, 1459-1464.

HARLEY, J. L. & BRIERLY, J. K. (1954). The uptake of
   phosphate by excised mycorrhizal roots of the
   beech. VI. Active transport of phosphorus from
   the fungal sheath into the host tissue.
   New Phytol., 53, 240-252.

HAYMAN, D. S. & MOSSE, B. (1971). Plant growth
   responses to vesicular-arbuscular mycorrhizae.
   I. Growth of *Endogone*-inoculated plants in phosphate deficient soils. *New Phytol.*, 70, 19-27.

HECHTER, O. (1965). Role of water structure in the
   molecular organisation of cell membranes.
   Fed. Proc., 24, 2-3: S91-S102.

HIGINBOTHAM, N., ETHERTON, B. & FOSTER, R. J. (1967).
   Mineral ion contents and cell transmembrane elec-
   tropotentials of pea and oat seedling tissue.
   Pl. Physiol., Lancaster, 42, 37-46.

HOPKINS, H. T. (1956). Absorption of ionic species of
   orthophosphate by barley roots; effects of 2:4
   dinitrophenol and oxygen tension. *Pl. Physiol.,
   Lancaster*, 31, 155-161.

KORNBERG, A., KORNBERG, S. R. & SIMS, E. S. (1956).
   Metaphosphate synthesis by an enzyme from *Escherichia coli*. *Biochim. biophys. Acta.*, 20, 215-227.

KUNDIG, W., & ROSEMAN, S. (1971). Sugar transport.
   J. biol. Chem., 246. I. 1393-1406, II. 1406-1418.

LAMPORT, D. T. A. (1965). The protein component of
   primary cell walls. *Adv. bot. Res.*, 2, 151-218.

LEHNINGER, A. L. (1970). *Biochemistry: the molecular
   basis of cell structure and function*. Worth Inc.

LEWIS, D. H. & SMITH, D. C. (1967). Sugar alcohols
   (Polyols) in fungi and green plants. I. Distribution, physiology and metabolism. *New Phytol.*,
   66, 143-184.

LEWIS, D. G. & QUIRK, J. M. (1965). Diffusion of phosphate to wheat plant roots. *Nature, London,* 205, 765-766.

MOTHES, von K. (1960). Über das Altern der Blätter und die Möglichkeit ihrer Wiederverjungung. *Naturwissenschaften,* 47, 337-351.

MARCHESI, V. T., TILLACK, T. W., JACKSON, R. L., SEGREST, J. P. & SCOTT, R. E. (1972). Chemical characterisation and surface orientation of one major glycoprotein of the human erythrocyte membrane. *Proc. natn. Acad. Sci. U.S.A.,* 69, 1445-1450.

MEDWECZKY, N. & ROSENBERG, H. (1969). The binding and release of phosphate by a protein isolated from *Escherichia coli. Biochim. biophys. Acta,* 192, 369-371.

MEDWECZKY, N. & ROSENBERG, H. (1970). The phosphate-binding protein of *Escherichia coli. Biochim. biophys. Acta,* 211, 158-168.

NEU, H. C. & CHOU, J. 1967. Release of surface enzymes in Enterobacteriacea by osmotic shock. *J. Bacteriol.,* 94, 1934-1945.

PALMER, J. M. (1967). The effect of some plant growth substances on the induction of enzymatic activities in thin slices of plant tubers. In *Biochemistry and physiology of plant growth substances.* Eds. Wightmann, E. & Setterfield, G., p401-415. Runge Press, Ottawa.

PINTO DA SILVA, P. & BRANTON, D. (1970). Membrane splitting in freeze-etching. *J. Cell Biol.,* 45, 598-605.

RADDA, G. K. & VANDERKOOI, J. (1972). Can fluorescent probes tell us anything about membranes? *Biochim. biophys. Acta,* <u>265</u>, 509 - 549.

ROTHSTEIN, A. & VAN STEVENINCK, J. (1966). Phosphate and carboxyl ligand of the cell membrane in relation to uphill and downhill transport of sugars in the yeast cell. *Ann. N.Y. Acad. Sci.,* <u>137</u>, 606-623.

RUSSELL, R. S. & MARTIN, R. P. (1953). A study of the absorption and utilisation of phosphate by young barley plants. *J. exp. Bot.,* <u>4</u>, 108-127.

SANDERS, F. E. & TINKER, P. B. (1971). Mechanism of absorption of phosphate from soil by *Endogone* mycorrhizas. *Nature, London,* <u>233</u>, 278-279.

SANDERS, F. E. & TINKER, P. B. (1973). Phosphate flow into mycorrhizal roots. *Pestic. Sci.,* <u>4</u>, 385-395.

SHARON, N. & LIS, H. (1972). Lectins: cell agglutinating and sugar specific proteins. *Science, N.Y.,* <u>177</u>, 949-959.

SIMON, E. W. (1974). Phospholipids and plant membrane permeability. *New Phytol.,* <u>73</u>, 377-420.

SINGER, S. J. & NICOLSON, G. L. (1972). The fluid mosaic model of the structure of cell membranes; cell membranes are viewed as two-dimensional solutions of oriented globular proteins and lipids. *Science, N.Y.,* <u>175</u>, 720-731.

SPATZ, L. & STRITTMATTER, P. (1971). A form of cytochrome b5 that contains an additional hydrophobic sequence of 40 amino acid residues. *Proc. natn. Sci. U.S.A.,* <u>68</u>, 1042-1046.

VAN STEVENINCK, J. & BOOIJ, H. L. (1964).   The role
of polyphosphates in the transport mechanism of
glucose.   *J. gen. Physiol.*, <u>48</u>, 43-60.

VAN STEVENINCK, J. & ROTHSTEIN, A. (1965). Sugar
transport and metal binding in yeast.   *J. gen.
Physiol.*, <u>49</u>, 235-246.

VAN STEVENINCK, J. (1968). Transport and transport-
associated phosphorylation of 2-deoxy-D-glucose in
yeast.   *Biochim. biophys. Acta*, <u>163</u>, 386-394.

VASEY, E. H. & BARBER, S. A. (1963). Effect of
placement on the absorption of Rb^{86} and P^{32} from
soil by corn roots.   *Proc. Soil Sci. Soc. Amer.*,
<u>27</u>, 193-197.

WEIGL, J. (1967). Beweis für die Beteiligung von
beweglichen transportstrukturen (Trägern) beim
ionen-transport durch pflanzliche membranen und die
kinetik des Anionentransports bei *Elodea* im Licht
und Dunkeln.   *Planta*, <u>75</u>, 327-342.

WEIGL, J. (1968). Austausch-mechanismus des ionen-
transports in pflanzen am beispiel des phosphat-
und chloridtransports bei maiswurzeln.   *Planta*,
<u>79</u>, 197-207.

WOOLHOUSE, H. W. (1969). Differences in the properties of the acid phosphatases of plant roots and
their significance in the evolution of edaphic
ecotypes.   In *Ecological aspects of the mineral
nutrition of plants*.   Ed. Rorison, I. H.
Symp. Br. ecol. Soc., <u>9</u>, 357-380.   Blackwell,
Oxford.

PHOSPHATE PHYSIOLOGY OF
VESICULAR-ARBUSCULAR MYCORRHIZAS

G. D. BOWEN, D. I. BEVEGE[1], AND B. MOSSE[2]

*Division of Soils, CSIRO, Glen Osmond,
South Australia, 5064.*

INTRODUCTION

There is mounting evidence that much of the plant growth response to endomycorrhizal infection is due to increased uptake of slowly diffusing ions such as phosphate as a result of increased soil exploration by hyphae (eg. Sanders & Tinker, 1971; Hattingh *et al.*, 1973).   Something also needs to be known about the physiology of the mycorrhizal system: What are the relative absorbing powers of fungus and root?   Does the mycorrhizal plant use its absorbed nutrients in the same ways as the uninfected plant? What is the partitioning between root and shoot? What is the biochemical fate of absorbed phosphate? Do endomycorrhizas behave similarly to ectomycorrhizas and act as a phosphate storage sink?

In previous studies with radioactive isotopes endomycorrhizas greatly increased phosphate and zinc absorption from solution (Gray & Gerdemann, 1967;

---

[1]Forest Biology Laboratory, Indooroopilly, Queensland, Australia.

[2]Rothamsted Experiment Station, Harpenden, Herts., U.K.

Bowen, G. D. *et al.*

Morrison & English, 1967; Bowen & Mosse, 1969; Bowen *et al.*, 1974). We have extended the use of radioactive isotopes to answer some of the questions raised above.

ABSORPTION OF PHOSPHATE

Clover and onion

Absorption of phosphate from solution was studied in short term uptake experiments with subterranean clover (*Trifolium subterraneum* L. var. Bacchus Marsh) and onion grown in a sterilised, phosphate deficient soil (Mt. Burr sand, Stephens *et al.*, 1941), inoculated with sporocarps of *Endogone mosseae*. Non-mycorrhizal seedlings received spore or sporocarp washings. At harvest the roots were detached, carefully washed free of adhering soil, mycorrhizal roots separated from non-mycorrhizal on infected plants, and equilibrated for $\frac{1}{2}$-1 hr in 0.5 mM $CaSO_4$. They were then placed in a fresh $CaSO_4$ solution containing ^{32}P (100-300 μM KH_2PO_4 (pH 6.5) at 20°C. This concentration of phosphate approximates that which may occur in soil solution. After absorption roots were washed free of unabsorbed phosphate in running water for 7 min., and lightly blotted. The radioactivity of digested roots was measured by GM counting or by liquid scintillation counting using the Cerenkov method.

Mycorrhizal roots of both onion and clover took up more than twice as much P as roots from non-mycorrhizal plants (Table 1). Differences in uptake between mycorrhizal and non-mycorrhizal roots on the same plant were much smaller and were not significant in the second experiment with onions.

Uptake into non-mycorrhizal roots was apparently stimulated by the presence of mycorrhizas on the same plant. Gray and Gerdemann (1969) obtained a similar result.

Table 1.   Absorption of phosphate by mycorrhizal and non-mycorrhizal roots of clover and onion.

$$\text{counts s}^{-1}\text{ mg}^{-1}$$

	Mycorrhizal plants		Non-mycorrhizal plants	Statistical significance
	mycorrh. roots a	uninfected roots b	c	
clover°	436 ± 59$^\Delta$		177 ± 31	a >>> c
onion†	89 ± 10	63 ± 8		a > b
	26 ± 0.3	23 ± 0.3	11 ± 0.1	a,b >>> c

* Different at >10%, >>1% and >>>0.1% probability level
° 30 min uptake from 10µM KH_2PO_4; ^{32}P at 200 µc l^{-1} 14 replicates
† 1 hr uptake from 5 µM KH_2PO_4; ^{32}P at 200 µc l^{-1} 12 replicates
Δ Mean ± S.E. of the mean.

Hoop pine

Plants were raised in either unsterilized nursery soil which carried a mixed inoculum including *E. araucareae* (Bevege, 1971) *E. mosseae* and *E. macrocarpa*, or in a sterilised peat/sand mixture inoculated with *E. araucareae* and watered initially with a complete nutrient solution. Bevege (1971) described the endomycorrhizal infection of hoop pine in detail. Briefly the hoop pine roots form many small, approx. 1 mm long laterals often incorrectly referred to as nodules. These quickly become heavily suberized, but growth may be renewed in subsequent seasons (Plate III, la & b.) The laterals may become infected (prior to suberization) with *Endogone* which forms coils, vesicles and arbuscules. The uptake

procedure was as for clover and onion, except that
intact plants and a 15 min. uptake period were used.
After uptake and washing, roots bearing the short
infected laterals were attached to chromatography
paper and dried.   They were then classified as
mycorrhizal or non-mycorrhizal using a dissecting
microscope to detect extra-matrical hyphae (confirmed
later by clearing and staining), and dissected individually for counting of radioactivity.   At least 40
individuals were counted in each replication if
possible.   Infected and uninfected laterals were of
similar size and the results refer to the radioactivity per lateral.

The results from three experiments are summarized in Table 2.   Unsuberized mycorrhizal roots
absorbed 1.6-1.8 times more ^{32}P than similar uninfected roots.   Absorption was metabolically mediated,
being inhibited by some 95-100 per cent by 1×10^{-3}M
KCN or temperatures of 2°C;   thus the data represent
true absorption.   The ^{32}P in suberized mycorrhizal
roots was 1.8-2.4 times that of uninfected suberized
roots but studies with KCN showed that most of the
"absorption" by uninfected suberized roots was not
metabolically mediated and micro autoradiographic
studies showed very great activity on the suberin
layer.   We therefore consider a high proportion of
the activity in the nonmycorrhizal suberized root to
be *adsorption* not true uptake.   The low *absorption*
by such roots is consistent with results of earlier
workers with other tree species (Addoms, 1946;
Kramer & Bullock, 1966).   Morrison and English
(1967) found a 50 percent reduction of "absorption"
of phosphate at 1°C (compared with 25°C) for mycorrhizal and non-mycorrhizal roots in the closely
related tree species *Agathis australis* and some adsorption by suberized tissue may have occurred in
that instance also.

Table 3 demonstrates some marked effects of
previous nutritional history on absorbing power of
hoop pine mycorrhizas.   Seedlings in peat/sand were

Table 2.   Absorption of phosphate by endomycorrhizas of hoop pine
^{32}P counts s^{-1}/lateral

Experiment	Mycorrhizal Plants		Non-mycorrhizal Plants	Statistical Significance*
	Mycorrhizal roots	Uninfected roots		
(i)				
Suberized roots	3.69 ± 0.26	1.89 ± 0.26		a>>b
(ii)				
Suberized roots	1.76 ± 0.46	1.0 ± 0.45	0.73 ± 0.10	a>>c
Unsuberized roots	2.23 ± 1.13	1.28 ± 0.10	0.90 ± 0.13	a>c,b
				b>>c
(iii)				
Suberized roots	12.71 ± 2.80	3.8 ± 0.17	5.9 ± 1.00	a>>>b,c
Unsuberized roots	8.40 ± 1.40	4.6 ± 1.30	3.6 ± 0.60	a>c

* Different at >5%, >>1% and>>>0.1% probability level.

(i)   Plants 18 months old, grown in soil, uptake at 20°C

(ii)   Plants 33 months old, grown in peat/sand, uptake at 26°C

(iii) Plants 32 months old, grown in peat/sand, uptake at 23°C

Table 3.   Phosphorus and nitrogen uptake by hoop pine mycorrhizas

| | Nutrient levels before uptake | | | |
| | N1 | | N2 | |
	P1	P2	P1	P2
PLANT DATA – (MEANS OF 2 PLANTS)				
Plant Dry Wt. (mg)				
Mycorrhizal plants (M)	382	345	422	453
Uninfected plants (U)	311	297	445	408
Total P (µg/plant)				
M	388	320	400	400
U	246	158	261	291
Total N (µg/plant)				
M	1376	1043	2223	2322
U	1448	1242	2525	2118
Shoots				
P (% of D.W.)				
M	0.12	0.10	0.11	0.11
U	0.08	0.06	0.06	0.08
N (% of D.W.)				
M	0.45	0.37	0.72	0.69
U	0.58	0.52	0.82	0.65
Roots				
P (% of D.W.)				
M	0.09	0.09	0.08	0.06
U	0.07	0.05	0.05	0.06

Table 3. (Contd.)

	Nutrient levels before uptake			
	N1		N2	
	P1	P2	P1	P2
N (% of D.W.)				
M	0.29	0.23	0.29	0.29
U	0.30	0.30	0.33	0.32

SHORT TERM ^{32}P UPTAKE COUNTS PER SECOND PER LATERAL*

	P1	P2	P1	P2
M Mycorrhizas (a)	4.20±0.63°	14.60±3.33	10.63±1.65	13.28±2.58
Uninfected roots (b)	2.97±0.77	5.31±1.10	3.47±1.67	4.27±1.80
U Uninfected roots (c)	3.43±0.63	7.07±1.77	6.35±0.92	6.02±0.78
Statistical analysis*		a>>>b	a>>>b	a>>>b,c
		a>>c	a>>c	
			c>b	

* Differences significant at >5%, >>1%, and >>>0.1% probability level.
° Mean±standard error of mean calculated from 40-100 individual laterals.
  Plants were 33 months old, grown in peat/sand mixture and uptake was at
  23°C.
  N1 = 1, N2 = 25 ppm N as NH$_4$NO$_3$ in peat/sand mixture.
  P1 = 0.5, P2 = 12.5 ppm P as Na$_2$HPO$_4$ in peat/sand mixture.

raised in 2 x 2 factorial combinations of nitrogen and phosphorus. Final concentrations in the medium were 1 and 25 ppm (N_1 and N_2) and 0.5 and 12.5 ppm (P_1 and P_2).

In each nutrient regime the effect of mycorrhizas on plant growth was slight and there was no large mycorrhizal effect on total N absorbed, or on the concentration of N in the shoots or roots. However, the total P absorbed by mycorrhizal plants was 1.4-2 times that of corresponding non-mycorrhizal plants and they had a higher P concentration in roots and shoots. The ^{32}P uptake of mycorrhizas was 1.5-3 times that of uninfected laterals on the same root. In contrast to the onion experiments, uptake by uninfected roots on mycorrhizal plants in N_1P_2, N_2P_1 and N_2P_2 treatments were 0.5-0.7 that of corresponding roots on non-mycorrhizal plants (significant at 5% probability level for N_2P_1 plants). Mycorrhizas and uninfected roots of plants grown in low P - low N absorbed less ^{32}P, 0.3-0.4 and 0.5-0.6 respectively, than roots from other nutrient treatments but had similar N and P concentrations.

P ACCUMULATION BY HYPHAE AND ROOT CELLS

Macro-autoradiographs of labelled mycorrhizal roots of onion, clover and hoop pine from the above experiments were made, using X-ray film ("Kodirex" - Kodak Australia, Pty. Ltd., Melbourne). Roots were fixed to a board and either dried before exposure to the X-ray film, or exposed in a frozen state. Young uninfected root tips had high phosphate uptake which decreased sharply with distance from the apex, except where mycorrhizal infection occurred. Here, radioactivity was also high (Fig. 1).

Microautoradiographs were carried out on pine, clover and onion roots to examine whether the increased uptake by mycorrhizal tissue was primarily due to uptake by the hyphae or to fungus stimulation of

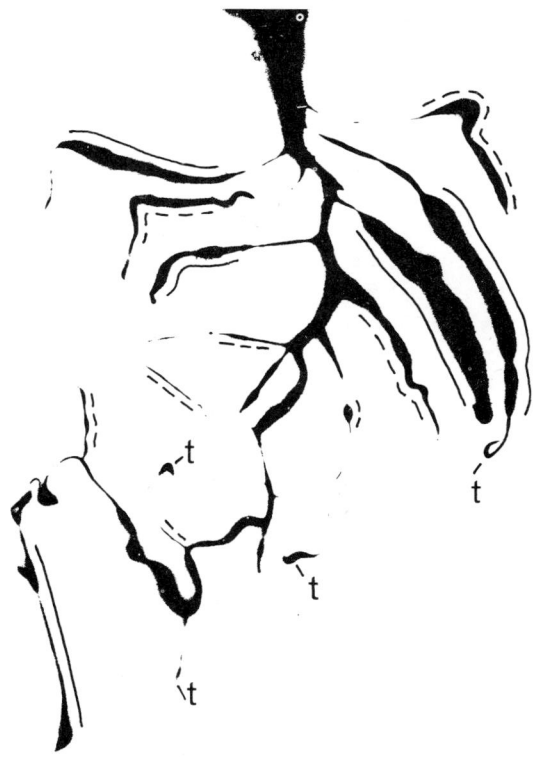

Figure 1. ^{32}P accumulation at tips and *Endogone*-infected portions of subterranean clover after 1 hr uptake from 5μM KH_2PO_4. ———— , heavy infection; ------ , light infection.

uptake by the root cells.   Stripping film was used (AR-10, Kodak Ltd., London) and the tissue preparation methods included (i) conventional methods of embedding and sectioning in paraffin wax and applying an experimentally determined correction factor for ^{32}P loss during fixation of mycelium and root cells and (ii) methods designed to retain water soluble

compounds e.g. freezing followed by freeze substitution or freeze drying and then wax infiltration.
Plate III, 2a demonstrates the accumulation of ^{32}P
by extra-matrical hyphae (frozen preparation) and
Plate III, 2b, of material prepared by method (i),
demonstrates the accumulation of ^{32}P in arbuscules
(and vesicles) of onion with very little accumulation
in neighbouring uninfected cortical cells.   A factor
of 2.5 must be applied to the uninfected cortical
cells to compensate for higher loss from these cells
than from mycelia during fixation in the formalin
(2.2%) – water (45%) solution.   Autoradiographs of
matched material using the freeze-drying-wax infiltration method of Branton and Jacobson (1961) i.e. no
loss of water soluble compounds, gave similar
results.   In such preparations transects across 3
uninfected cortical cells gave silver grain counts of
20, 8 and 12 per cell transect (adjusted for a background of 16) while transects across three neighbouring cells of similar size but with arbuscules gave
counts of 52, 56 and 62, mainly in the arbuscle, i.e.
arbuscules had some 4 times the activity of uninfected cells.

     In other uptake studies, roots from which we
detached the external hyphae took up much less P.
Phosphate was also taken up by germ tubes arising
from *Endogone* spores and this uptake was inhibited by
1°C.

     We conclude that uptake by hyphae external to
the root is a primary cause for increased absorbing
power of mycorrhizas.   In a soil situation phosphate
is translocated to arbuscules (and vesicles) inside
the root, and moves from them into the higher plant.

PHOSPHATE STORAGE AND DISTRIBUTION

     The possibility that the fungus in endomycor-
rhizas acts as a phosphate storage pool, as in ectomycorrhizas (Harley, 1969), was examined in two ways.

(a) *Partitioning of phosphate.*   Six week old mycorrhizal and non-mycorrhizal onion plants were washed
free of soil and placed in a large volume of 5μM
KH_2PO_4 (plus ^{32}P at 100 μc/1) in 0.5 μM $CaSO_4$ for 18
hr, in a growth cabinet at 20°C with light at 20,400
lumen m^{-2}.   After washing and digestion of the material the ^{32}P content of external mycelium, roots and
shoots was measured.   Table 4 shows that mycorrhizal
and non-mycorrhizal plants had similar root weights
but the shoot growth of mycorrhizal plants was double
that of non-mycorrhizal, thus leading to a significantly lower root/shoot ratio (P<05).   The hyphae
were approximately one-sixth the weight of the roots.
It is difficult to remove all organic material from
the mycelium and to obtain a reliable fresh weight.
That of the mycelium in Table 4 is rather higher than
reported for clover (6% of the infected roots)
(Bevege, Bowen & Skinner, 1975), but it is unlikely
that this has greatly affected the conclusions
reached.

Table 4.   Partitioning of absorbed phosphate
in mycorrhizal and non-mycorrhizal onion

	Mycorrhizal Plants	Non-Mycorrhizal Plants
F.Wt. Shoots (g)	$0.510^{+}\pm0.13$	0.227 ± 0.07
F.Wt. Roots (g)	0.240 ± 0.01	0.261 ± 0.08
F.Wt. Mycelium (g)	0.038 ± 0.007	
Root/Shoot ratio	0.52 ± 0.11	1.19 ± 0.22
^{32}P Uptake (counts/s^{-1})		
^{32}P Total uptake/mg root	2.83 ± 0.32	2.95 ± 0.62
^{32}P /mg root	2.37 ± 0.40	2.58 ± 0.61
^{32}P /mg mycelium	3.65 ± 0.42	
% ^{32}P transported to shoot	$10.6 \pm$	13.3 ± 2.0
% ^{32}P in mycelium	14.7 ± 0.45	

[+] Mean ± S.E. of mean (3 replicates)

There was little difference between mycorrhizal and non-mycorrhizal plants in the percentage of absorbed phosphate transported to the shoot.   The ^{32}P in the mycelium accounted for almost 15 percent of the total label and the mycelium had a somewhat greater ^{32}P concentration than the roots.   However too much should not be made of such a difference as a large proportion of the weight of roots may not be able to accumulate ^{32}P eg. xylem vessels.   The data here, and that of the microautoradiographs indicate that the fungus has a slight phosphate storage capacity, but because of its much smaller mass, the total stored is small compared with that of ectomycorrhizas.   A corollary of the similar transport to the tops of endomycorrhizal and non-mycorrhizal plants is that transfer probably occurs continually without the need for arbuscule degeneration.

(b) *Redistribution of phosphate in roots*.   Subterranean clover seedlings were planted in sterilized Mt. Burr sand in 2.5 cm wide polyethylene tubing, with phosphate (as KH_2PO_4) added at 23 µg/plant ("Low P" plants).   Half the plants received an extra 30 µg/plant 12 days later ("High P" plants).   *Endogone mosseae* inoculum (as sporocarps) was placed in a layer either 1 cm below the surface ("shallow") or 7 cm below the surface ("deep").   After 18 days under controlled environment conditions (20°C, 16 hr day with light at 20,400 lumen m^{-2}) infection was well developed with many arbuscules and a few vesicles.   The roots and tubes were then severed 1 cm below the deep inoculum layer and were stood in a 2 cm deep layer of 0.5 µM $CaSO_4$ containing 5µM KH_2PO_4 with ^{32}P at 1 mc/1.   After 4 hr and 24 hr roots were removed, unabsorbed phosphate was removed by washing and the ^{32}P content measured in roots from the 1 cm zone of soil corresponding to the positions of deep and shallow inoculum.   Six replicate plants were used for each mycorrhizal placement and fertilizer treatment.   The ^{32}P counts gave no indication of a high phosphate storage in the "deep" mycorrhizal

inoculum layer (in the ^{32}P solution) nor of any preferential redistribution of ^{32}P to the "shallow"
mycorrhizal infection, 5 cm away.   Almost no redistribution of ^{32}P occurred to untreated parts of the
roots.   The ^{32}P translocated to the tops of high and
low P plants was 72 and 58 percent respectively of
the total absorbed, with no difference between mycorrhizal and non-mycorrhizal plants.   It is concluded
that phosphate is neither translocated preferentially
towards mycorrhizas, nor do they store it.

BIOCHEMICAL INCORPORATION

Onion.   Following uptake from 5µM KH_2PO_4 (^{32}P at 300
µc/1) solution for 1 hr, the acid soluble fraction of
the excised roots was removed by freezing and thawing
in 0.2 N perchloric acid.   In an experiment with 4
replicates and another with 7, the percentage incorporation into the acid soluble fraction was identical
in mycorrhizal and non-mycorrhizal roots (90 ± 1.5
percent in one experiment and 86.5 ± 1.6 percent in
the other).   However in a similar experiment on
mycelia alone there was 79 percent incorporation of
^{32}P into acid soluble phosphates.   This and the
greater extraction of ^{32}P from roots than from mycelia by formalin-alcohol-acetic acid fixation (see
above) suggests some differences in labelling pattern
between mycelia and roots, but the greater mass of
roots masks this.

Hoop pine.   Roots of 39 month hoop pine grown with
0.5 or 2.5 ppm P as NaH_2PO_4 were given a 15 min
uptake from 5 µM KH_2PO_4 with ^{32}P at 5 mC l^{-1}.Unabsorb-
ed phosphate was removed and the plants transferred
to a large volume of a solution of the same composition but omitting the ^{32}P.   Fractionation of ^{32}P was
made by the Ogur-Rosen extraction method for acid
soluble phosphate, nucleic acids, phospholipids and
phosphoproteins (Martin & Morton, 1956).   At time 0

and after 4 hrs there were only minor differences
between roots of mycorrhizal and non-mycorrhizal
plants in the distribution of label; consequently
only the shoots were fractionated at 26 hr.   The
relative incorporation of ^{32}P into the nucleic acid
and protein fractions in shoots of non-mycorrhizal
plants was 1½-2 times that of corresponding mycorrhizal plants (Table 5), suggesting (i) larger
amounts of these in non-mycorrhizal plants, (ii)
greater metabolic turnover of phosphate in non-mycorrhizal plants or (iii) equilibration with a smaller
pool of acid soluble phosphates in the non-mycorrhizal plants.   To a very large extent the labelling
pattern reflected phosphate status of the leaf
(except for the high P, non-mycorrhizal) and the last
interpretation is the most likely.

Table 5. ^{32}P incorporation into shoots
of mycorrhizal and non-mycorrhizal hoop pine

	Mycorrhizal Plants		Non-Mycorrhizal Plants	
	Low P*	High P*	Low P*	High P*
% Total P in shoot - D.W. basis	0.14	0.09	0.04	0.10
%^{32}P incorporation at 26 hr.				
Acid soluble	80.9	71.5	61.5	49.9
Nucleic acids	12.3	18.1	23.6	31.7
Phospho lipid	0.9	1.2	2.1	0.6
Phospho protein	5.9	9.3	12.8	17.8

* One plant per treatment.   Low P plants
were grown with 0.5 ppm P and High P
with 2.5 ppm P.

DISCUSSION

The results here show that although vesicular arbuscular mycorrhizas can absorb several times more phosphate from solution than uninfected roots, sometimes very much smaller increases in absorbing power (Nye, 1966) occur, especially if non-mycorrhizal and mycorrhizal roots of the same plant are compared (Table 1 and 2). The autoradiographic studies indicate that in mycorrhizas much of the increase in uptake is due to uptake and translocation by the fungal hyphae external to the root, rather than to a fungal stimulation of ion uptake by uninfected cells. However non-mycorrhizal roots on mycorrhizal plants can have appreciably greater phosphate absorbing power (from solution) than those on non-mycorrhizal plants. (The experimental methods precluded the possibility of redistribution of absorbed phosphate from mycorrhizal to non-mycorrhizal roots on the same plant). Correlations occur between absorbing power of roots and assimilate supply (Rovira & Bowen, 1973); increased growth and more assimilates in mycorrhizal plants may thus increase the absorbing power of non-mycorrhizal roots in such plants. Although hyphal growth into soils is apparently the main reason for the greater uptake by mycorrhizas of poorly mobile ions from soil (Bowen & Rovira 1969; Sanders & Tinker, 1971) increases in absorbing power of uninfected roots may well affect also the uptake of more mobile ions e.g. nitrate and potassium (Nye, 1966).

Increases in phosphate absorbing power from solution, and phosphate uptake from soil, are affected primarily by the extent of external hyphal growth, which can vary widely depending on soil conditions (Daft & Nicolson, 1966; Mosse, 1973) and the fungus-host combination (Bevege and Bowen, these proceedings). Much remains to be found out concerning the penetration of soil by hyphae, the longevity of absorption by individual hyphae and possible differences in nutrient absorption characteristics of the different fungi concerned. Assuming equal specific gravity, 1 mg of hyphae of 10 μm diameter has the same length as 1600 mg of root of 400 μm diameter or 1-4 mg of root hairs. Compared to roots, hyphae are therefore a highly efficient and energy conserving mechanism for the absorption of poorly mobile ions. This is particularly important for many trees and shrubs which have much lower rooting densities than herbs and grasses (Barley, 1970) and under most nutritional conditions must be an important competitive factor in mixed communities of trees. In low nutrient conditions the growth of hyphae into soil may, of course, also be important to nutrition of grasses and herbs. Bowen *et al.*, (1974) have shown that in phosphate deficient wheat plants growth of main root axes and first order laterals is favoured over that of finer, lower order laterals which give efficient soil exploitation. Happily, the infection of grass roots by vesicular arbuscular mycorrhizas offsets this scarcity of fine laterals under low phosphate conditions.

A further feature of endomycorrhizal infection, particularly evident with trees such as hoop pine but possibly present also to a lesser extent in grasses and herbaceous plants, is the longer functional life of older parts of roots. Mycorrhizal infection of hoop pine enables the short laterals to take up nutrients even after they are suberized. Fungi in the infected part subsequently infect tissue arising from renewed growth of the beads (Plate 1b), thus making them efficient absorbing organs of nutrients and water for the next 2-3 years.

While the mycorrhiza clearly affected phosphate uptake by clover, onion and hoop pine, the rate of transport of the absorbed P into the shoot, which can be stimulated by some rhizosphere micro-organisms (Bowen & Rovira, 1969), was little affected. That the mycorrhizal endophytes neither store appreciable amounts of absorbed phosphate, nor obtain them preferentially from the plant is shown by the ^{32}P content of the mycelium which was only a little greater than that of roots of non-mycorrhizal plants (Table 4) and by the non-accumulation of ^{32}P in infected parts of the root when uninfected parts were supplied with ^{32}P- phosphate.   We suggest that a consequence of the poor P storage capacity must be a rapid transfer to the host plant, *via* the relatively large surface area of the arbuscules, without a prior need for arbuscule degeneration.

Rhizosphere micro-organisms can enhance incorporation of absorbed phosphate into nucleic acid fractions of roots and shoots (Bowen & Rovira; 1969) but from our results, there appears to be no evidence of this with VA mycorrhizas.   With the exception of shoots of the high phosphate, non-mycorrhizal hoop pines where phosphate incorporation into the nucleic acid fraction was higher, not lower, than in mycorrhizal plants, labelling patterns reflected phosphate concentration of the shoots.

REFERENCES

ADDOMS, R. M. (1946). Entrance of water into suberized roots of trees.   *Pl. Physiol.*, Lancaster, 21, 109-111.

BARLEY, K. P. (1970). The configuration of the root system in relation to nutrient uptake.   *Adv. Agron.*, 22, 159-201.

BEVEGE, D. I. (1971). Vesicular arbuscular mycorrhizas of *Araucaria:* Aspect of their ecology and physiology and role in nitrogen fixation.   Ph.D. Thesis, Univ. of New England, Armidale, Australia, 369 pp.

BEVEGE, D. I., BOWEN, G. D. & SKINNER, M. F. (1975). This symposium, p. 149.

BEVEGE, D. I. & BOWEN, G. D. (1975). This symposium, p. 77.

BOWEN, G. D. (1973). Mineral nutrition of ectomycorrhizae. In *Ectomycorrhizae,* (eds. Marks, G. C. and Kozlowski, T. T.) pp. 151-205, Academic Press, New York.

BOWEN, G. D., CARTWRIGHT, B. & MOONEY, J. R. (1974). Wheat root configuration under phosphate stress. In *Mechanisms of Regulation of Plant Growth,* eds. Bielieski, R. L., Ferguson, A. R. & Cresswell, M. M. pp. 121-125. Bul. 12, Roy. Soc. N.Z., Wellington, New Zealand.

BOWEN, G. D. & MOSSE, B. (1969). In Bowen, G. D. & Rovira, A. D. (1969), see below.

BOWEN, G. D., SKINNER, M. F. & BEVEGE, D. I. (1974). Zinc uptake by mycorrhizal and uninfected roots of *Pinus radiata* and *Araucaria cunninghamii.* *Soil Biol. Biochem.,* 6, In Press.

BOWEN, G. D. & ROVIRA, A. D. (1969). The influence of microorganisms on root growth and metabolism In *Root Growth,* (ed. Whittington, W. J.) pp. 170-201, Butterworths, London.

BRANTON, D. & JACOBSON, L. (1961). Freeze-drying of plant material. *Exp. Cell Res.,* 22, 559-568.

DAFT, M. J. & NICOLSON, T. H. (1966). Effect of *Endogone* mycorrhiza on plant growth. *New Phytol.,* 65, 343-350.

GRAY, L. E. & GERDEMANN, J. W. (1967). Influence of vesicular-arbuscular mycorrhiza on the uptake of phosphorus-32 by *Liriodendron tulipifera* and *Liquidamber styraciflua.* *Nature, London,* 213, 106-107.

GRAY, L. E. & GERDEMANN, J. W. (1969). Uptake of phosphorus-32 by vesicular-arbuscular mycorrhizae. *Pl. Soil*, <u>30</u>, 415-422.

HARLEY, J. L. (1969). *The Biology of Mycorrhiza*, Leonard Hill, London.   334 pp.

HATTINGH, M. J., GRAY, L. E. & GERDEMANN, J. W. (1973). Uptake and translocation of ^{32}P-labelled phosphate to onion roots by endomycorrhizal fungi. *Soil Sci.*, <u>116</u>, 383-387.

KRAMER, P. J. & BULLOCK, H. C. (1966). Seasonal variations in the proportions of suberized and unsuberized roots of trees in relation to the absorption of water.   *Am. J. Bot.*, <u>53</u>, 200-204.

MARTIN, E. M. & MORTON, R. K. (1956). The chemical composition of microsomes and mitochondria from silver beet.   *Biochem. J.*, <u>64</u>, 221-235.

MORRISON, T. M. & ENGLISH, D. A. (1967). The significance of mycorrhizal nodules of *Agathis australis*. *New Phytol.*, <u>66</u>, 245-250.

MOSSE, B. (1973). Advances in the study of vesicular-arbuscular mycorrhiza.   *A. Rev. Phytopath.*, <u>11</u>, 171-196.

NYE, P. H. (1966). The effect of the nutrient intensity and buffering power of a soil; and the absorbing power, size and root hairs of a root, on nutrient absorption by diffusion.   *Pl. Soil*, <u>25</u>, 81-105.

ROVIRA, A. D. & BOWEN, G. D. (1973). The influence of root temperature ^{14}C assimilate profiles in wheat roots.   *Planta*, <u>114</u>, 101-107.

SANDERS, F. E. & TINKER, P. B. H. (1971). Mechanism of absorption of phosphate from soil by *Endogone* mycorrhizas.   *Nature, London,* <u>233</u>, 278–279.

STEPHENS, C. G., CROCKER, R. L., BUTLER, B. & SMITH, R. (1941). A soil and land use survey of the Hundreds of Riddock, Hindmarsh, Grey, Young and Nangwarry, County Grey, South Australia.   *Coun. scient. ind. Res. Aust. Bull.,* <u>142</u>, 55 pp.

THE EFFECT OF FOLIAR-APPLIED PHOSPHATE ON THE MYCORRHIZAL INFECTIONS OF ONION ROOTS

F. E. SANDERS

Department of Plant Sciences,
University of Leeds, U.K.

INTRODUCTION

It is now generally accepted that large growth responses can be the result of the development of vesicular-arbuscular mycorrhizas on the root systems of plants growing in phosphorus deficient soils. These growth responses are usually associated with improved phosphorus nutrition of the host plant.

Sanders and Tinker (1971, 1973) showed that this improved phosphorus nutrition resulted from an increased efficiency of phosphorus uptake from the soil. This and subsequent work (Hattingh *et al.*, 1974; Pearson & Tinker, 1975) have shown beyond reasonable doubt that the external hyphae of the endophyte are able to absorb phosphate from the soil and translocate it into the host root, providing a pathway for transport in parallel with the normal diffusion pathway. There is little evidence to suggest that mycorrhizas have access to forms of soil phosphate unavailable to non-mycorrhizal roots (Sanders & Tinker, 1971; Hayman & Mosse, 1972).

The efficiency of phosphate uptake by roots may be measured as the rate of uptake per unit length of root, or *inflow*. Inflows may be calculated from growth analysis data (Brewster & Tinker, 1972,

Sanders & Tinker, 1971, 1973).   Using this technique
phosphate inflows to developing mycorrhizal root
systems can readily be calculated and compared with
inflows to non-mycorrhizal root systems.   Inflows to
mycorrhizal root systems are substantially larger
than those to non-mycorrhizal root systems where a
mycorrhizal growth response is observed.

In pot experiments, when onion seedlings are
inoculated at the early seedling stage, if the percentage of root length infected (% infection) is
plotted against time, the resulting curve is typically sigmoid (Tinker, 1975; Sanders, Tinker &
Palmerley, in preparation).   The process of development of the mycorrhizal root system during vegetative
growth of the host may therefore be described in
terms of three phases; an establishment phase, a
phase where the % infection increases rapidly, and an
equilibrium phase where the % infection tends to a
constant value.

The degree to which root systems become infected
by VA endophytes is related to the availability of
phosphorus in the growth medium (see Daft & Nicolson,
1969: Mosse, 1973; Sanders & Tinker, 1973).
Sanders & Tinker's work suggested that the final
equilibrium % infection was related to the quantity
of phosphorus fertiliser added to the soil.   It is
therefore necessary to ask how phosphorus supply
regulates the spread of mycorrhizal infection.   There
are two possible ways in which soil phosphate status
could affect the spread of mycorrhizal infection.
1) The external phase of the fungus could be adversely affected by high concentrations of phosphate in
the soil solution.   In the work mentioned above,
(Sanders & Tinker, 1973), high soil solution phosphate concentrations resulting from large additions
of KH_2PO_4 may have had a direct inhibitory effect on
external hyphal growth.   This would lead to a slower
rate of increase of % infection.   2) The phosphorus
supply to the host roots influences either directly
or indirectly the susceptibility of the roots to

infection by the mycorrhizal fungus.

A choice between these two hypotheses cannot be
made in experiments in which phosphate is supplied to
mycorrhizal root systems from the soil.   Therefore
the effects of foliar application of phosphate on
mycorrhizal onion roots were examined, since in this
way phosphorus supply to the host could be controlled
without altering soil phosphate status.

MATERIALS & METHODS

Soil - The soil used was a sandy loam from
Woburn Experimental Farm which contained 170 μg g^{-1}
isotopically exchangeable phosphorus and had a pH of
6.2.   The soil was collected from the same site as
soil No. 12 of Hayman & Mosse (1970).   The soil was
sieved, given a partially sterilising dose of γ-irra-
diation (0.8 M rad) to eliminate indigenous mycorrhizal fungi, and stored moist for several months.
The soil received 0.5 g KNO_3 per kg. immediately
before use.

Plants - Onion seedlings (*Allium cepa* L., F_1
hybrid var. Mustang, chosen for seedling uniformity),
germinated on moist filter paper, were planted out in
6 cm pots.   Seedlings were either inoculated with
freshly wet-sieved spores and mycelium of *Glomus
mosseae* (Nicol. & Gerd.) Gerd. & Trappe, or inoculum
washings to ensure that both mycorrhizal and non-
mycorrhizal treatments received the same contaminating microorganisms.

Growth conditions - After planting, the pots
were randomised in trays in a growth chamber maintained at a constant temperature of 18°C.   The
photoperiod was 16 hours at 20 000 lux.   Pots were
watered from below with deionised water.

Harvesting - three replicates of each of the
non-mycorrhizal (C) and mycorrhizal (M) treatments
were harvested at intervals.   Fresh and dry weights
of roots and shoots were determined after samples

had been taken for infection assessment.   External
mycelium was stripped from mycorrhizal roots and wet-
sieved from the soil in which they had been growing.
The mycelium was cleaned of all debris and weighed
dry.   Infection was assessed in the mycorrhizal root
samples, which had been taken at random, by examination of sixty 1 cm pieces after clearing and staining (10% KOH followed by trypan blue/lactophenol).
The pieces were mounted on microscope slides and the
% length of each piece carrying infection was recorded.   A final % infection was then calculated.

Samples of the dry shoot and root material were
analysed for phosphorus content (wet ashing followed
by phosphorus estimation by the molybdenum blue
method).   Phosphorus % in the dry matter, and total
phosphorus contents of the roots and shoot of each
plant, were calculated.

Root lengths were measured by ruler in some
samples and related to fresh weight.   Root length
for the remaining plants were calculated from fresh
weight.   This is not a very accurate method, but
sufficiently so for these purposes.

Foliar feeding - at 19 days after inoculation
the seedlings had a fully expanded first leaf and a
growth response to mycorrhizal infection was just
becoming visible.   Three replicates each of control
(C) and mycorrhizal (M) plants were harvested at this
time.   A further nine mycorrhizal and three control
plants were then selected at random, and foliar
phosphate application started.   Phosphate was
injected into the hollow leaves with a microsyringe
in the form of a solution containing 10 mg cm^{-3} P.
Higher concentrations led to leaf damage while the
use of lower concentrations required the injection of
inconveniently large volumes.   Details of the
injection schedule are given in Fig. 1.   As far as
could be seen, all injected solution remained within
the leaf.   The schedule was designed so that the
injected plants would receive phosphorus at a rate
just equal to the expected rate of phosphorus uptake

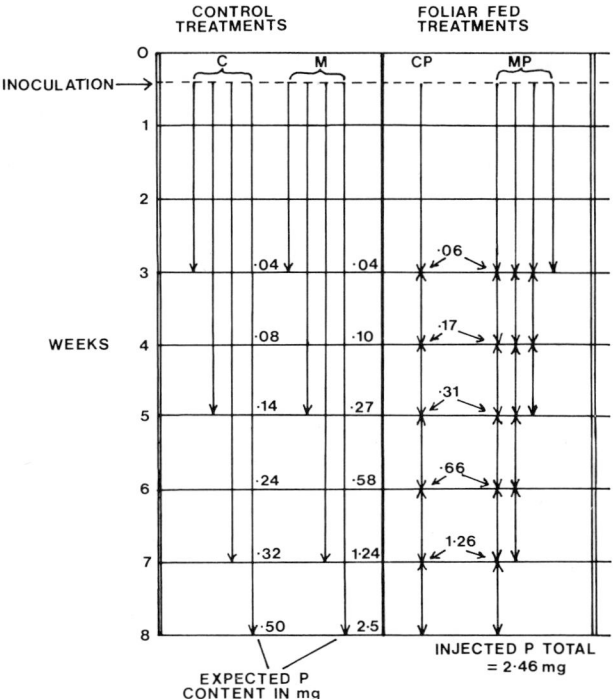

HARVEST AND P INJECTIONS FOR FOLIAR FEEDING
EXPERIMENT (EACH LINE REPRESENTS 3 REPLICATES)

Figure 1.   Harvest and phosphate injection
schedule for foliar feeding experiment.
Each vertical line represents three replicates.   V, harvest; X injection time.
The numbers against the control treatment
lines are the *expected* phosphorus contents
(mg) at weekly intervals (see text).   The
numbers against the foliar fed treatments
are the quantities of phosphorus (mg)
injected at weekly intervals.   The amount
injected at the beginning of each week
equals the *expected* uptake during that week.

by the untreated mycorrhizal plants.   The results
of a previous serial harvest experiment using the
same soil-plant-endophyte combination and the same
growth conditions were used to calculate this expected uptake rate.   Plants were therefore injected at
weekly intervals with steadily increasing doses of
phosphate as shown in Fig. 1, where times of harvest
are also shown.

RESULTS AND DISCUSSION

Effect on growth and phosphorus content

     Fig. 2 shows the way in which dry matter per
plant increased with time for the four treatments.
There was a growth response to foliar-applied phosphate by both control and mycorrhizal plants (MP and
CP treatments).   M plants showed the expected
increased phosphorus uptake compared to the C plants
as usual (Fig. 3).   At the last harvest M plants
contained on average 2.1 mg of phosphorus compared
to the expected phosphorus content of 2.5 mg.   The
foliar fed mycorrhizal plants (MP) which had received
2.5 mg P by injection contained 3.4 mg at the last
harvest, while the foliar-fed non-mycorrhizal plants
(CP) contained 3.3 mg.

Table 1.   % phosphorus in dry matter of
root and shoot for mycorrhizal and non-
mycorrhizal plants with and without foliar
feeding.   Data from last harvest.

| | Treatments | | | |
| | MP | M | CP | C |
	%	%	%	%
Shoot	0.47	0.28	0.46	0.16
Root	0.33	0.36	0.39	0.21

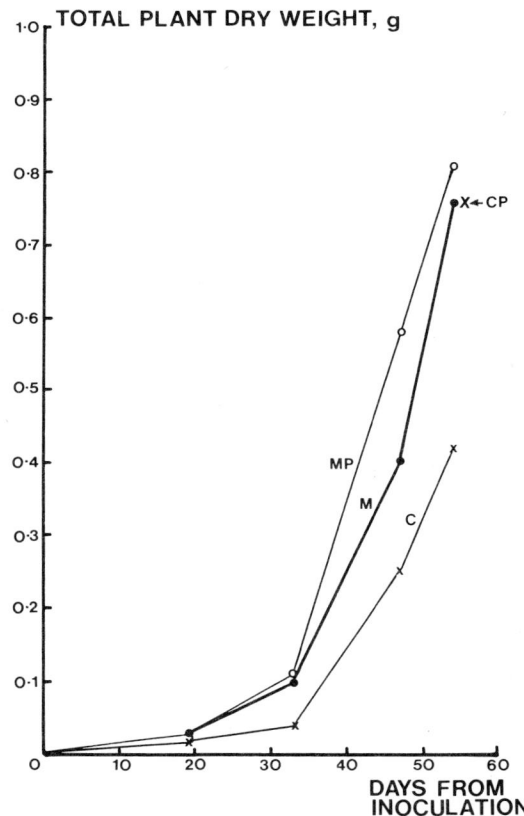

Figure 2.   Change in plant dry weight with
time for the four treatments.

The non-mycorrhizal control plants (C) showed a
growth response to foliar-applied phosphorus nearly
equal to the mycorrhizal growth response.   This was
accompanied by a near doubling in root % phosphorus
(Table 1) showing that phosphorus was absorbed and
translocated from shoot to root.   Retention of
phosphate as solid, unabsorbed salt inside the leaf
cylinder was therefore unlikely.   The relatively

Sanders, F. E.

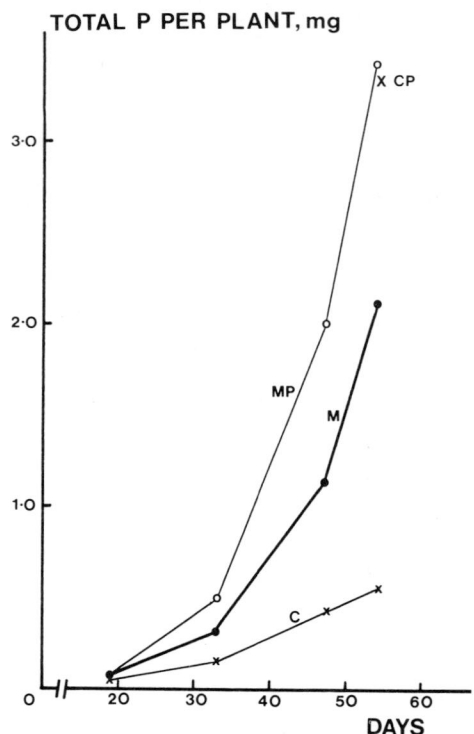

Figure 3. Change in total plant phosphorus with time for the four treatments.

large % phosphorus in the roots of M plants was probably due to the storage of phosphorus in the fungus and associated host cytoplasm, and is commonly found.

Root length data are shown in Fig. 4. The M and C plants had similar root lengths until day 47 after inoculation, whereas the mycorrhizal response in terms of dry matter and phosphorus content became apparent between 20 and 30 days. It seems to be

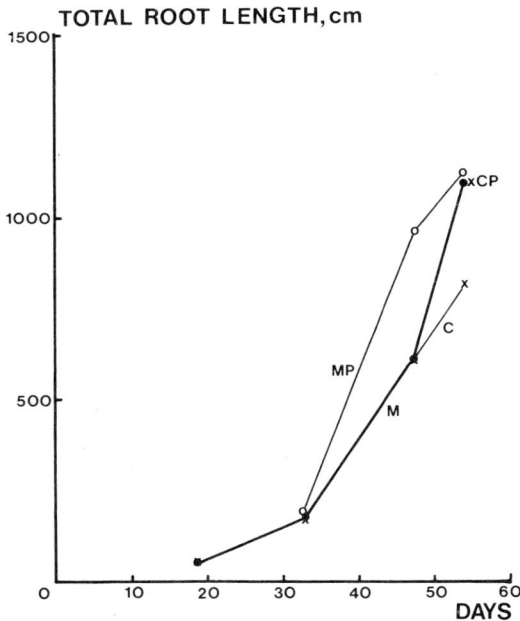

Figure 4.   Change in root length with time
for the four treatments.

normal that the response to mycorrhizal infection is
expressed last in terms of root length, and this
implies of course that the root:shoot ratio of the M
plants diverged from that of the C plants in response
to increased phosphorus supply (Table 2).

Effect on infection

     % infection showed the usual sigmoid increase
(Fig. 5).   At the last harvest (day 54) the MP
plants had a 15% lower mean % infection than the M
plants.   The agreement between replicates was good,
and another experiment gave similar results, so the

Sanders, F. E.

Table 2.   Root : shoot ratios at the first and last harvests.

| | Treatment | | | |
	MP	M	CP	C
Harvest 1 (Day 19)	0.4	0.4	0.4	0.4
Harvest 4 (Day 54)	0.25	0.25	0.3	0.4
Means of three replicates				

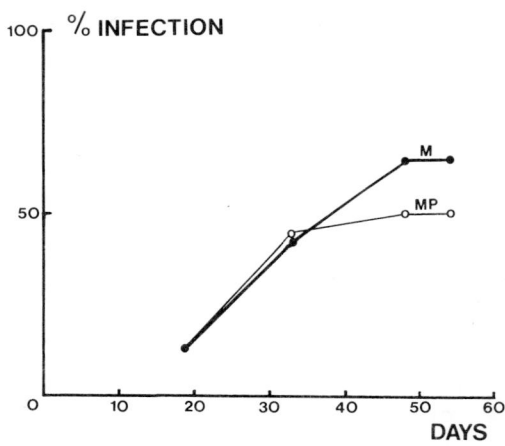

Figure 5.   Change in % infection with time for M and MP treatments.

difference is probably a true effect.   The weight of external mycelium per cm of infected root remained nearly constant for the M plants (Fig. 6), but for the MP plants, the rate of increase in mycelium weight was slower, so that by the last harvest (day 54), the MP roots had 2 mg of external mycelium per 500 cm of infected root, compared to 3.5 mg associated with the M roots.

There was also an effect of foliar-applied phosphate on the anatomy of the internal infection, similar to that found by Mosse (1973) for soil-

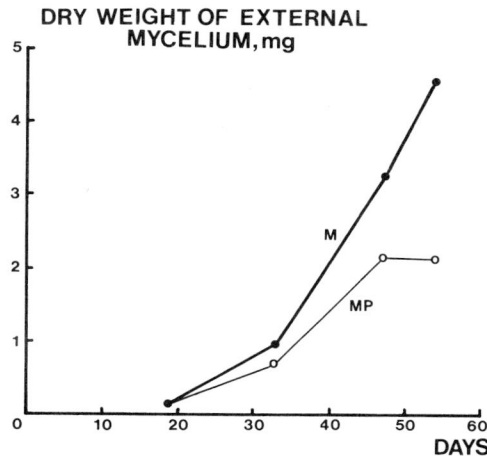

Figure 6.   Change in dry weight of external
mycelium with time for M and MP treatments.

applied phosphate.   Compared to the M roots, MP
roots had a lower intensity of infection and the
hyphae were thinner and stained less heavily than
normal internal hyphae, giving the impression that
they had thinner walls.   Such observations are very
difficult to quantify, and this was not attempted.
     The effect of foliar-applied phosphate on the
development of mycorrhizal infection therefore
appeared to be as follows:

1)   an effect on the rate of spread of the endophyte
within the root in relation to the rate of increase
in total root length;
2)   an effect on the production of external mycelium
per cm of infected root length;
3)   an effect on the intensity of infection of the
root cortex.

The associated changes in the ability of the roots to
take up phosphate can now be considered.

Effect on phosphate uptake from the soil

If the quantity of injected phosphorus is subtracted from the total content of the plants at
harvest, figures are obtained for uptake from the
soil (Fig. 7).   The effect of the MP treatment was

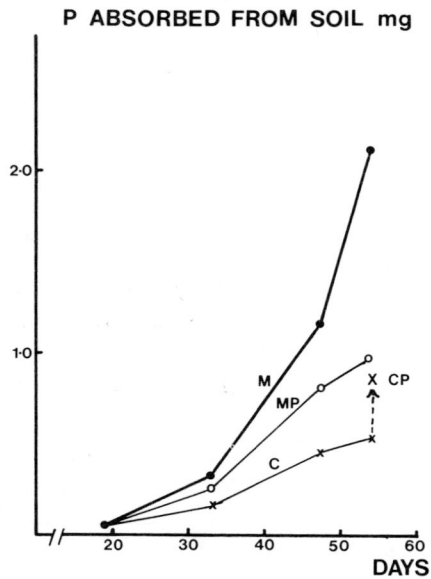

Figure 7.   Phosphorus uptake from the soil
at the four harvests, for M, MP, C and CP
treatments.

to depress phosphorus uptake to about half of that of
the M plants by the final harvest (Table 3).   In the
non-mycorrhizal case, however, uptake by the CP
plants is actually greater than that of the C plants.
The phosphorus balance figures for the MP and CP
plants are similar.   Since the uptake from soil was
about the same, the growth response of the MP plants
compared to the C plants must have been due largely

Table 3.   Phosphorus balance for plants
harvested at day 54.

Treatment	Total P	Injected P	P uptake
MP	3.4 mg	2.5 mg	0.9 mg
M	2.2	0	2.2
CP	3.3	2.5	0.8
C	0.5	0	0.5

to the injected phosphate, the mycorrhizas playing a
much less important role than in the case of the M
plants.

Phosphorus uptake can also be expressed as
inflows.   Inflows were calculated from the data
given in Fig. 7, using the formula:

$$I = \frac{(N_2 - N_1) \, \ln(L_2/L_1)}{(L_2 - L_1)(T_2 - T_1)}$$

where I is the average inflow between times 1 and 2
(T), and L and N are the root lengths and phosphorus
contents at the two times.   The results of calculations using this formula are only approximate but
sufficient for this purpose (Table 4).

Table 4.   Mean inflows for plants of the
four treatments over the time interval,
days 19 - 54.

Treatment	Inflow moles phosphorus $cm^{-1}s^{-1}$
M	6.2×10^{-14}
MP	2.7×10^{-14}
C	2.1×10^{-14}
CP	2.7×10^{-14}

MP, C, & CP treatments have phosphorus inflows that
are probably not significantly different from each
other, while the M inflow is much greater. The
difference between the C and the CP inflow figures
may reflect the fact that C plants are acutely phosphorus deficient, which may have reduced the absorbing power of the roots below that of the CP plants.
MP and CP plants clearly behaved similarly all
through the experiment. Inflows attributable to the
hyphal pathway (using CP inflow as a control figure)
were therefore MP, O moles $cm^{-1}s^{-1}$; M, 3.5 x 10^{-14}
(see Sanders & Tinker, 1973). This suggests that
the mycorrhizal infection in the MP plants was
completely non-functional as an aid to host phosphorus uptake, even though external mycelium was in the
ratio MP:M = 0.46.

CONCLUSIONS

The effect of phosphate applied to the leaves
of mycorrhizal plants has been a) to reduce the
rate of spread and intensity of the mycorrhizal
infection, b) to reduce the weight of external
mycelium associated with each cm of infected root,
and c) to depress the supply of phosphorus to the
host via the mycorrhizas, reducing the status of the
endophyte to that of a benign parasite. Hypothesis
2, that the control of the spread and of mycorrhizal
infection is linked to the phosphate status of the
host, seems most acceptable on the basis of these
results. The exact mechanism by which the spread
of the fungus is regulated is still unknown.

ACKNOWLEDGMENTS

I wish to acknowledge the invaluable assistance
of Mrs. S. Palmerley in the carrying out of this
work, and Professor P. B. Tinker for much helpful
discussion .

REFERENCES

BREWSTER, J. L. & TINKER, P. B. H. (1972). Nutrient
flow rates into roots. *Soils, Ferts.*, 35, 355–
359.

DAFT, M. J. & NICOLSON, T. H. (1969). Effect of
Endogone mycorrhiza on plant growth. II. Influence
of soluble phosphate on endophyte and host in maize.
New Phytol., 72, 127–136.

HATTINGH, M. J., GRAY, L. E. & GERDEMANN, J. W.
(1974). Uptake and translocation of ^{32}P-labelled
phosphate to onion roots by endomycorrhizal fungi.
Soil Sci., 116, 383–387.

HAYMAN, D. S. & MOSSE, B. (1971). Plant growth
responses to vesicular-arbuscular mycorrhiza.
I. Growth of *Endogone*-inoculated plants in phosphate-deficient soils. *New Phytol.*, 70, 19–27.

HAYMAN, D. S. & MOSSE, B. (1972). Plant growth
responses to vesicular-arbuscular mycorrhiza.
III. Increased uptake of labile P from soil.
New Phytol., 71, 41–47.

MOSSE, B. (1973). Plant growth responses to vesicular-
arbuscular mycorrhiza.   IV. In soil given additional phosphate. *New Phytol.*, 72, 127–136.

PEARSON, V. & TINKER, P. B. (1975). This symposium,
p. 277.

SANDERS, F. E. & TINKER, P. B. (1971). Mechanism of
absorption of phosphate from soil by *Endogone*
mycorrhizas. *Nature, London,* 233, 278–279.

SANDERS, F. E. & TINKER, P. B. (1973). Phosphate flow
into mycorrhizal roots. *Pestic. Sci.*, 4, 385–395.

TINKER, P. B. (1975). Effects of vesicular-arbuscular mycorrhizas on higher plants. *Symp. Soc. exp. Biol.*, <u>29</u> (in press).

MEASUREMENT OF PHOSPHORUS FLUXES IN THE EXTERNAL HYPHAE OF ENDOMYCORRHIZAS

VIVIENNE PEARSON AND P. B. TINKER

Department of Plant Sciences,
University of Leeds, U.K.

INTRODUCTION

Vesicular-arbuscular mycorrhizas can have very striking effects on plant growth and phosphate uptake. Recent work (Sanders & Tinker, 1971; 1973; Mosse, Hayman & Arnold, 1973; Hattingh, Gray & Gerdemann, 1973) supported the simplest theory of the mycorrhizal effect, i.e. that the external hyphae of the endophyte provide the host plant with an additional, well-distributed absorbing surface which absorbs phosphate from outside the depletion zone around the root. If this is so, this phosphate must reach the mycorrhizal root via the hyphal entry points. Using this hypothesis, Sanders and Tinker (1973) calculated the very high mean value for phosphate flux through entry point hyphae of 3.8×10^{-8} mol. cm^{-2} sec^{-1}. The objective of the present work was (1) to demonstrate the transport of phosphorus within the hyphae and (2) to measure the steady state flux of this transport, to test whether the high calculated rates could occur. This paper reports the development of a suitable system, and preliminary results obtained with it.

Pearson, V. & Tinker, P. B.

EXPERIMENTAL METHODS AND RESULTS

Aseptic culture of external hyphae of endomycorrhizas

Spores of *Glomus mosseae* (yellow vacuolate endophyte), maintained on onion plants grown in γ-irradiated soil (0.9 Mrads), were surface sterilised with
Chloramine 'T' (Mosse, 1959) and germinated on water
agar.   Seed of *Trifolium repens* L. (white clover)

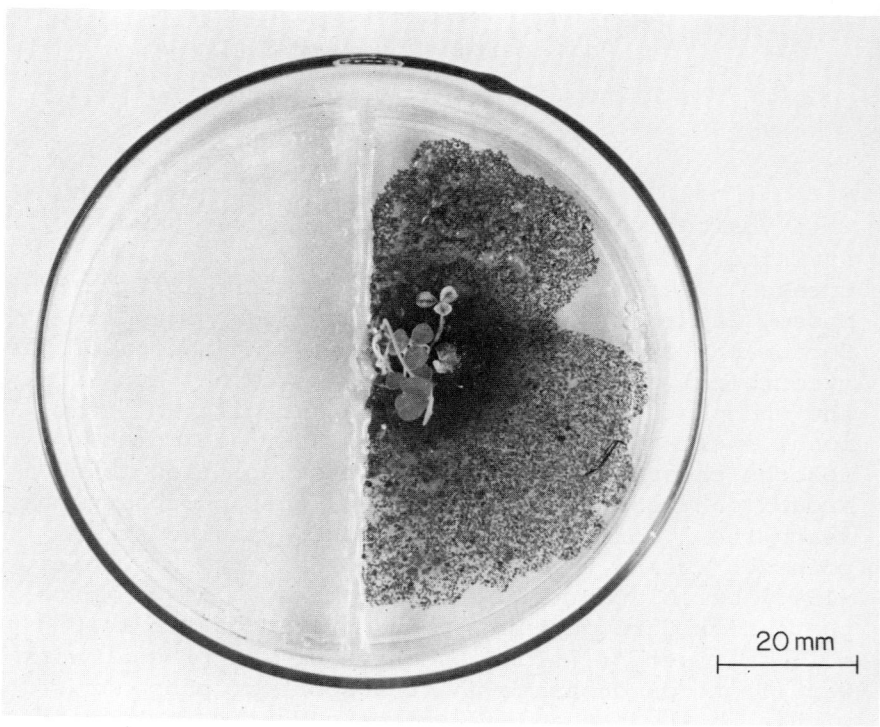

20 mm

Figure 1. Split-plate technique used for
the study of phosphate fluxes within endomycorrhizal hyphae.

was germinated on sucrose agar until the radicle had emerged.

External hyphae of a mycorrhizal seedling were grown in aseptic culture (Fig. 1) by a modification of the split-plate technique used by Pearson and Read (1973) in studies of ericaceous mycorrhizas.   Pre-sterilised plastic petri dishes divided into two halves by a 6 mm high partition were used, one half of each dish containing 0.5% tap water agar and the other half containing 0.5% tap water agar with 0.08 g 1^{-1} KNO_3 + 0.13 g 1^{-1} Ca $(NO_3)_2$, and about 2 g sterile soil (Woburn A, see Sanders & Tinker, 1973).

A sterile seedling of *T. repens* was aseptically transferred to the soil agar in each dish, close to the partition, and inoculated with pre-germinated spores of *G. mosseae*.   The dishes were kept in a growth cabinet with 16 hours light (2200 lux) at 19°C.   After 1 - 2 months hyphae ramified through the soil agar, and stained samples of the roots (Phillips & Hayman, 1970) showed normal vesicular-arbuscular infections (Figs. 2 & 3).   External hyphae only spread in the medium after infection of the host root had occurred.   After a further period the hyphae also grew over the partition and into the water agar.

Transport of phosphorus within the external hyphae

After the number of hyphae crossing the partition had been counted and their diameters measured, lanol-in, which had been sterilised by heating to 100°C for 24 hours, was cooled and dispensed from a sterile syringe on to the partition to form a barrier to diff-usion around the hyphae.   1.5 ml of 0.2% water agar containing 5 µCi ^{32}P as orthophosphate and 4.5 µg phosphorus as KH_2PO_4 were evenly pipetted over the water agar side of the dish.   After diffusion throughout the agar, this was calculated to give a final phosphate concentration of 5 x 10^{-6} M, near the average for the soil solution phosphate in the exper-iments of Sanders and Tinker (1973).   The procedure

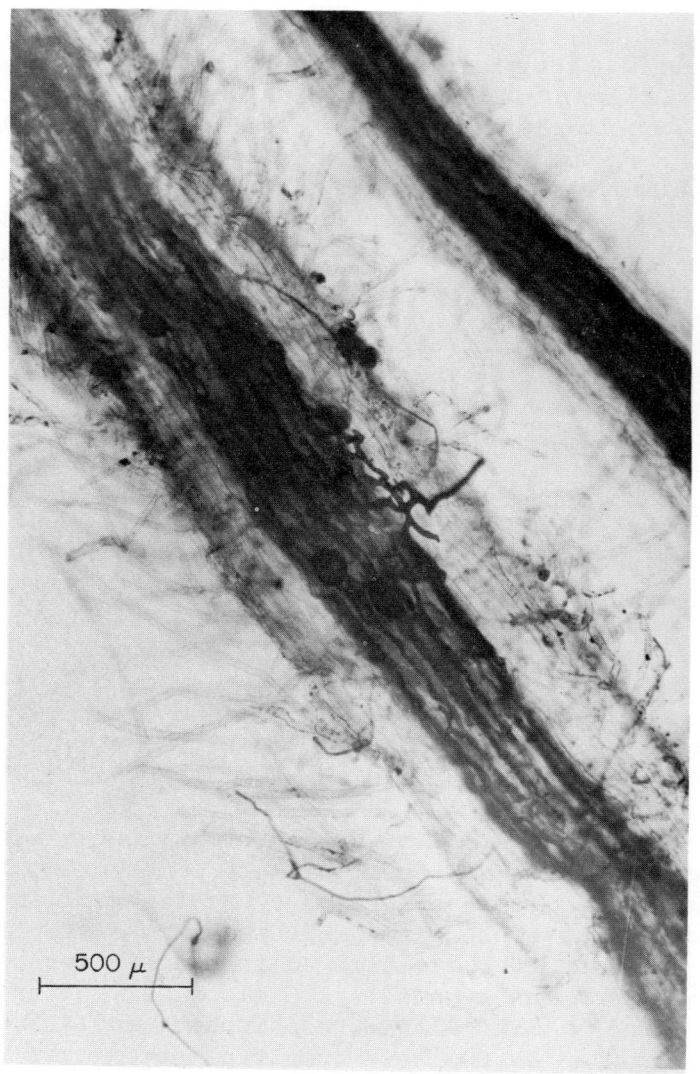

<u>Figure 2.</u>   Aseptically produced mycorrhizal roots of white clover (*Trifolium repens*) showing an entry point, internal hyphae and vesicles.

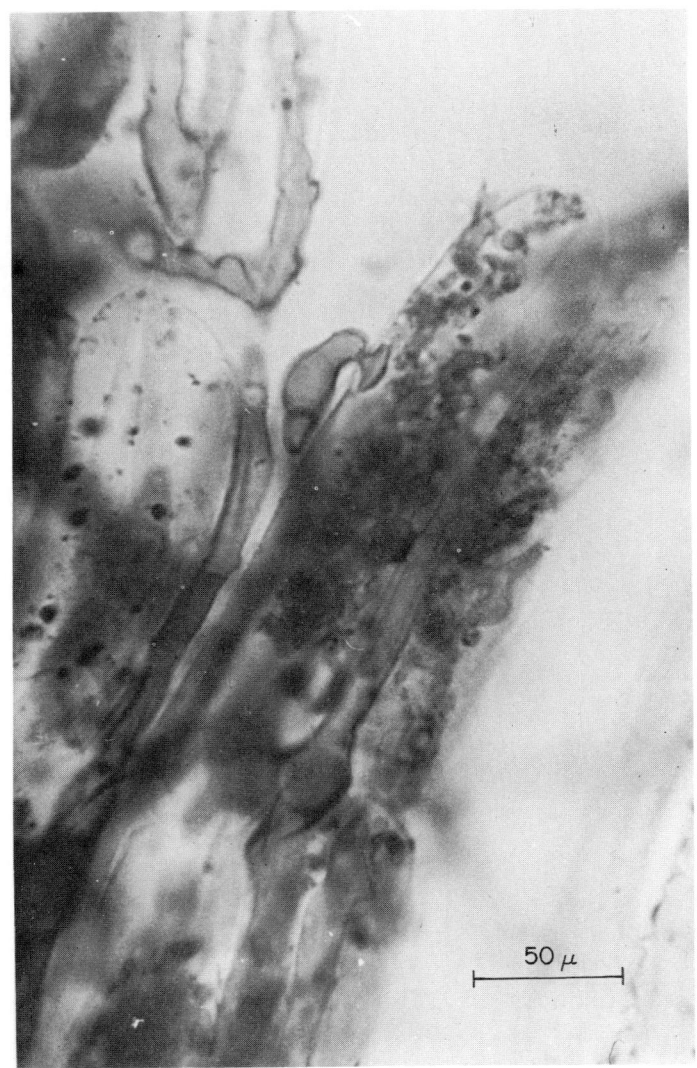

Figure 3.   Root cortical cells of white clover (*Trifolium repens*) containing hyphae and arbuscules.

was repeated with control dishes containing uninoculated seedlings.

The appearance of ^{32}P in the shoots of the
clover seedlings was detected with a specially designed β-ray counter, the detector head consisting of
a cylinder of plastic scintillator with a 1.5 cm hole
(Appleby, in preparation)*.   This was lowered down
to enclose the seedling shoot (Fig. 4).   The plastic
scintillator was integral with a photomultiplier,
which was connected to a Nuclear Enterprises counting
system.   A lead plate was placed around the base of

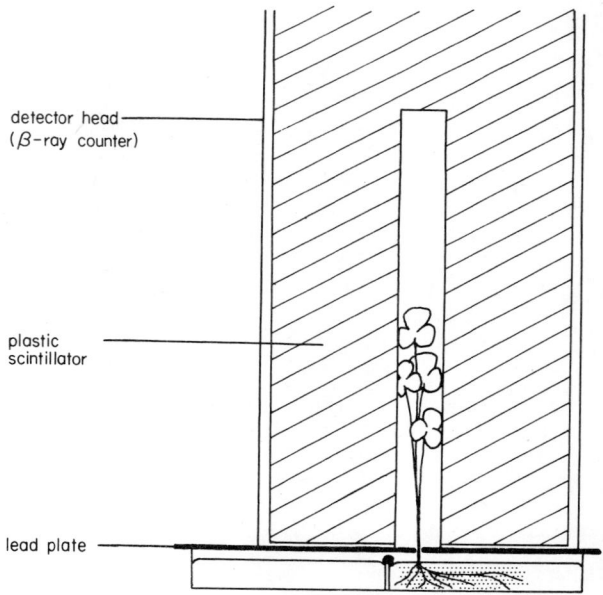

detector head
(β-ray counter)

plastic
scintillator

lead plate

Figure 4.   A system for studying steady
state nutrient fluxes in vesicular-arbuscular
mycorrhizas.

---

* This detector was designed and built by Mr. D.
Appleby of the Medical Physics Department of Leeds
University.

the seedling shoot to screen off the petri dish contents, and the whole system was placed inside a lead box to reduce background radiation to a minimum. This counter was calibrated approximately by inserting strips of filter paper, of roughly the same size as the shoot, which had absorbed varying amounts of ^{32}P solution and then been dried.

During the experimental period of 6 days, there was a steady increase of ^{32}P in the shoots of the mycorrhizal seedlings; representative results are in Fig. 5. The count rate with the uninoculated control seedlings was similar to that obtained when the

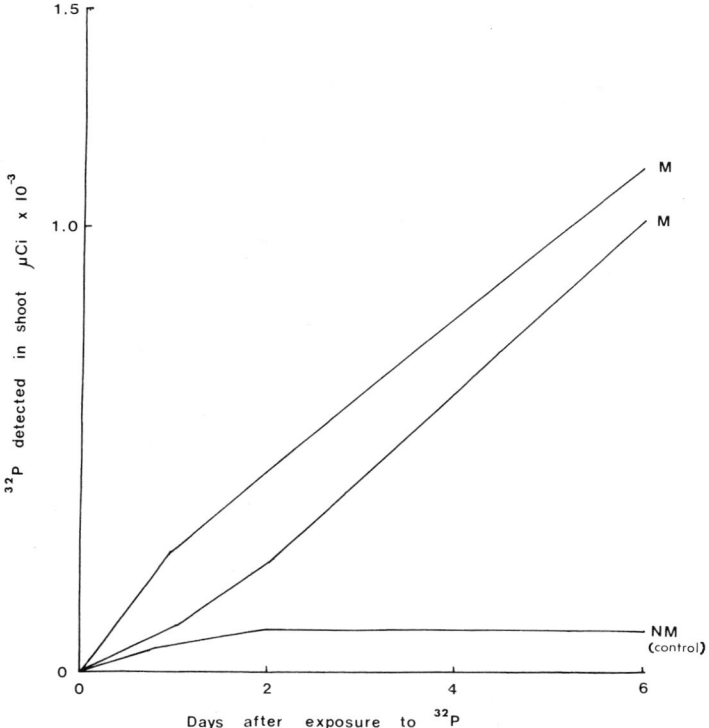

Figure 5. Appearance of ^{32}P in shoots of clover seedlings with vesicular-arbuscular mycorrhizas.

detector head was placed above a similar dish containing the same quantity of ^{32}P in agar, but without a seedling.   These results demonstrated translocation of phosphorus, and release from the fungus into the host.

The total phosphorus from the seedling shoot, root and surrounding soil agar was extracted by wet-acid digestion using conc. HNO_3 and a triacid mixture (conc. HNO_3 : 80% $HClO_4$: conc. H_2SO_4), until no radioactivity was detected on the material remaining. All digests were reduced to near dryness, made up to a volume of 5 mls and counted on a Nuclear Chicago liquid scintillation counter using Cerenkov radiation.

The approximate quantity of ^{32}P in the shoot measured by the 'whole plant counter' was much smaller than the total found after wet digestion. Only about 5% of the ^{32}P was found in the shoot digest, and it is uncertain whether the remaining 95% was in the roots, hyphae or soil agar.   Some more recent results (Cooper, pers. comm.) indicate that there may be a lag in the transfer of ^{32}P to the shoot, since measurements over longer periods show a larger fraction of the ^{32}P in the shoot.

The total amount of phosphate transported per unit time as found by wet digestion, when divided by the total cross-sectional area of the hyphae gave the flux, the largest of four values being 1.0×10^{-9} and the smallest 0.3×10^{-9} mols $cm^{-2}s^{-1}$.   It is assumed that all the hyphae were living; if any were dead, the flux values would be correspondingly larger.

DISCUSSION

This work confirms that hyphae of *Glomus mosseae* transport phosphorus from some distance (perhaps 1 - 2 cm) into the host root, as found by Hattingh, Gray and Gerdemann (1973).   The minimum hyphal flux of phosphorus was between 0.3 and 1.0×10^{-9} moles

cm^{-2} sec $^{-1}$, which is much less than the value of
3.8×10^{-8} moles cm^{-2} sec^{-1} previously calculated by
Sanders and Tinker (1973).   However, their value for
flux was for entry point hyphae, whilst the experiments reported here measured the phosphorus flux in
finer hyphae some distance from the root.   Since
these hyphae converge into the thick hyphae at the
point of entry into the root, it seems reasonable to
expect that the flux of phosphorus across the latter
will be greater.   Bearing this in mind, and the very
different conditions of culture of the host plants, a
difference between the fluxes obtained in these two
ways is to be expected.   It was not possible to
measure the total number of fine hyphae which could
be considered to lead into a single entry point,
which might have clarified this point.   It would
have been desirable to estimate the total length of
hyphae absorbing the ^{32}P, since it would have yielded
a minimum value for the absorbing power of the hyphae
per unit length, but the hyphae could not be followed
in the agar.

The retention of most of the ^{32}P in the roots or
hyphae was unexpected.   Possibly the very low transpiration of the plants in the petri dishes delayed
translocation from the root to the shoot.

The mechanism of transport of nutrients by fungal hyphae is poorly understood.   Cowan, Lewis and
Thain (1972) studied the transport of potassium in
Phycomyces blakesleeanus and concluded that all
transport could be explained by passive movement of
ions within the hyphae along a gradient of concentration and of electrical potential.   In order to
explain the largest flux of 1.0×10^{-9} moles cm^{-2}
sec^{-1} obtained here, a concentration gradient of 0.1
moles $litre^{-1}$ cm^{-1} would be necessary, assuming that
the effective diffusion coefficient of $H_2PO_4^-$ is
equal to that in water.   0.1 M phosphate solution
contains 0.32% P, and direct analysis of external
hyphae indicated a total P concentration of about
0.3% in the contents of the hyphal lumen.   Diffusion

alone is therefore scarcely credible as a transport
mechanism over distances of centimetres, and becomes
even less so for the higher flux of Sanders and
Tinker (1973).   It seems unlikely that electrical
potential gradients could account for the discrepancy
and bulk flow or protoplasmic streaming appear likely
alternatives (Tinker, 1975).

ACKNOWLEDGMENTS

We thank Mr. D. Appleby for the design and con-
struction of the whole-shoot counter and for much
valuable advice and help.   We thank the Agricultural
Research Council for financial support, and for a
fellowship for Vivienne Pearson.

REFERENCES

COWAN, M. C., LEWIS, B. G. & THAIN, J. F. (1972).
Mechanism of translocation of potassium in mycelium
of *Phycomyces blakesleeanus* in an aqueous environment.   *Trans. Br. mycol. Soc.*, 58, 103-112.

HATTINGH, M. J., GRAY, L. E. & GERDEMANN, J. W. (1973)
Uptake and translocation of ^{32}P-labelled phosphate
to onion roots. *Soil Sci.*, 116, 383-387.

MOSSE, B. (1959). The regular germination of resting
spores and some observations on the growth requirements of *Endogone. Trans. Br. mycol. Soc.*, 42,
274-286.

MOSSE, B., HAYMAN, D. S. & ARNOLD, D. J. (1973).
Plant growth responses to vesicular-arbuscular
mycorrhiza.   V.   Phosphate uptake by three plant
species from P-deficient soils labelled with ^{32}P.
New Phytol., 72, 809-815.

PEARSON, V. & READ, D. J. (1973). The biology of mycorrhiza in the Ericaceae II.   The transport of carbon and phosphorus by the endophyte and the mycorrhiza.   *New Phytol.*, 72, 1325–1331.

PHILLIPS, J. M. & HAYMAN, D. S. (1970). Improved procedures for clearing roots and staining parasitic and vesicular-arbuscular mycorrhizal fungi for rapid assessment of infection.   *Trans. Br. mycol. Soc.*, 55, 158–161.

SANDERS, F. E. & TINKER, P. B. (1971). Mechanism of absorption of phosphate from soil by *Endogone* mycorrhizas.   *Nature, London,* 233, 278–279.

SANDERS, F. E. & TINKER, P. B. (1973). Phosphate flow into mycorrhizal roots.   *Pest. Sci.*, 4, 385–395.

TINKER, P. B. (1975). This symposium, p. 353.

UPTAKE OF ^{32}P-LABELLED PHOSPHATE BY ENDOMYCORRHIZAL ROOTS IN SOIL CHAMBERS

M. J. HATTINGH

Department of Plant Pathology,
University of Stellenbosch,
Stellenbosch, South Africa.

INTRODUCTION

The simplest theory to account for increased phosphate uptake by endomycorrhizas is that the mycelium outside the root constitutes an additional, better distributed surface for absorbing phosphate from the soil solution (Mosse, 1973). Hattingh *et al.* (1973) presented direct evidence in support of this theory. A method was developed to demonstrate uptake of ^{32}P-labelled phosphate by mycorrhizal roots of onion in small soil chambers.

METHOD

The soil chambers (Fig. 1) and the method of obtaining mycorrhizal onion plants have been described in detail (Hattingh *et al.*, 1973). The chambers were kept at a 45° angle with the face sides down in a growth cabinet in order to confine root growth to the surface of the soil plane (Fig. 2). The root system developed without entering the central dish.

A network of fungal hyphae extended over the soil plane in the central dishes of chambers support-

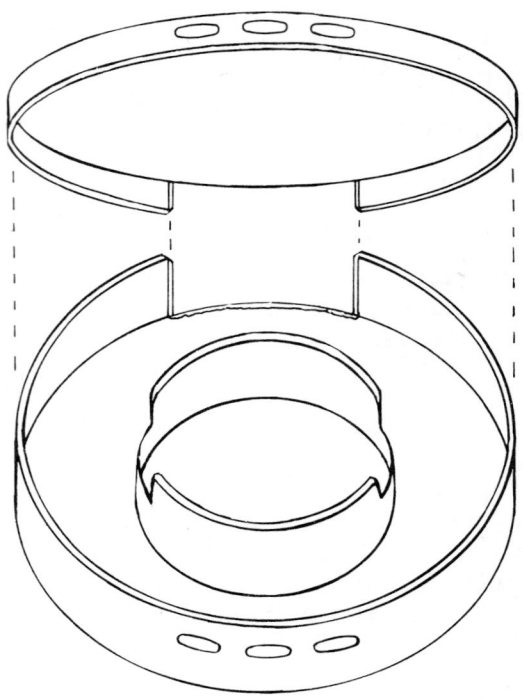

<u>Figure 1</u>.   Diagram of an empty soil chamber.
From Hattingh *et al.*   (1973), by permission
of The Williams & Wilkins Co.

ing plants, inoculated with *Glomus mosseae* or *G.
fasciculatus,* after a growth period of 70 days.   No
hyphae were present on the soil plane of control
chambers containing nonmycorrhizal roots. $H_3{}^{32}PO_4$
in 0.2 ml H_3PO_4 carrier solution was injected at soil
level in the centre of each chamber, which was returned to the growth cabinet for three days before
radioactivity of root segments was determined.   The
tracer was found in mycorrhizal roots, but no signif-

Figure 2.   A mycorrhizal onion plant after 70 days' growth in a soil chamber.   The lid and plastic film covering the soil have been removed.   From Hattingh *et al.* (1973), by permission of The Williams & Wilkins Co.

icant radioactivity was demonstrable in nonmycorrhiz-al roots.   Autoradiography indicated that the tracer did not reach the surface of mycorrhizal roots by diffusion.

DISCUSSION

Special containers have been used by several other workers to study phosphate uptake by plants (see Hattingh *et al.*, 1973). Recently Skinner and Bowen (1974) used boxes with removable perspex sides to investigate the uptake and translocation of ^{32}P-orthophosphate by mycelial strands of *Rhizopogon luteolus* to *Pinus radiata*.

An important feature of the method described by Hattingh *et al.* (1973) was that the tracer was not placed in contact with the root surface. Radio-activity could only reach the roots if the ^{32}P-labelled phosphate were taken up and translocated by the endomycorrhizal hyphae. This was confirmed when the hyphae extending from the roots over the soil plane were severed and tracer prevented from reaching the root tissue.

More information is required on the distance the external hyphae of endomycorrhizal fungi extend through soil away from the root surface. Mosse (1959) and Sanders and Tinker (1973) mentioned a figure of at least 1 cm. In our system (Hattingh *et al.*, 1973) the hyphae extended a distance from the root surface equalling at least the radius (3 cm) of the central dish. Rhodes and Gerdemann (personal communication) are using modified soil chambers to determine how far away from the root surface phosphate is taken up by endomycorrhizal fungi. Sanders and Tinker (1973) calculated that in a moderately phosphate-deficient soil there were 80 cm of endomycorrhizal hyphae/cm of infected onion root. The quantity of external mycelium and the distance to which the hyphae extend must both be important factors in the uptake of phosphate from soil by endomycorrhizal roots. However, results obtained in growth chambers on the spread of the mycelium away from the roots should be considered with caution. Conditions on the soil plane may induce more profuse hyphal growth than is possible under normal field conditions.

Bieleski (1973) suggested that autoradiographs should be used to compare depletion zones around mycorrhizal and nonmycorrhizal roots. Similarly the transport of tracer through the mycelium from a single point of application in a soil chamber could be demonstrated. Severing the hyphae near the root surface at known time intervals could provide information on the rate of phosphate uptake.

The study of microbial activity in soil surrounding ectomycorrhizas has led to the recognition of the "mycorrhizosphere" (Katznelson *et al.*, 1962; Neal *et al.*, 1964; Foster & Marks, 1967; Oswald & Ferchau, 1968). The microorganisms presumably stimulated to develop around endomycorrhizal fungi should also receive attention. Microorganisms could be isolated from the surface of hyphae of endomycorrhizal fungi from soil chambers, where they are easily accessible.

If endomycorrhizal fungi are to be used in practice to eliminate some problems which arise when plants are grown in steamed or fumigated soils, considerable attention will have to be directed to the selection of suitable inocula. Mosse (1972) indicated that plant responses to different strains of vesicular-arbuscular fungi vary considerably. Bowen (1973) suggested the selection of ectomycorrhizal fungi for tree inoculation on the organisms' ability to produce mycelial strands readily in a wide range of conditions. Similarly, by using the soil chamber method, it should be possible to recognise those endomycorrhizal fungi producing profuse external mycelial networks. Other selection criteria must be the ability of the spores to persist under adverse conditions, the production of long germ tubes to ensure contact with the root system, and aggressive infection of the host. It would also be advantageous if the selected fungi could be effectively established in natural soils in competition with indigenous endomycorrhizal fungi. The possibility has been demonstrated by Mosse *et al.* (1969).

REFERENCES

BIELESKI, R. L. (1973). Phosphate pools, phosphate transport, and phosphate availability. *A. Rev. Pl. Physiol.*, 24, 225-252.

BOWEN, G. D. (1973). Mineral nutrition of ectomy-corrhizae. In *Physiology and Ecology of Ectomy-corrhizae* (Eds. Marks, G. C. and Kozlowski, T. T.) p. 151-205. Academic Press, New York.

FOSTER, R. C. & MARKS, G. C. (1967). Observations on the mycorrhizas of forest trees. II. The rhizo-sphere of *Pinus radiata* D. Don. *Aust. J. biol. Sci.*, 20, 915-926.

HATTINGH, M. J., GRAY, L. E. & GERDEMANN, J. W. (1973). Uptake and translocation of ^{32}P-labelled phosphate to onion roots by endomycorrhizal fungi. *Soil Sci.*, 116, 383-387.

KATZNELSON, H., ROUATT, J. W. & PETERSON, E. A. (1962). The rhizosphere effect of mycorrhizal and non-mycorrhizal roots of yellow birch seedlings. *Can. J. Bot.*, 40, 377-382.

MOSSE, B. (1959). Observations on the extra-matrical mycelium of a vesicular-arbuscular endophyte. *Trans. Br. mycol. Soc.*, 42, 439-448.

MOSSE, B. (1972). Effects of different *Endogone* strains on the growth of *Paspalum notatum*. *Nature, London*, 239, 221-223.

MOSSE, B. (1973). Advances in the study of vesicular-arbuscular mycorrhiza. *A. Rev. Phytopath.*, 11, 171-196.

MOSSE, B., HAYMAN, D. S. & IDE, G. J. (1969). Growth
responses of plants in unsterilized soil to inoculation with vesicular-arbuscular mycorrhiza.
Nature, London, <u>224</u>, 1031-1032.

NEAL, J. L., BOLLEN, W. B. & ZAK, B. (1964). Rhizosphere microflora associated with mycorrhizae of
Douglas Fir.   *Can. J. Microbiol.,* <u>10</u>, 259-265.

OSWALD, E. T. & FERCHAU, H. A. (1968). Bacterial
associations of coniferous mycorrhizae.   *Pl. Soil,*
<u>28</u>, 187-192.

SANDERS, F. E. & TINKER, P. B. (1973). Phosphate flow
into mycorrhizal roots.   *Pestic. Sci.,* <u>4</u>, 385-395.

SKINNER, M. F. & BOWEN, G. D. (1974). The uptake and
translocation of phosphate by mycelial strands of
pine mycorrhizas.   *Soil Biol. Biochem.,* <u>6</u>, 53-56.

ULTRASTRUCTURAL EVIDENCE RELATING TO HOST-ENDOPHYTE TRANSFER IN A VESICULAR-ARBUSCULAR MYCORRHIZA

GUY COX, F. E. SANDERS, P. B. TINKER AND J. A. WILD

Department of Plant Sciences, University of Leeds, Leeds, U.K.

INTRODUCTION

Investigations on the growth responses produced by vesicular-arbuscular (VA) mycorrhizal infection have shown that the most important effect is in improved phosphorus nutrition of the host plant (Mosse, 1973). It has been inferred, on theoretical grounds, that phosphorus must enter the root *via* the fungal hyphae (Sanders & Tinker, 1971) and experimental proof that this can occur has recently been obtained (Hattingh *et al.*, 1973; Pearson & Tinker, 1975). The transfer to the fungus of carbon derived from host photosynthate has also recently been demonstrated (Ho & Trappe, 1973), albeit only over a period of months.

Ultrastructural studies have hitherto concentrated on the structure of the host-endophyte interface and the life-history of the infection structures (Scannerini & Bellando, 1967, 1968; Scannerini, 1972;

THE FIGURE NUMBERS IN THIS PAPER REFER TO PLATES IV - VI.

Protsenko *et al.*, 1971; Kaspari, 1973; Cox &
Sanders, 1974; Scannerini, *et al.*, 1975; Kaspari,
1975). This study applies techniques of light and
electron microscopy to the problems of nutrient
transfer in onion mycorrhiza. We have investigated
two aspects of this:
1. The occurrence of granules which stain as polyphosphate. Polyphosphate granules are found in
many fungi, algae and bacteria (Keck & Stich, 1957)
and may well be important in the mycorrhizal situation.
2. The transfer of carbon from host to endophyte.
Lipid droplets have frequently been observed in VA
mycelium (Mosse, 1959; Cox & Sanders, 1974) and
particular attention has been given to their distribution.

MATERIALS AND METHODS

'Mustang', an Fl hybrid *Allium cepa* L., infected
with *Glomus mosseae* (Nicol. & Gerd.) Gerdemann &
Trappe, was grown at 18°C, under a 20,000 lux 16-hour
day regime.

Electron microscopy

Material was fixed and embedded according to the
schedule of Cox & Sanders (1974). Some material had
0.5% $PbNO_3$ added to the glutaraldehyde fixative as a
stain for polyphosphate. Sections were cut at 70-
120 nm for conventional electron microscopy, or at
1-5 μm for light microscopy or high-voltage electron
microscopy. Sections for electron microscopy were
either stained in Reynold's lead citrate and 2%
uranyl acetate, or examined unstained.
For conventional electron microscopy, an AEI EM
6B was used at 60kV. For high-voltage electron
microscopy, the AEI EM7 at Oxford was used at 1 MeV,

and the GEPSA microscope at CNRS Toulouse was used at 2.5 and 3 MeV.

Light microscopy

To examine lipid distribution, whole roots were cleared and stained in Trypan Blue/Sudan IV. (Cox & Sanders, 1974). To stain for polyphosphate, resin-embedded sections were stained in 1% Methylene Blue in 1% Borax or 0.05% Toluidine Blue in 0.1M phosphate buffer, pH 6.8, and differentiated in 1% H_2SO_4. Fresh material (external mycelium and sections of roots) was stained in Toluidine Blue as above, and differentiated in tap water or 1% H_2SO_4, and examined in temporary water mounts. Control samples were extracted with 10% trichloracetic acid to remove polyphosphate before staining (Keck & Stich, 1957). Fresh material was also stained with Ebel, Colas and Muller's (1958) lead sulphide stain for polyphosphate.

Autoradiography

Infected and control plants, grown as above, were exposed to 80 μC $^{14}CO_2$ for three hours. The plants were kept in the dark for 24 h, then returned to their normal light regime (above). Root systems (3 control; 5 mycorrhizal) were harvested after the 24 h dark period. Further harvests were taken 4, 10 and 17 days after labelling. Samples of each root system were fixed and embedded as described above. Some root systems from each harvest were used for macro-autoradiography of the entire root system prior to selecting samples for fixation. External mycelium was sieved from the pots and treated in the same way as the roots.

Kodirex X-ray film was used for the macroauto-radiography. For microautoradiography, 1-2 μm sections were coated with Kodak AR 10 stripping film and exposed for 7 days.

Cox, G. *et al.*

Chemical analysis for polyphosphate

    Tissue samples were washed, blotted dry, weighed
and frozen.   They were extracted using a method
adapted from Bennett and Scott (1971).   The purified
extract, containing nucleic acids and polyphosphate,
was assayed for purity and RNA content by ultraviolet
spectroscopy.   Aliquots of the extract were characterised by acrylamide gel electrophoresis using 2.4%
gels with 5mA/tube at room temperature.   Gels were
scanned in the UV using a Joyee-Loebl U.V. Scanner
with PDAB filter, then stained in Toluidine Blue (1%
in 1% acetic acid) for 2 hours.   After differentiation overnight in 1% acetic acid at 0.4°C the gels
were optically scanned using various filters in a
Joyce-Loebl Chromoscan 200.

RESULTS

Light microscopy

Polyphosphate
    When fresh material (external mycelium or mycor-
rhizal roots) was stained in buffered toluidine blue,
hyphal contents showed an overall granular metachromatic red staining.   Some host cell walls also
stained red or purple.   Extraction with TCA prior to
staining completely removed the metachromatic reaction from fungal material, while not affecting the
staining of host cell walls.   Similar staining of
the hyphal contents was seen with Ebel, Colas and
Muller's (1958) lead sulphide stain for polyphosphate.   Although these results suggested the presence of polyphosphate, it was difficult to obtain
adequate resolution from fresh material.
    Sections of resin-embedded material presented
the possibility of obtaining more detailed information.   Toluidine blue staining was extremely slow,
though it gave the same metachromatic reaction as in

fresh material; methylene blue proved more suitable.
Three minutes staining followed by up to 5 minutes
differentiation showed small metachromatic granules,
near the limit of resolution of the light microscope,
within both intercellular hyphae and arbuscules
(Plate IV , Fig. 4).   Given longer staining, followed by much longer differentiation, the entire contents of the fungal vacuoles stained metachromatic-
ally, though the granules remained visible as darker
red grains (Plate IV , Fig. 3).

Lipid
         Abundant lipid droplets were seen in the fungal
mycelium after staining with Sudan IV.   These were
not uniformly distributed, but were more numerous in
the older parts of the infection unit, near the entry
point (see Cox & Sanders, 1974).   Lipid is abundant
near the entry point, both in intercellular hyphae,
the appressorium and 'infection wedge' and in the
external hyphae (Plate IV , Fig. 1).   Intracellular
structures are not visible in this region.   In the
region of senescent arbuscules (Plate IV , Fig. 2)
lipid droplets are smaller but still numerous in the
intercellular hyphae, and are also seen in the
arbuscule trunk.   Lipid droplets are scarce around
mature arbuscules, and we have not observed them (in
the light microscope) in arbuscules in the early
stages of formation or in intercellular hyphae in
advance of the region of arbuscule development.

Electron microscopy

         Both lipid droplets and the putative polyphos-
phate granules are readily observed in the electron
microscope.   Lipid is distinguished by its affinity
for osmium tetroxide; this makes it relatively
electron dense in an unstained section (Plate VI ,
Fig. 13) but less so in a section in which the contrast of other structures has been increased by post-
staining with lead citrate and uranyl acetate

(Plate IV , Fig. 7).   Lipid droplets are large and
numerous in some intercellular hyphae (Plate IV ,
Fig. 7) but almost or entirely absent from others
(Plate IV , Fig. 6).   Lipid droplets are not numerous in arbuscules, though they are seen occasionally
(Plate VI , Fig. 13) in the branches of mature arbuscules; they are more common in the arbuscule
trunk during senescence (see Fig. 14 of Cox &
Sanders, 1974).

'Polyphosphate' granules are visible in the
electron microscope whether or not lead nitrate is
included in the fixative.   The lead treatment increases their contrast, however, particularly in
sections which have not been post-stained.   This
agrees with other observations of polyphosphate granules in the electron microscope (Jensen, 1968;
Atkinson, *et al.,* 1974; Gezelius, 1974).

In thin sections the granules are seen as electron dense circular profiles 0.1 - 0.2 μm in diameter,
which often show a hollow centre (Plate IV , Fig. 5).
The hollow centre may result from volatilization in
the electron beam (Gezelius, 1974) but since it is
also sometimes seen in high-voltage micrographs of
thick sections (e.g. in some granules of Plate VI,
Fig. 14), incomplete penetration of stain or fixative
may be a more likely explanation.   The granules,
which are not themselves membrane-bounded, are always
seen in vacuoles of the fungus.   Polyphosphate granules occur in vacuoles in *Chlorella* (Atkinson *et al.,*
1974) and are usually in vacuoles in *Dictyostelium*
(Gezelius, 1974) though in other organisms the granules may occur in a wide variety of sites (Keck &
Stich, 1957).

The use of thick sections in the high-voltage
electron microscope permits an in-depth visualization
of structure, particularly when stereo-pair micrographs are taken.   The granules are seen to be very
numerous in both the arbuscule (Plate VI , Figs. 12,
14) and the intercellular hyphae (Plate VI , Fig. 12;
see also Plate IV , Fig. 6).   The granules seem to

occur regularly one in each vacuole, whether the
vacuoles are crowded together in the characteristic
'reticulate vacuolation' (Cox & Sanders, 1974) of the
intercellular hyphae and the arbuscule trunk (Plate
VI , Fig. 14) or strung out in line in the fine
branches (Plate VI , Fig. 13).   In a few fine
branches (Plate VI , Fig. 14) the vacuoles are devoid
of granules.   None of our high-voltage electron
micrographs has shown vacuoles without granules in
the arbuscule trunk and main branches, or in the
intercellular hyphae.

The high-voltage electron microscope also permits visualization of three-dimensional aspects of
arbuscule collapse amplifying the sequence given in
a previous paper (Cox & Sanders, 1974).   The early
stages of collapse seem to be notably localized
(Plate VI , Figs. 12, 14), a number of branches in
close proximity tending to collapse together.   The
high-voltage microscope confirmed our earlier conclusion that the trunk is the last part of the arbuscule to collapse, and that it is rich in lipid at
this stage.

Chemical analysis

The gels showed a number of relatively fast-
moving (ca. 2cm h^{-1}) bands, which stained blue. These
were larger in shoot extracts than root extracts, and
were interpreted as various RNA fractions.   A slower
-moving band (ca. 1cm h^{-1}), purplish-staining,
and much larger in mycorrhizal root extracts than
other extracts, probably contained polyphosphate.
Synthetic polyphosphate (Graham's salt), run under
the same conditions, gave a rather diffuse peak (as
might be expected from a wide range of chain lengths)
but the least mobile components were comparable in
metachromasy and mobility to the slow-moving band in
plant extracts.   However, scanning of the gels in
the ultraviolet prior to staining showed some absorption in the slow-moving band, indicating that nucleic

acids were also present.

Table 1.   Integrated area of the slow-
moving peak, fresh weight of tissue
extracted, and integrated area corrected
for fresh weight ($\int ODg^{-1}$) for mycorrhizal
and control roots and tops.

		inte-grated peak	fresh wt.(g)	$\int ODg^{-1}$
mycorrhizal roots)	means of 2 replicates	138	16.1	8.6
control roots)		22	19.7	1.1
mycorrhizal tops)	means of 3 replicates	18	14.4	1.2
control tops)		16	20.9	0.8

Table 1 shows the integrated area of the peaks
corresponding to the slow-moving, metachromatic band
from each extract, and corrects it for the fresh
weight of tissue used in each case.   It will be seen
that corrected figures give only small differences
between mycorrhizal tops, control roots and control
tops, but a clearly much larger value for mycorrhizal
roots.   This band also gave a much clearer metachromatic staining reaction than the others (though
the other bands were too small to assess clearly).
Although we have not achieved a perfect separation of
polyphosphate from nucleic acids, there is clearly a
component of mycorrhizal roots which is absent or
much reduced in non-mycorrhizal organs, and we feel
that this is most probably polyphosphate.

Autoradiography

     27h after commencement of ^{14}C labelling,

activity was found in the external mycelium (Plate
V , Figs. 10, 11) and in all internal fungal structures, as well as throughout the host root.   Earlier
harvests were not taken, but are clearly needed in
future work.   Cortical cells containing arbuscules
showed much higher activity than uninfected cortical
cells (Plate V , Figs. 8, 9) but the label could not
be assigned specifically to the fungus rather than
the investing host cytoplasm.   Label was related
specifically, but not quantitatively, to the fungus
in the arbuscule trunk and intercellular hyphae
(Plate V , Fig. 9).   Lipid-rich intercellular
hyphae, distinguished by osmium staining, contained
more activity than lipid-poor hyphae (Plate V ,
Fig. 9).   Vesicles, which are also rich in lipid,
showed a high activity (Plate V , Fig. 8).

DISCUSSION

Phosphorus translocation

The evidence presented above suggests that the
metachromatic, lead-staining bodies found in the
fungal vacuoles are polyphosphate granules.   The
vacuoles as a whole show a weaker, overall metachromasy (Plate IV ,Fig. 3).   This suggests that the
granules may exist in equilibrium with a vacuolar sap
saturated with polyphosphate.   The apparent lack of
a membrane around the granules supports this conten-
tion (see also Keck & Stich, 1957;   Atkinson *et al.*,
1974).   It may be, therefore, that the small vacuoles so characteristic of VA endophytes can be regarded as 'packets' of polyphosphate.   The granules may
occupy up to 3% of the volume of the fungus (our
preliminary calculation);   if there is also polyphosphate dissolved in the vacuolar sap, polyphosphate
must represent a substantial proportion of the inorganic phosphate content of the fungus.   These
'packets' of polyphosphate could constitute a means

of translocation of phosphorus in the fungus.

Bidirectional cytoplasmic streaming at the rate
of several cm.h^{-1} has been observed in germ tubes of
various VA endophytes and in other external mycelium
(Mosse, 1959 and our unpublished observations).
This could provide a mechanism for transporting vacuoles containing polyphosphate through the fungal
hyphae. It seems unlikely that polyphosphate would
be released directly to the host, and therefore at
some point it must undergo hydrolysis. Micrographs
such as Plate VI, Fig. 14 suggest that the vacuoles
lose their polyphosphate granules after they reach
the fine branches. These may be the site at which
polyphosphate is hydrolysed, and perhaps the site of
transfer of phosphorus compounds to the host.

Carbon translocation

Smith, Muscatine & Lewis (1969) showed that
transfer of carbohydrate from the host to the fungus
in both fungal parasites and ecto-mycorrhizas, involved the conversion by the fungus of host-derived glucose into other sugars (such as trehalose) and polyols (mannitol) which could not be metabolized by the
host plant. So far, neither trehalose or mannitol
have been detected in VA mycorrhizal fungi (Hepper
& Mosse, 1972; Hayman, 1974; and our unpublished
work). However, the synthesis of lipid by the fungus could provide an alternative sink for photosynthate, and it seems possible that the presence of
numerous lipid globules in the fungal mycelium is the
result. Lipid globules could also be involved in
carbon translocation; they could be moved by cytoplasmic streaming in the same way that we have postulated for polyphosphate-containing vacuoles. The
translocation of lipid into the external mycelium
could help to maintain volume balance in the mycelium
if there is an overall movement of vacuoles into the
root; some proportion of the lipid could also represent recycled lipid from the membranes of these

PLATE SECTION

Nicolson, T. H.

PLATE I   One of the Plates from Kidston and
               Lang's (1921) paper which shows hyphae,
               spores and vesicles associated with
               stems of *Asteroxylon* and *Rhynia* species.

Figs. 12-14.   Large thick-walled resting spores
               up to 500 μm diameter.   The wall
               is shown in Fig. 14 and displays
               stratification with distinct wall
               layers.

Figs. 15 and 16.   Smaller spores with diameters of
                   some 120-150 μm.

Figs. 17 and 18.   Spores of diameters of about 60
                   μm with thick walls.

Fig. 19.   Hyphae, vesicles and spores of a
               fungus in stems of *Rhynia* species。

Figs. 20-24   Hyphae and vesicles of a fungus
               which occurred "characteristically
               between the cells of the inner
               cortex of many of the rhizomes of
               Asteroxylon".   The hyphae range
               in diameter 4-20 μm and the
               vesicles up to 80 μm。

(This plate is reproduced by kind permission of the
Royal Society of Edinburgh).

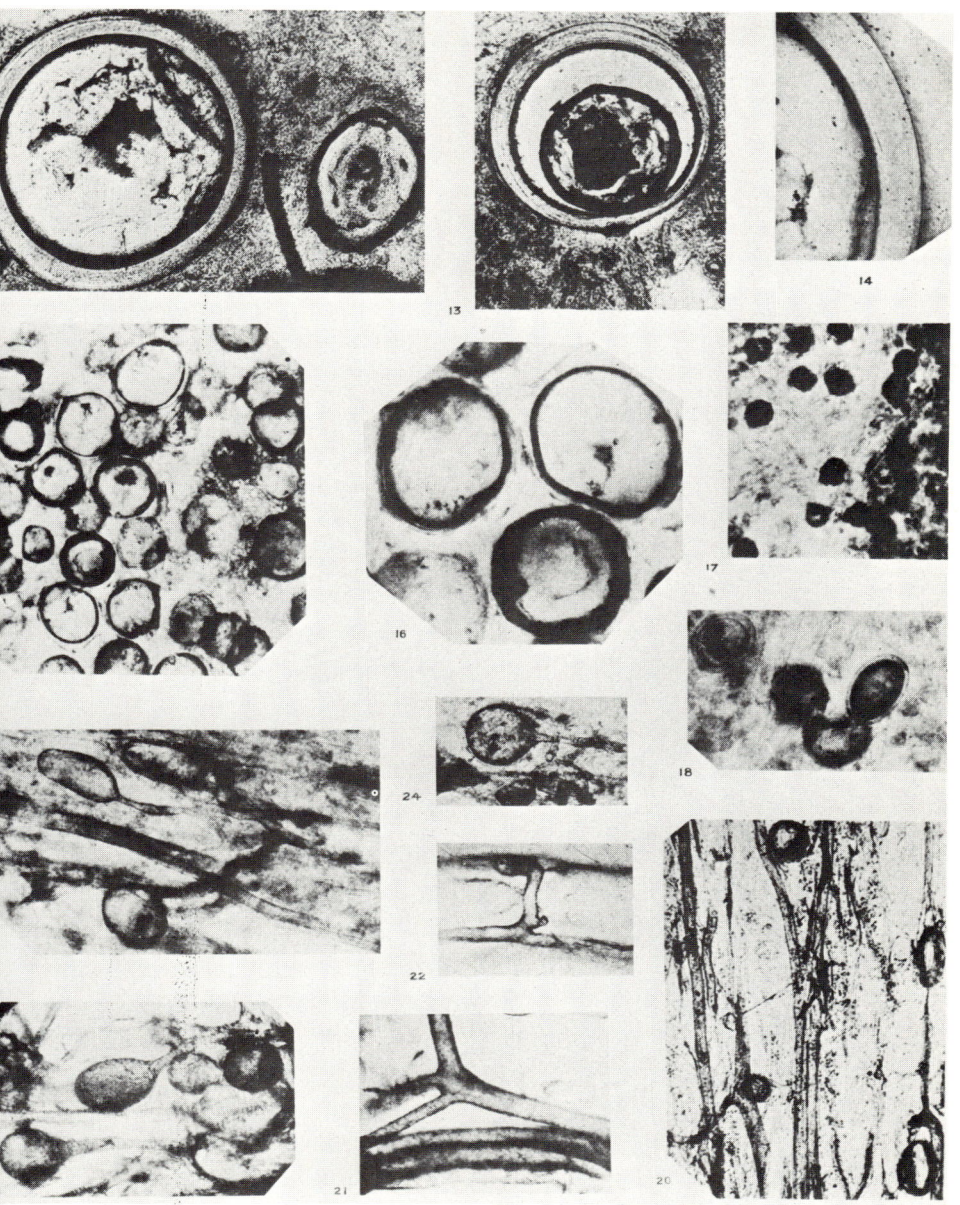

PLATE I

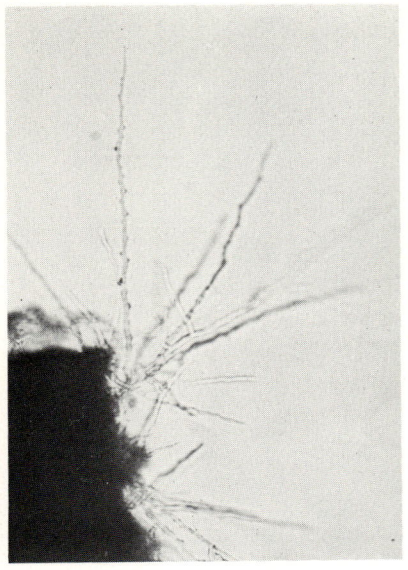

Plate II,1   Hyphae of
Endogone eucalypti growing
from a root segment of
Eucalyptus regnans.

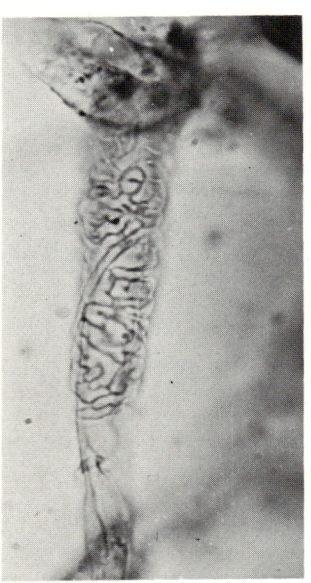

Plate II,2   Portion of a sheath
of *E. eucalypti* on a root hair
of *Eucalyptus regnans*

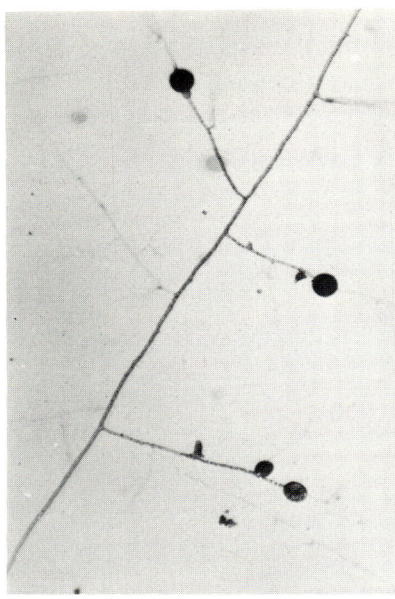

Plate II,3
Hyphal swellings of
E. eucalypti
in agar culture.

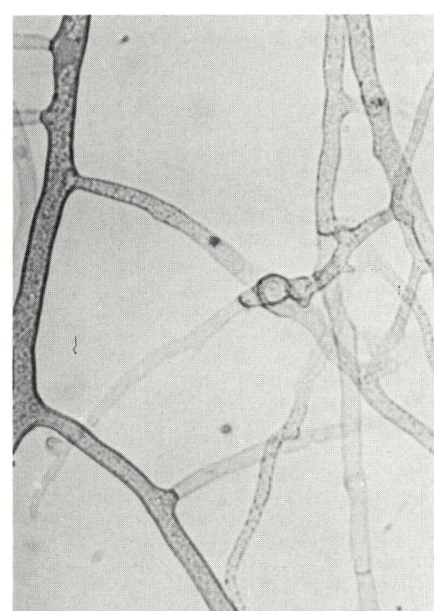

Plate II,4   Anastomosis
of hyphae of *E. euc-
alypti* in agar culture.

Explanation of Plate III
Infection of *Araucaria cunninghamii* by *Endogone araucareae*

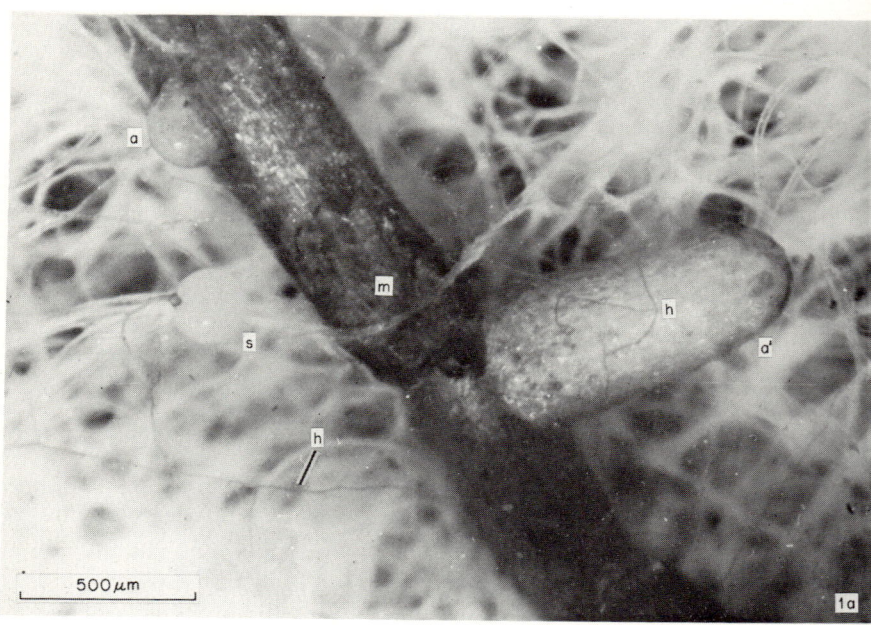

1a   Portion of suberised mother root (m) carrying
     unsuberised short roots (a). Hyphae (h) and a
     spore (s) of *E. araucareae* are attached to the
     unsuberised infected lateral, a'.

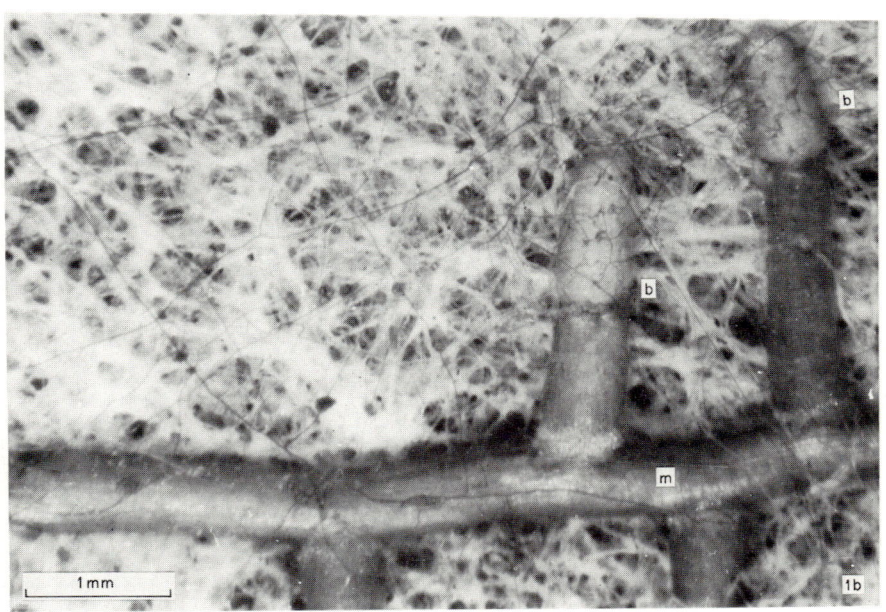

PLATE III

1b Unsuberised infected "beads" (b) growing out
    from older, suberised laterals. Note the
    extramatrical hyphae and infection points on
    both beads and mother root (m).

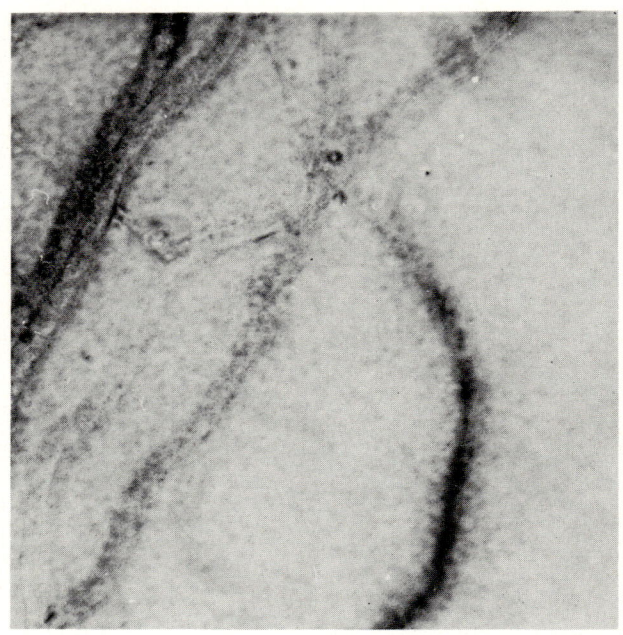

PLATE III

2a ^{32}P accumulation in hyphae attached to clover roots after 1 hr uptake from 5μM KH$_2$PO$_4$ with ^{32}P at 1 μc/ml. Hyphae covered with AR10 stripping film and dried.

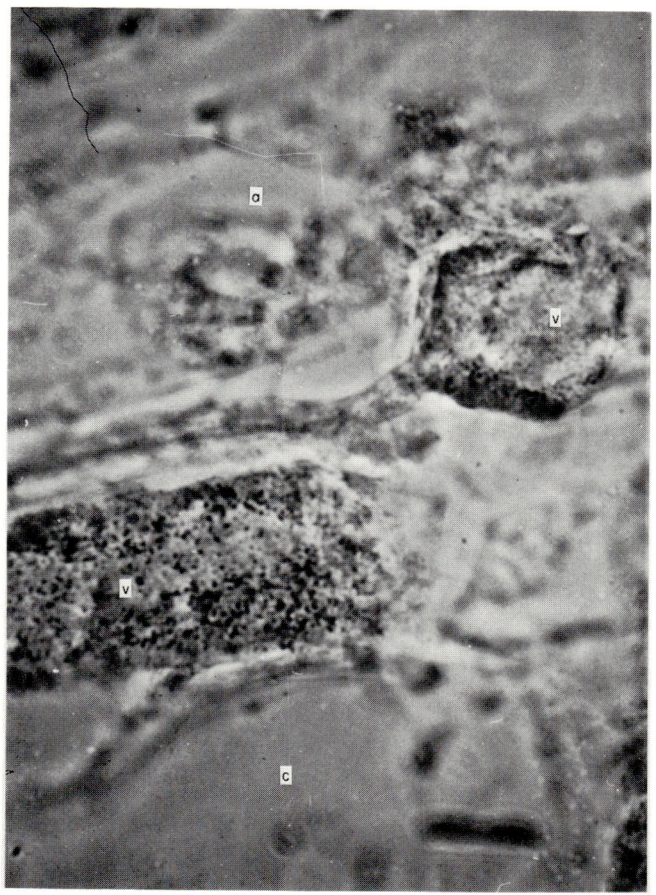

Plate III

2b   Autoradiograph of accumulation of ^{32}P in
      arbuscules (a) and vesicles (v) in onion with
      little label in adjoining uninfected cortical
      cells (c). Uptake as for (b) above, material
      fixed in formalin-alcohol-acetic, embedded in
      paraffin wax and sectioned at 15μm.

PLATE IV

1 & 2. Light micrographs of cleared, squashed roots
stained with trypan blue and Sudan IV.   Lipid
droplets (some of which are arrowed), stained red,
appear as grey or black in these black and white
prints.   eh - external hypha;   ep - entry point;
ih - internal hypha;   ac - collapsed arbuscule.
1. Region of the entry point.   2. Region of collapsed arbuscules in the same infection unit.
3. Light micrograph of a resin-embedded section,
stained in methylene blue.   This has been given a
long staining, followed by a long differentiation,
and the entire contents of the fungal vacuoles have
stained red (and hence appear black in this print).
The small branches of the arbuscules (arb) appear
uniformly dense, but more detail can be seen in the
arbuscule trunks (at) and intercellular hyphae (ih).
4. Light micrograph at higher magnification of a
resin-embedded section, stained with methylene blue.
A shorter staining period renders only the granules
visible (arrowed);   the vacuole contents remain
unstained.   ih - intercellular hyphae;
h - intracellular hypha.
5. Electron micrograph of two fine arbuscule branches,
one sectioned longitudinally, the other transversely.
Granules (arrowed) are seen in the vacuoles.   Lead
nitrate stain in the fixative;   section post-stained
with uranyl acetate and lead citrate.
6. Electron micrograph of an intercellular hypha.
Granules (arrowed) are seen in some of the vacuoles
(v).   No lipid droplets are visible.   Treatment as
in 5.
7. An intercellular hypha, showing vacuoles (v) and
lipid droplets (l).   Treatment as 5.

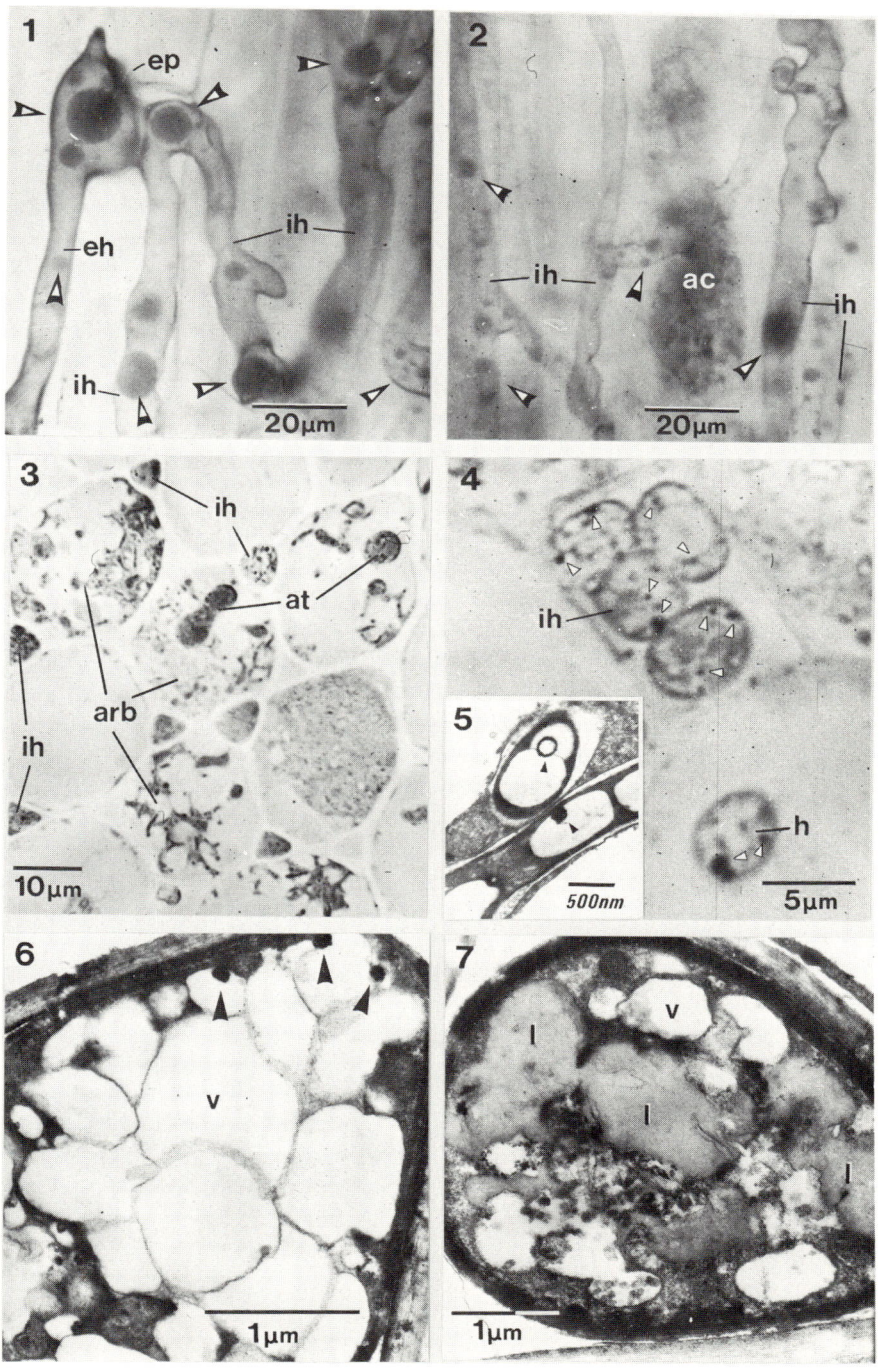

PLATE V

8 - 10 are stripping film autoradiographs of resin-embedded sections.   All show material harvested 24h after the end of the incubation period.
a - focussed on the section
    (phase contrast illumination)
b - focussed on the silver grains
    (bright field illumination)
8. More silver grains are seen over the lipid-rich vesicle (ve) than over the arbuscule (arb), but both have more grains than the surrounding uninfected cortical cells.
9. An arbuscule-containing cell.   Many silver grains are seen over the arbuscule as a whole; silver grains can be related specifically to the fungus over two large branches (arrowed).   More grains are seen over a lipid-rich intercellular hypha (lih) than a lipid-poor intercellular hypha (ih).
10. External hypha (eh) running along the root surface.   Many grains are seen over the hypha, and very few over the uninfected host epidermal (e) or cortical (c) cells.
11. Part of a macroautoradiograph of an entire root system (cut into segments).   Heavy incorporation of label is seen in the root tip (large black arrow), rather less a short distance from the tip (large white arrow).   Uninfected roots, outside the elongating zone, show very little incorporation (small arrows); rather higher incorporation is seen in mycorrhizal regions (double small arrows).   Some label has been incorporated into external mycelium (circled).

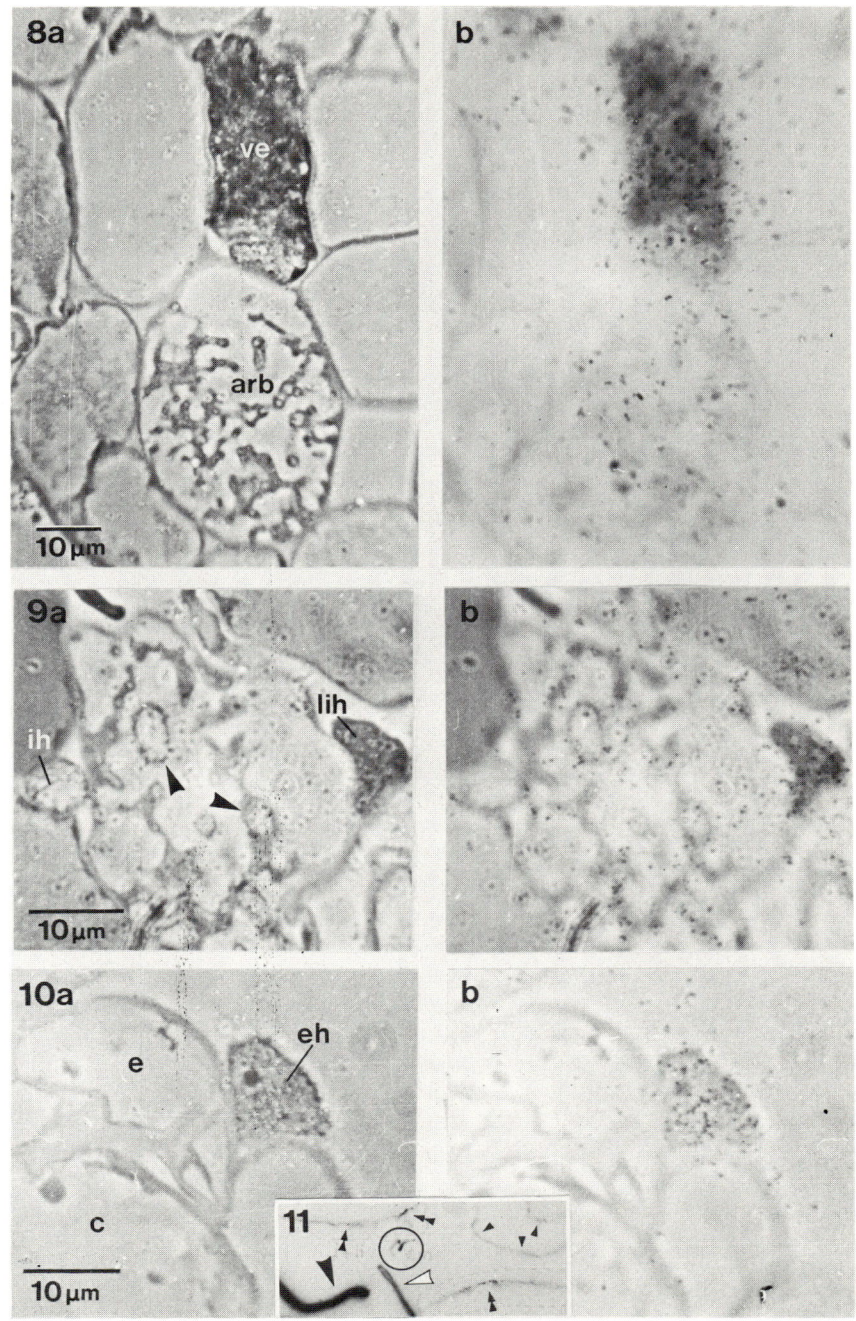

PLATE VI

Stereo pair electron micrographs of 4 μm sections,
taken in the Toulouse high-voltage electron microscope at 2.5 MeV. Material was fixed in glutaraldehyde + lead nitrate, postfixed in osmium tetroxide
but the sections were not post-stained.
12. Low-magnification micrograph of an entire cell.
Granules are numerous in both large and small
branches, and in intercellular hyphae (top r & l).
A small, localised, region of collapse (arrowed) can
be seen.
13. Higher-magnification micrograph of fine branches,
showing the regular occurrence of one granule per
vacuole. Lipid droplets (l) have a clearly different appearance.
14. Large and small branches, generally showing one
granule per vacuole. However, in a few fine
branches (hollow arrow) vacuoles are seen devoid of
granules. A region of collapse is also visible
(solid arrow).

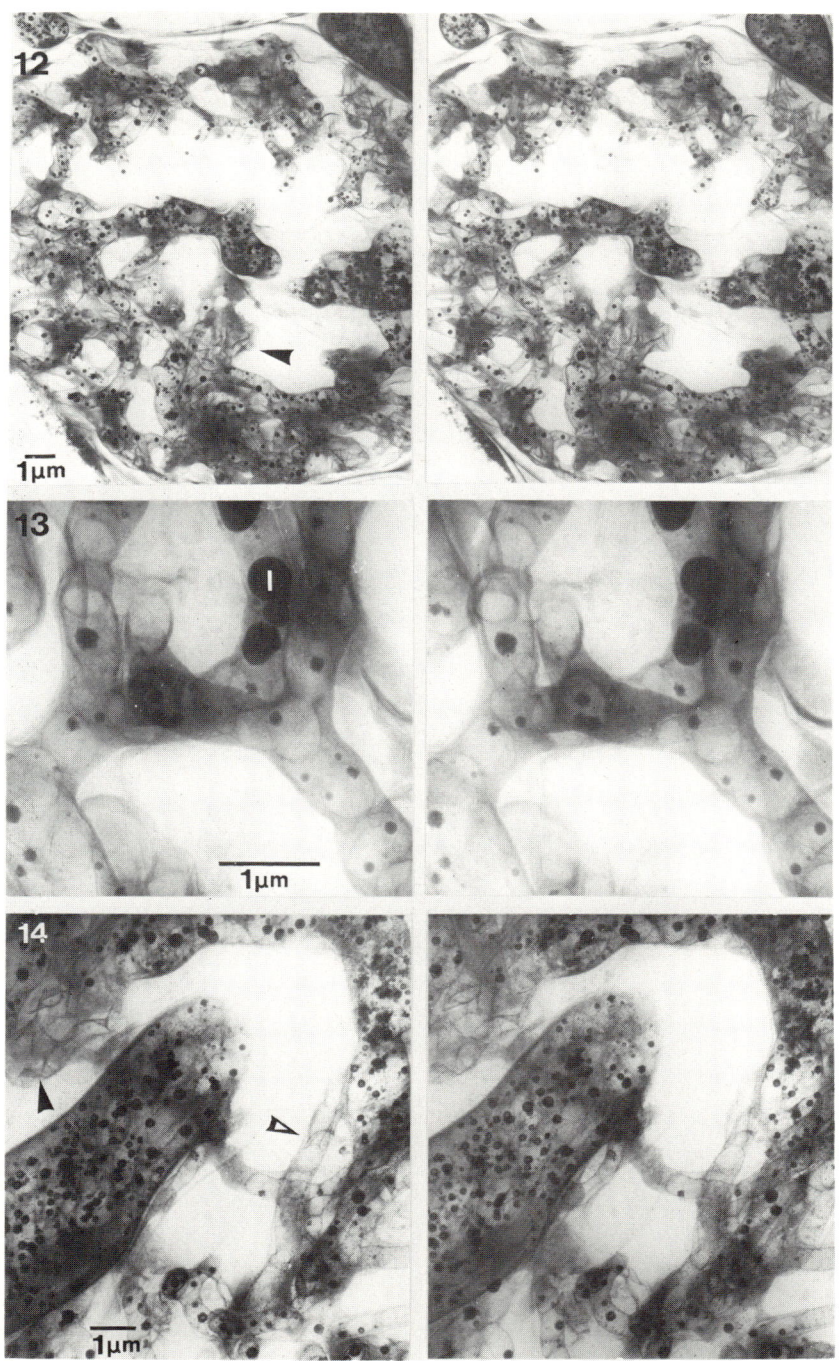

Scannerini, S. *et al.*

<u>Plates VII - XI, key to figure lettering.</u>

A	:	arbuscule
B	:	bacterium
CW	:	cell wall
D	:	dictiosome
Dfm	:	digested fungal material
Dg	:	dense globule
ER	:	endoplasmic reticulum
FW	:	fungal wall
H	:	hypha
HC	:	host cell
Lz	:	layered zone
M	:	mitochondrion
Ma	:	dense matrix
ML	:	middle lamella
N	:	nucleus
Nuc	:	nucleoid
P	:	proplastid
S	:	septum
Sh	:	sheath
Shm	:	sheath membrane
T	:	tonoplast
V	:	vacuole
Ve	:	vesicle

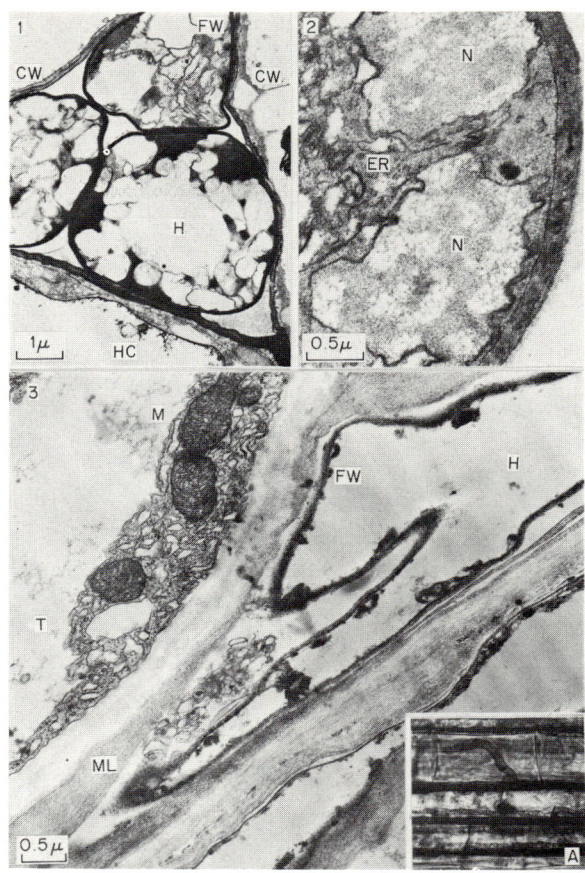

PLATE VII

Figure 1. Cross section of *Endogone* intercellular cords. Intact walls at the interface between the host and the endophyte.

Figure 2. A detail of a structured intercellular hypha. Note the finely layered wall.

Figure 3. Branching of an intercellular hypha, showing an appressorium like structure.

Figure A. *Endogone* intercellular cords and a young haustorium.

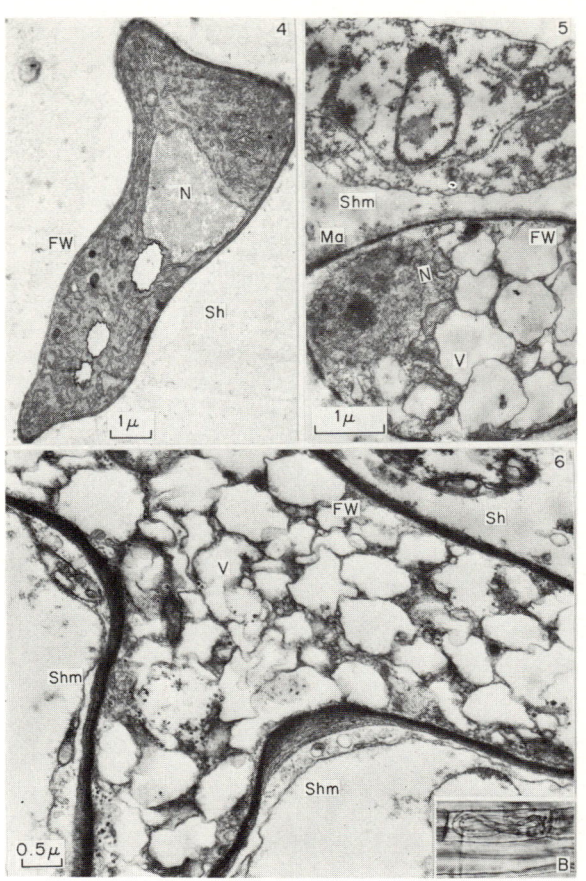

PLATE VIII

Figure 4.   Young haustorium of *Endogone*.

Figure 5.   A vacuolated haustorium in a stage possibly preluding to a hyphal coiling.

Figure 6.   A further stage of vacuolation and
branching of the coiled hyphae.

Figure B.   *Endogone* coil in a host cell.

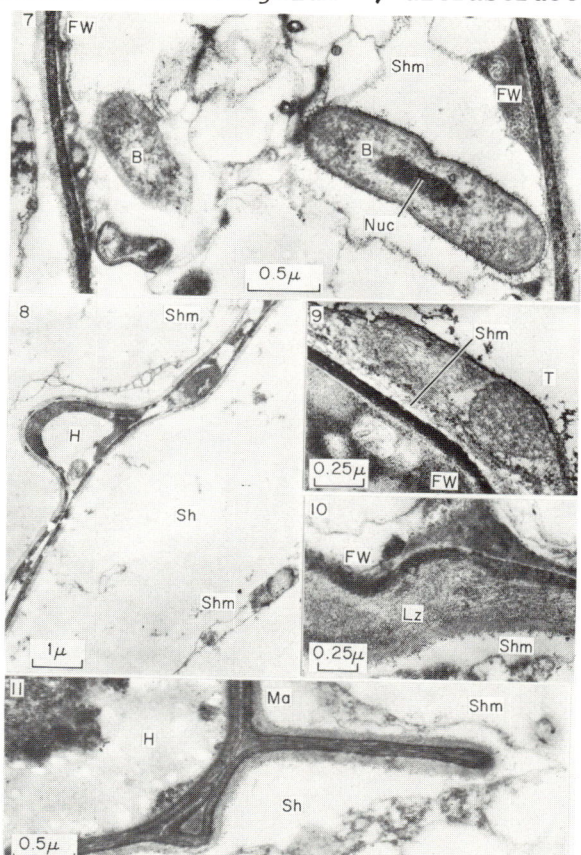

PLATE IX

Figure 7. Coiled hypha containing some bacteria. The bacteria, both living or undergoing initial lysis, are surrounded by an interface of IT 24 type. Note a bacterial division.

Figure 8. Sheath showing a transparent hypertrophic area.

Figure 9. Reduced sheath between the coiled hypha and the host protoplasm.

Figure 10. Layered zone showing a dense hypertrophic matrix.

Figure 11. Swelling and collapsing of some peripheral branches of the coil.

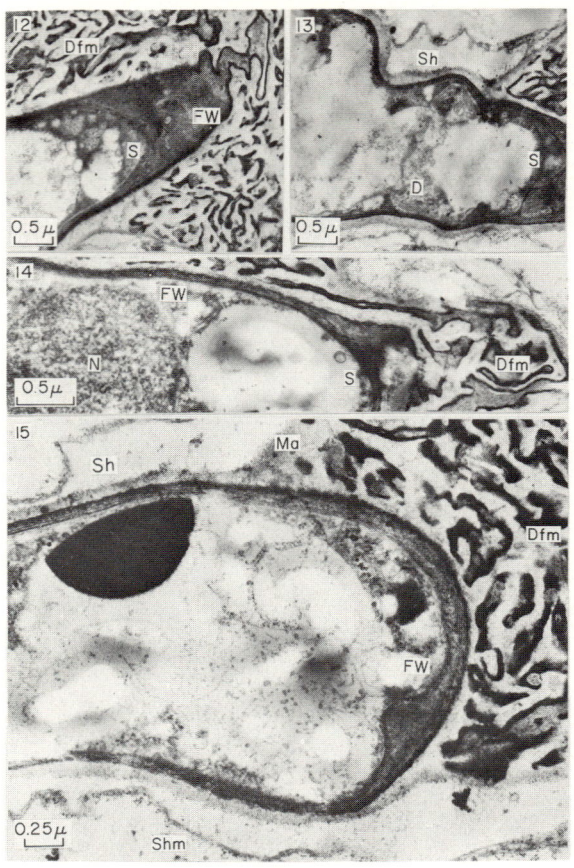

PLATE X

<u>Figure 12.</u>   Developing septum, initial steps in fungal lysis and arbuscule formation.

<u>Figure 13.</u>   Development of a secondary septum behind the septum isolating the hypha from the digested material。

<u>Figure 14.</u>   A nucleated hypha near the digested material, exhibiting an almost complete septum.

<u>Figure 15.</u>   Final stage in the process leading to the separation between the living hypha and the digested material.

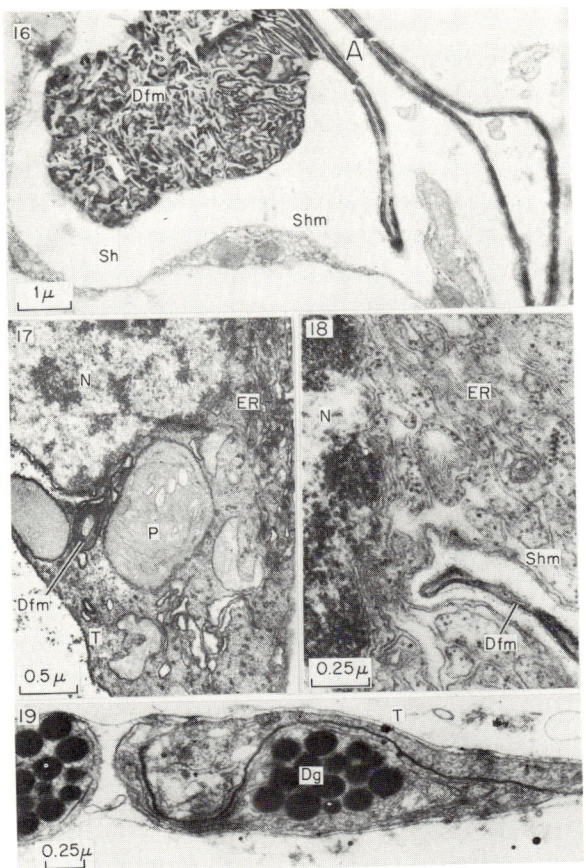

PLATE XI

Figure 16.   Arbuscule, with swollen hyphae and digested fungal material.

Figure 17.   Slender fungal outgrows in proximity to the host nucleus and normal proplastids.

Figure 18.   Hypertrophic endoplasmic reticulum of the host around the digested fungus.

Figure 19.   Degenerated plastids containing dense globules.

Abbreviations

A   =   arbuscule
D   =   dictyosome
E   =   cisterna of the endoplasmic reticulum
EMA =   extra-haustorial matrix
EME =   extra-haustorial membrane
FC   =   fungus cytoplasm
FL   =   lumen of the dead hypha
FP   =   fungal plasmalemma
FV   =   fungal vacuole
FW   =   fungal wall
H   =   intracellular hypha
HC   =   host cytoplasm
HP   =   host plasmalemma
HV   =   host cell vacuole
HW   =   host wall
M   =   mitochondrion
N   =   nucleus
O   =   oil body
P   =   papilla
S   =   septum
T   =   tonoplast
Ve   =   vesicle

All material was fixed in $KMnO_4$.

Fine structure of tobacco mycorrhiza

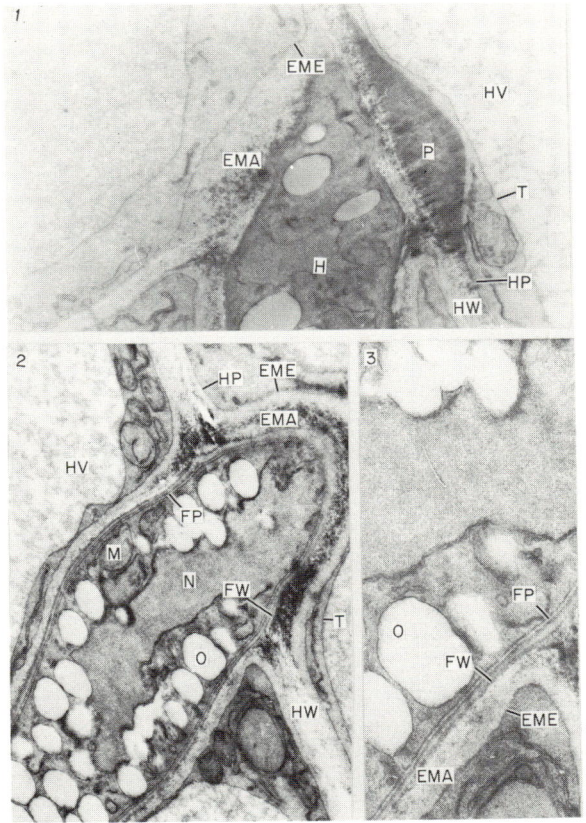

PLATE XII

<u>Figure 1</u>. Formation of papilla near an intra-cellular hypha penetrating the host cell wall. (X *11700*). (Kaspari, 1973).

<u>Figure 2</u>. Penetrated host cell wall; remains of the penetrated papilla (arrow) can be seen in the extra-haustorial matrix. (X *11700*). (Kaspari, 1973)

<u>Figure 3</u>. Invaginations of the fungal plasmalemma (arrow) showing some relationship to oil bodies of the intracellular hypha. (X *23400*). (Enlargement of part of Fig. 2)

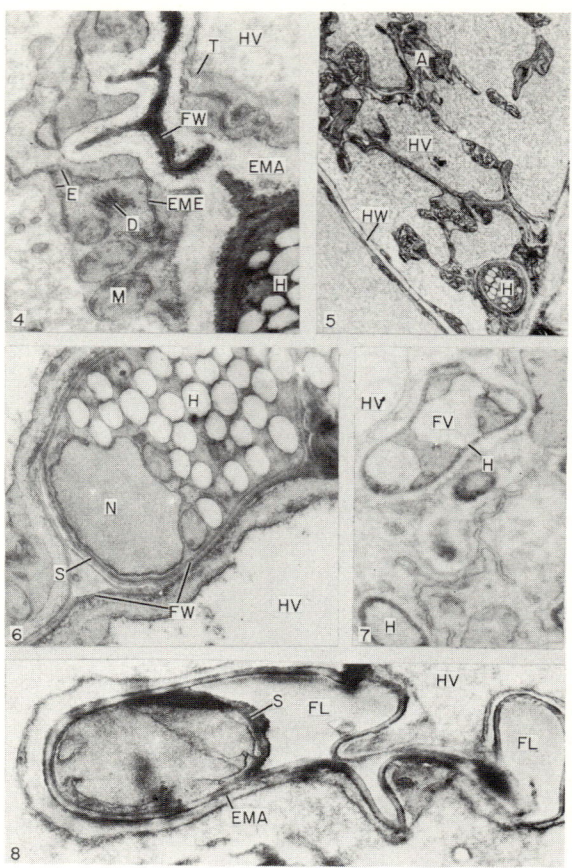

PLATE XIII

Figure 4.   Intracellular hypha and collapsed hyphal segment with accumulation of organelles in the host cytoplasm.   (X *11700*).

Figure 5.   Host cell with intracellular hypha and stages in arbuscule development.   (X *23400*).

Figure 6.   Intracellular hypha with a complete septum cutting off a collapsed hyphal segment, and an incomplete septum (arrow).   (X *12400*).

Fine structure of tobacco mycorrhiza

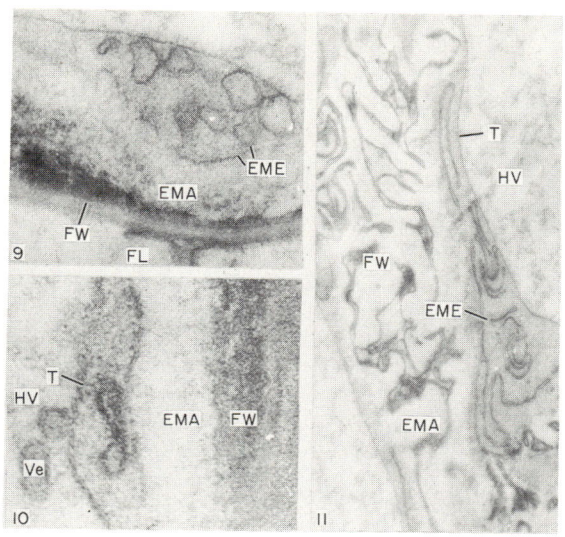

PLATE XIV

Figure 9. Advanced stage of arbuscule development. The hypha only contains membrane fragments (arrow) and the extra-haustorial membrane is badly preserved. (X 36400).

Figure 10. Hyphal wall and matrix material form compact masses in the host cytoplasm. (X 12400).

Figure 11. Tonoplast invagination and vesicle in the host vacuole near a collapsed hyphal segment. (X 71500).

PLATE XIII

Figure 7. Vacuoles in the hyphae show the increasing degeneration of the fungus during arbuscule formation. (X 12400).

Figure 8. Part of an arbuscule. A complete septum separates a dead from a moribund part of the hypha. (X 12400).

Key to lettering used on Plates   XV - XXII
Figs. 1-38.

	Fungus structures			Host structures	
c	- collapsed hyphae		cw	- cell wall	
d	- dolipore septum		cyt	- cytoplasm	
dh	- disorganized hyphae		e	- encasement material	
dp	- digested (lysed) peloton		er	- endoplasmic reticulum	
h	- normal, healthy hyphae		g	- golgi body	
l	- lipid material		hl	- lipid body	
m	- mitochondrion		hm	- mitochondrion	
mat	- matrix of material of obscure origin in digested pelotons		hn	- nucleus	
			hp	- amyloplast	
n	- nucleus		p	- plasmalemma	
pv	- partially vaculoate hyphae		pb	- paramural body	
			re	- reticulate encasement material	
r	- ribosomes				
up	- undigested peloton		sp	- space	
v	- vacuole		t	- tonoplast	
ves	- vesicles		vac	- vacuole	
w	- wall of hypha		ve	- vesicles	
			x	- crystal	

Captions to Figures 1 - 38.   All material is *Dactylorhiza purpurella* infected with *Rhizoctonia* sp. isolate T, unless otherwise stated.

Hadley, G.

PLATE XV

Figs. 1 - 3. Individual hyphal clusters (pelotons) teased out from infected host cells (all X *132*).

Fig. 1. Coiled hyphae beginning to produce lateral branches.

Fig. 2. Further branching and anastomosis of laterals to form a complex network (a peloton).

Fig. 3. Densely packed peloton, with a small digested peloton (dp) adjacent.

Fig. 4. Oblique thin section through protocorm stained with toluidine blue, showing distribution of "host cells", mostly in outer cortical parenchyma, containing healthy pelotons (up) and "digestion cells" containing massed remains of digested pelotons (dp). Epidermal cells, many of them collapsed, are not usually infected. (Light micrograph, X *55*).

Fig. 5. Phase contrast micrograph of thin section of host cell infected with *Rhizoctonia solani (Thanatephorus cucumeris)* isolate W 82, showing secondary infection. Densely staining hyphae, rich in cytoplasm, are intermixed with sparsely-filled, vacuolate hyphae which do not take up much stain. Part of a digested peloton (dp) remains from a primary infection. (Section and micrograph by B. Williamson).

Fig. 6. Light micrograph of thin section stained with toluidine blue, c.f. Fig. 4. Densely staining hyphae full of cytoplasm form healthy pelotons (up) in the outer cortical cells. Other hyphae are partially vacuolate (pv) or collapsed (c). Clumps or hyphal remains undergoing digestion (dp) are present in some cells.

Fine structure of orchid mycorrhiza

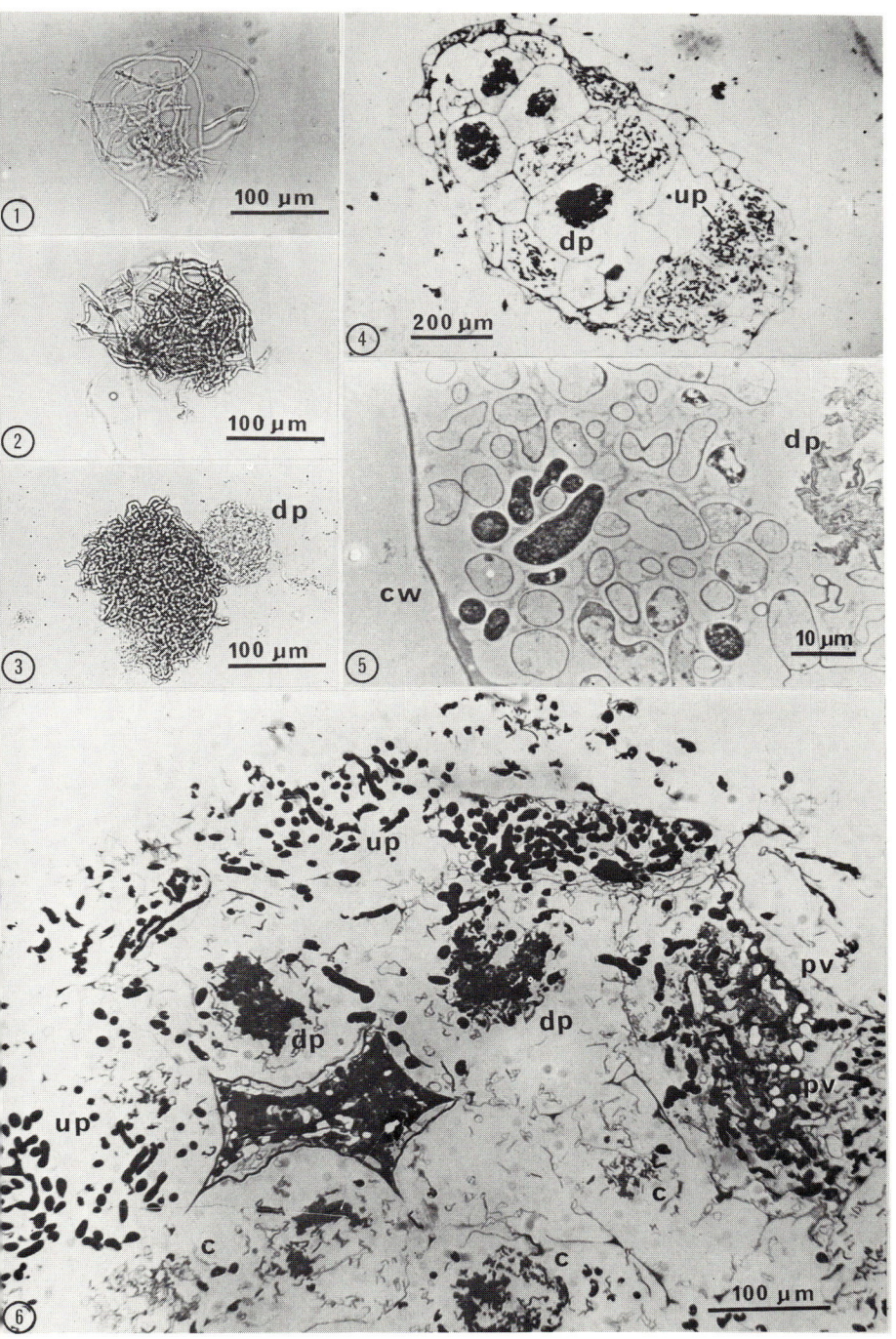

Hadley, G.

PLATE XVI

Fig. 7.   General view of distribution of hyphae in
host cell.   Some hyphae appear healthy and contain
normal cell organelles (h), others appear to have
disorganized contents and densely staining walls of
variable thickness (dh).   Host cytoplasm (cyt) is
mostly as a thin film around hyphae.

Fig. 8.   Section through intracellular hypha showing
nucleus (n), mitochondria (m), ribosomes (r) and
plasmalemma of fungus.   The hyphal wall (w) is surrounded by a granular encasement layer (e).   Host
cytoplasm contains mitochondria (hm) and some endoplasmic reticulum (er).   The host tonoplast (t)
appears to be adjacent to the plasmalemma (p) where
the cytoplasm forms a very thin layer.   A paramural
body or lomasome (pb) can be seen but the spaces (sp)
between plasmalemma and encasement may be an artefact.
From Hadley, Johnson & John, (1971).

Fig. 9.   Part of a hypha showing a complex of vesicles (ves) which appear to be connected to the plasmalemma by a system of canals.   The host plasmalemma
(p) forms irregular invaginations or undulations over
protuberances of the encasement layer.
From Hadley, Johnson & John (1971).

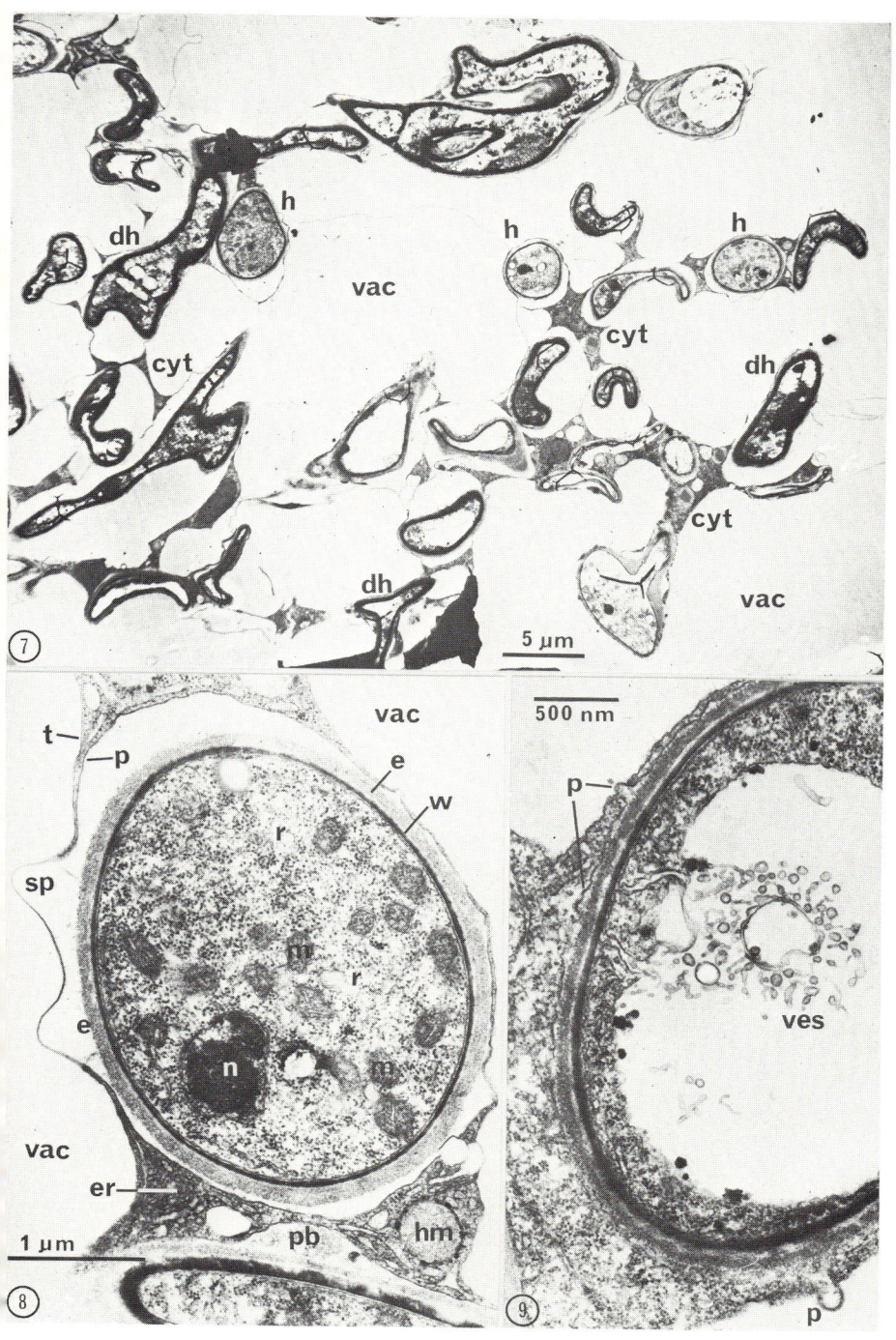

PLATE XVII

<u>Fig. 10.</u> Section through hyphae of *Rhizoctonia* sp.
in agar culture. Hyphal walls (w) appear to consist
of a single electron-dense layer. A second, transparent layer separates the adjacent walls of the
hyphae.

<u>Fig. 11.</u> Vacuolated ageing hyphae of *Rhizoctonia*
sp. from agar culture. An electron-transparent
material separates the adjacent walls (c.f. Fig. 10).

<u>Fig. 12.</u> Section of hypha enclosed by encasement
layer (e) and host cytoplasm (cyt) containing nucleus
(hn), mitochondrion (hm) and an amyloplast (hp).
The host tonoplast (t) is lobed in a complex manner.

<u>Fig. 13.</u> Hyphae of *Rhizoctonia* sp. isolate Rs 16,
an aggressive strain, in host cell. The encasement
layer (e) is inconspicuous or absent and the host
cytoplasm lacks an organized structure.

<u>Fig. 14.</u> Hyphae in host cytoplasm. The encasement
layer (e) is thicker and more granular around the
flattened, disorganized hypha (dh) than the normal
hypha (h). Endoplasmic reticulum (er) is conspicuous near the flattened hypha. Paramural bodies
(pb) occur between plasmalemma and encasement.
A crystal (x) and a mitochondrion (hm) can be
seen.   Inset:- detail of endoplasmic
reticulum and encasement layer.

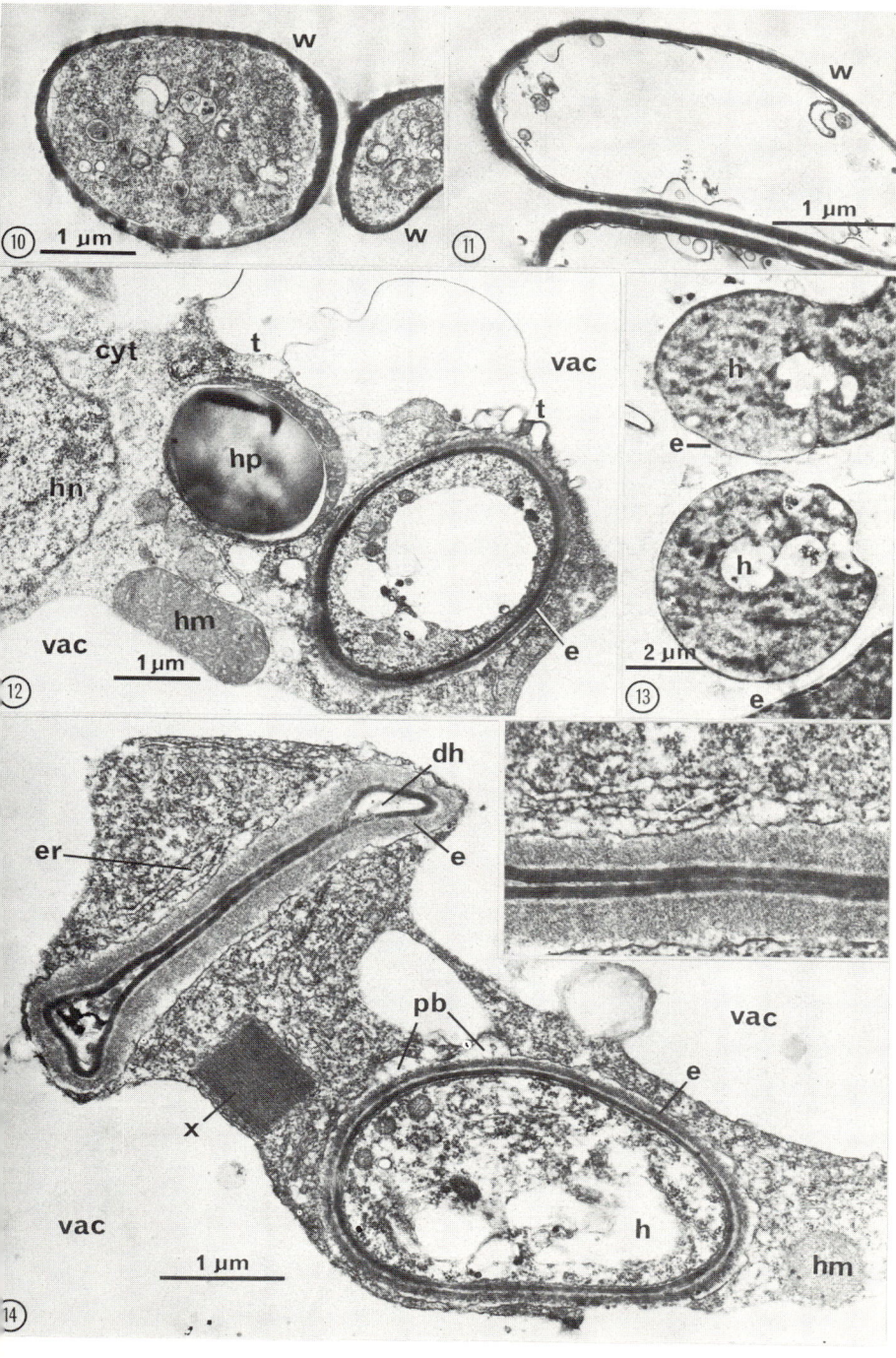

PLATE XVIII

Fig. 15.   Intracellular hypha near host cell wall
(cw) with a very thick encasement layer (e).   The
host plasmalemma (p) is partially separated from the
encasement by a space (sp) and appears to be forming
lobes or vesicles in some areas (arrows).   Endoplasmic reticulum (er) is present.

Fig. 16.   Intracellular hypha with a well defined
encasement layer (e) which is continuous with the
adjacent host cell wall (cw).

Fig. 17.   Part of two adjacent intracellular hyphae
(h) surrounded by a continuous encasement layer (e).
There appear to be vesicles (ve) between the host
plasmalemma and the encasement.

Fig. 18.   Two adjacent intracellular hyphae with a
broad band of encasement material (e) forming a
connecting layer between them.

Fig. 19.   Part of an intracellular hypha surrounded
by host plasmalemma (p) which is forming undulations
over the encasement layer (e) and invaginating into
the host cytoplasm.   Golgi bodies (g) and a mitochondrion (m) can be seen.

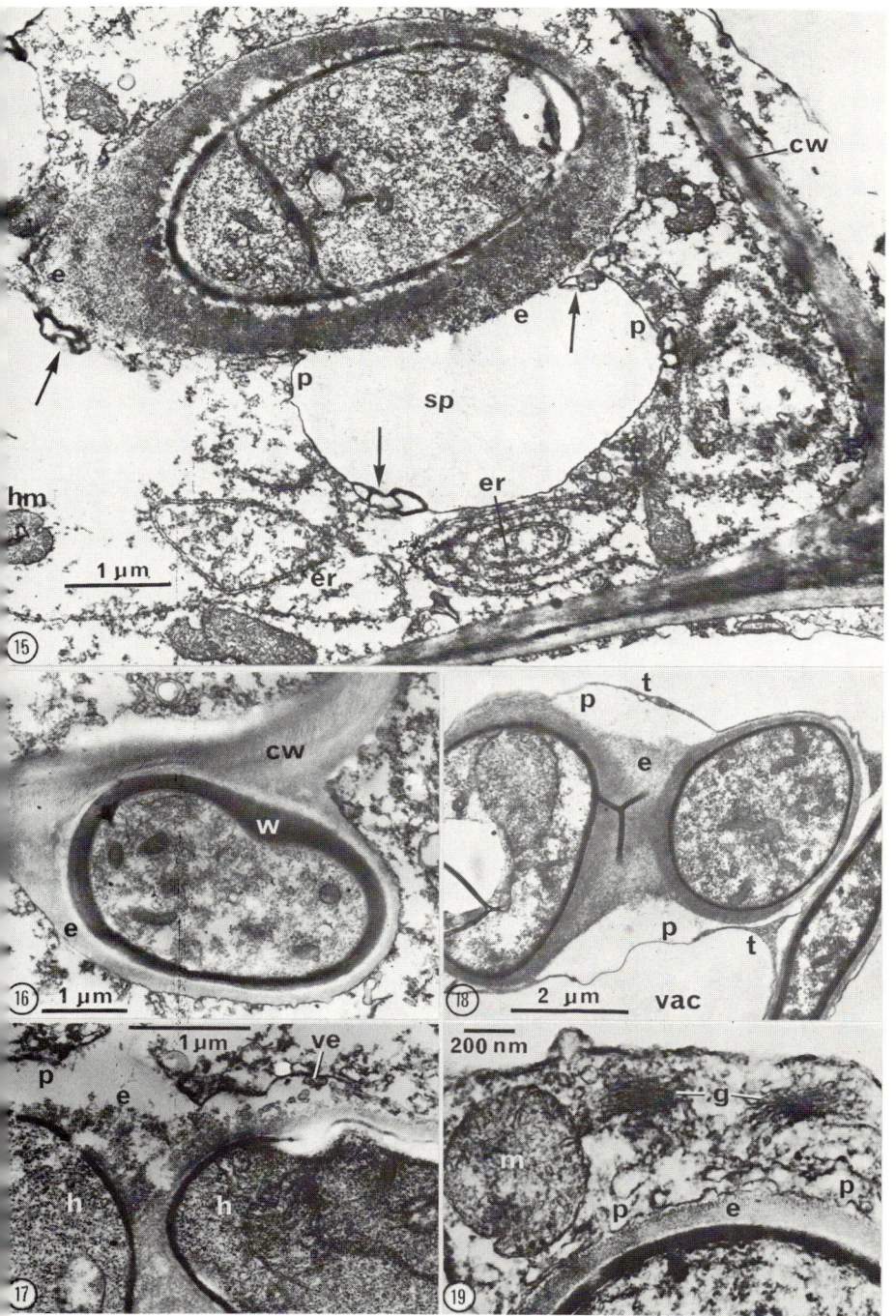

PLATE XIX

<u>Fig. 20.</u>   Host cell plasmalemma (p), endoplasmic
reticulum (er) and tonoplast (t) invaginating into
the vacuole (vac) adjacent to the wall of an intracellular hypha (h).

<u>Fig. 21.</u>   Part of an intracellular hypha surrounded
by host plasmalemma (p) which is invaginating into
the host vacuole (vac).   Paramural bodies (pb)
appear to be present between the plasmalemma and the
encasement layer (e).

<u>Fig. 22.</u>   Host plasmalemma and tonoplast separating
from the encasement layer (e) around a hypha and
invaginating into the host cell vacuole (vac).
Cytoplasmic inclusions (1) may be lipid or other
reserve material.

<u>Fig. 23, 24.</u>   Detail from Fig. 22.

<u>Fig. 25.</u>   Part of host cell showing intracellular
hyphae in the first stages of disorganization.   Some
are collapsed (c) and others are vaculoate (v) without a clearly defined tonoplast.

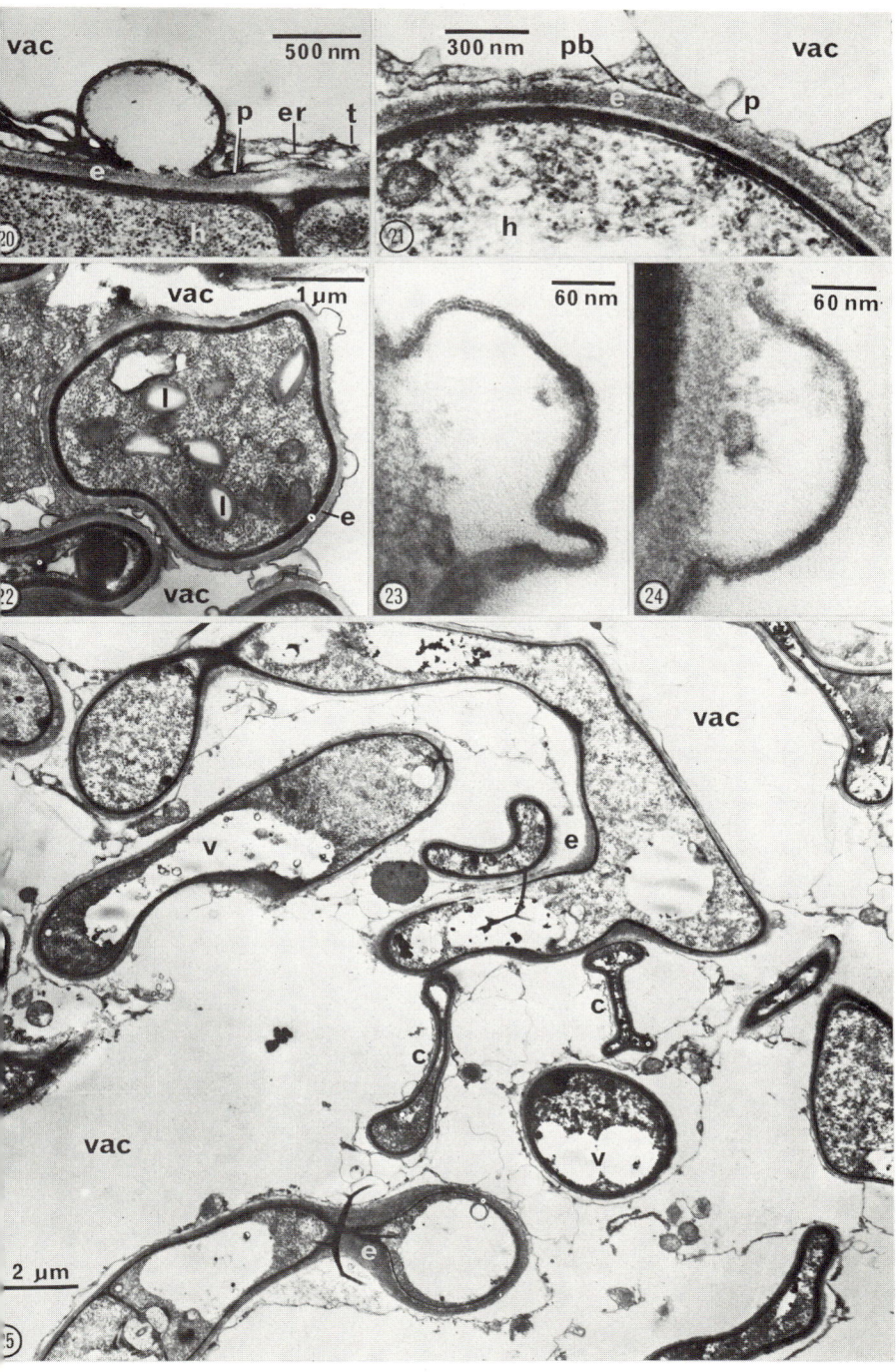

Hadley, G.

PLATE XX

Fig. 26. Two intracellular hyphae, one of which (h) appears healthy, containing a nucleus and mitochondria. The other has an irregular, thickened wall (w), lacks an organized cytoplasmic structure and appears to be degenerating. A dolipore septum (d) can be seen.

Fig. 27. A disorganized, collapsed hypha bounded by a thick granular encasement layer (e) and host cytoplasm (cyt).

Fig. 28. Remains of a collapsed hypha (c) adjacent to a digested peloton (dp). The encasement layer (e) appears to become reticulate (re) around totally disorganized hyphal remains.
Inset:- Detail showing the host plasmalemma (p) forming lobes or vesicles around hyphal remains.

Fig. 29. A single hypha in an advanced stage of degeneration with a reticulate encasement layer (re) and material of undetermined nature (mat) forming a matrix inside the wall. The length of wall (approx. 12 µm) corresponds to the circumference of a healthy hypha.

Fig. 30. Remains of a collapsed hypha (c) being drawn into the edge of a digested peloton (dp). The continuity of the reticulate encasement material can be seen. Host plasmalemma and ER are not clearly defined. Lipid droplets (hl) occur at the boundary of the host vacuole, adjacent to the tonoplast.

Fine structure of orchid mycorrhiza

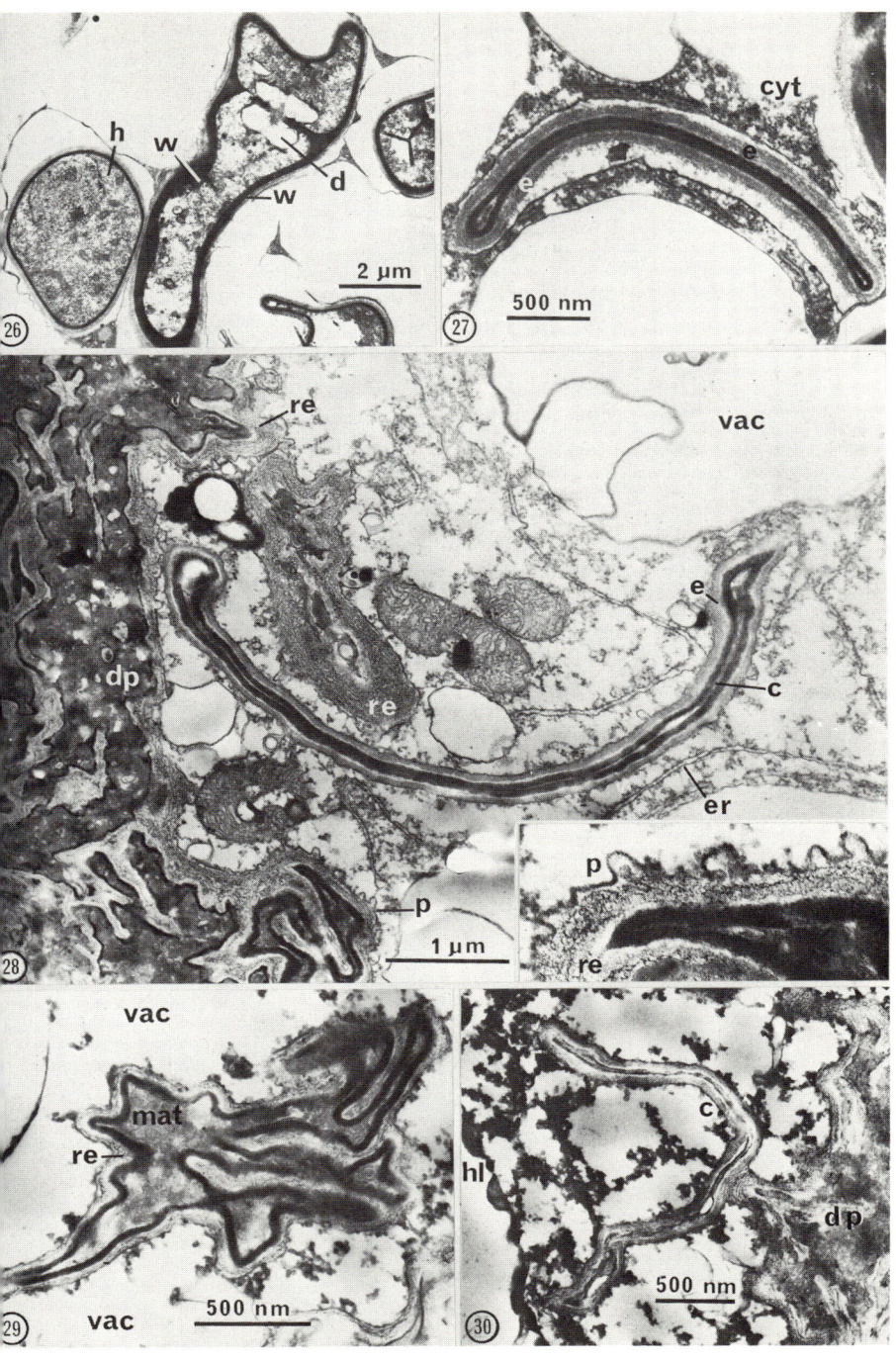

PLATE XXI

Fig. 31.   Complex series of host membranes near the edge of a digested peloton.   There are lipid droplets (hl) associated with the tonoplast.

Fig. 32.   As in Fig. 31, but the membranes are more closely associated into an endoplasmic reticulum with many ribosomes and a mitochondrion.   Lipid material (hl) is associated with the tonoplast.

Fig. 33.   Portions of double membrane present in the matrix material (mat) of a digested peloton, along with the remains of hyphal walls.

Fig. 34.   Part of the central region of a digested peloton.   The electron dense strands appear to be the remains of hyphal walls, together with residual encasement material (e) and matrix material (mat) of uncertain origin.

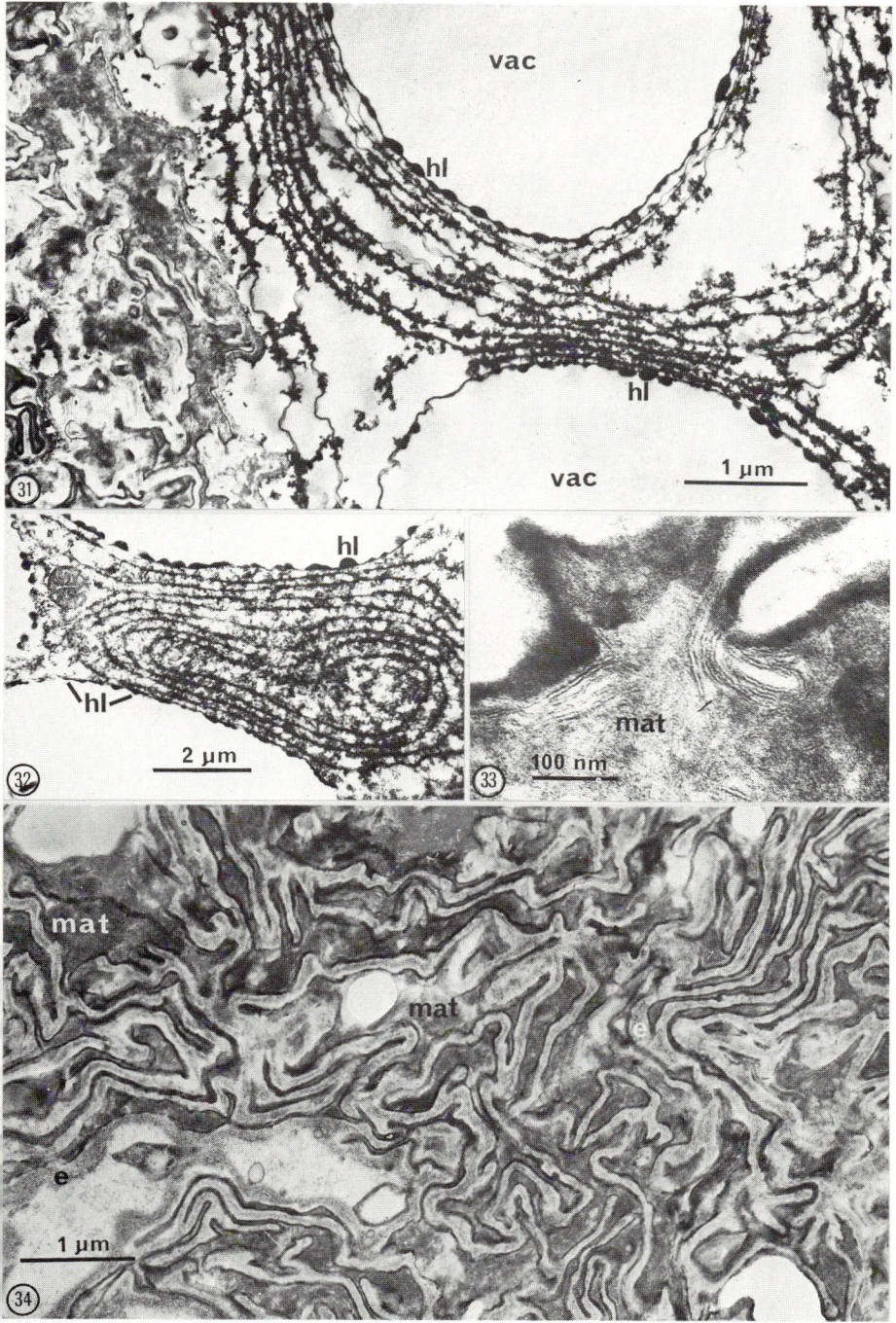

Hadley, G.

PLATE XXII

Figs. 35 - 38 show *Rhizoctonia solani* isolate Rs 16
which infects in a compatible fashion but may become
aggressive in host cells.

Fig. 35. Part of a digested peloton containing a
complex vesicle-like structure of unknown origin.
Electron dense layers are assumed to be the remains
of hyphal walls.

Fig. 36. Hyphae of isolate Rs 16 in disorganized
host cell. Encasement material (e) forms a sparse
layer or is absent and there is no host plasmalemma
or recognizable cytoplasm. In the adjacent host
cell the plasmalemma (p) appears normal.

Fig. 37. Hypha of isolate Rs 16 alongside the wall
(cw) of a host cell which lacks plasmalemma or other
recognizable structures. There is no encasement
layer. Plasmalemma (p) occurs in the adjacent host
cell.

Fig. 38. Hypha of isolate Rs 16 in intercellular
space of host, showing a normal structure. The
hyphal wall (w) is not encased and appears to be
closely pressed against the host cell wall (cw).

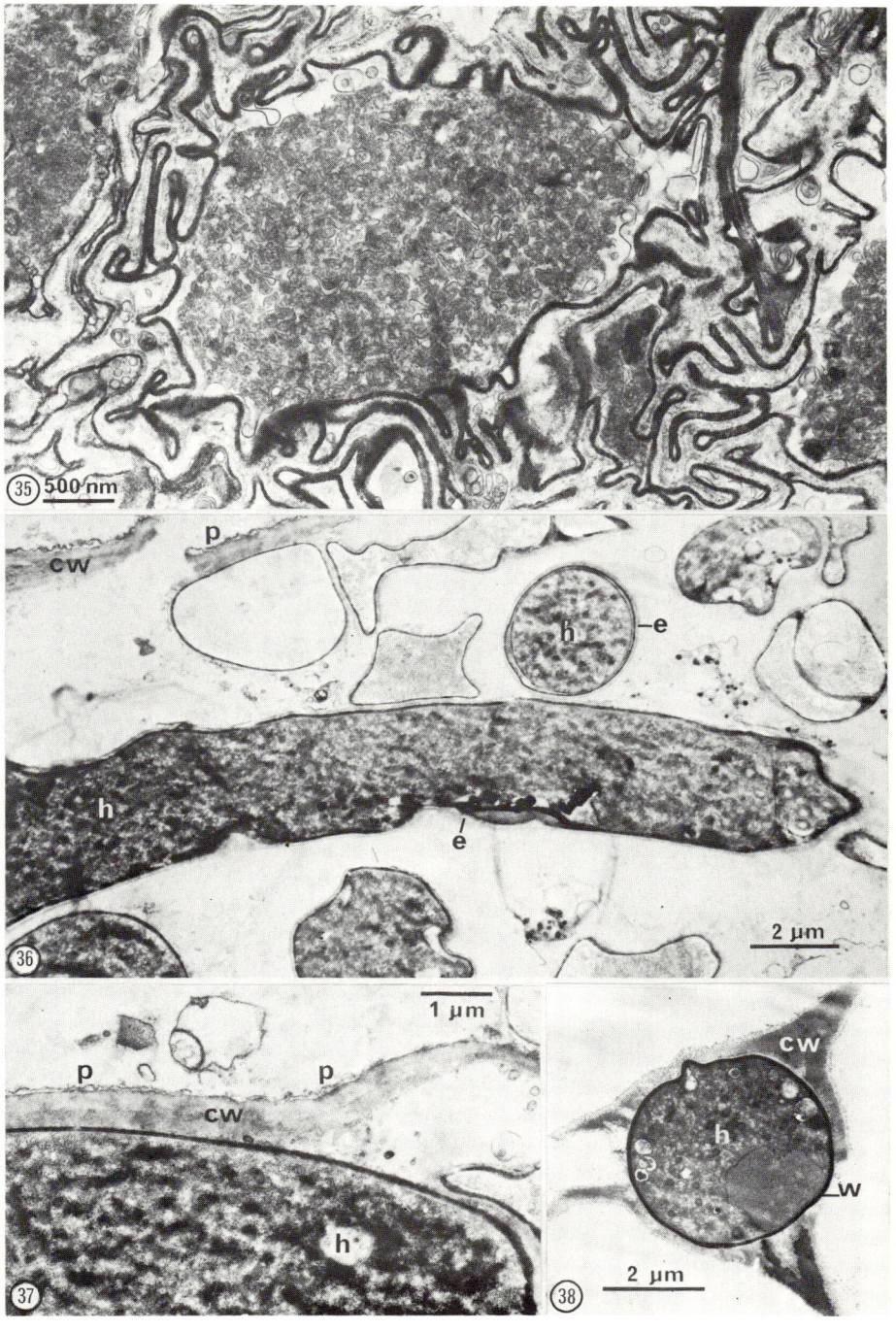

PLATE XXIII

1. Colonization of the upper slopes at the
   Devon coal tip showing vegetated and
   bare areas.
2. Dense stand of Rosebay willowherb and
   thistle on a stable slope at the Devon tip.
3. Success of hand planted trees on the
   northern slope of pile number 1 at Nanty Glo.

Arbuscular mycorrhizas in coal spoils

PLATE XXIV

4. Mass of *Glomus* hyphae on roots of *Robinia hispida.* (X *96*).
5. Echinulate clusters of *Gigaspora gigantea* on roots of *Coronilla varia*. (X *240*).
6. Large azygospores of *G. gigantea* from roots of *Coronilla varia* growing at Tamaqua. (X *24*).
7. Irregular accessory vesicles of *Gigaspora calospora,* sieved from roots of *Dactylis geomerata* grown on the Devon tip. (X *600*).
8. Vesicles produced within the root cortex of *Alnus glutinosa* (*Glomus* sp., X *240*).
9. Section of a root nodule from *Robinia hispida* showing bacteroids(b), vesicle(v), and hypha(h) of fungal endophyte. (X *60*).

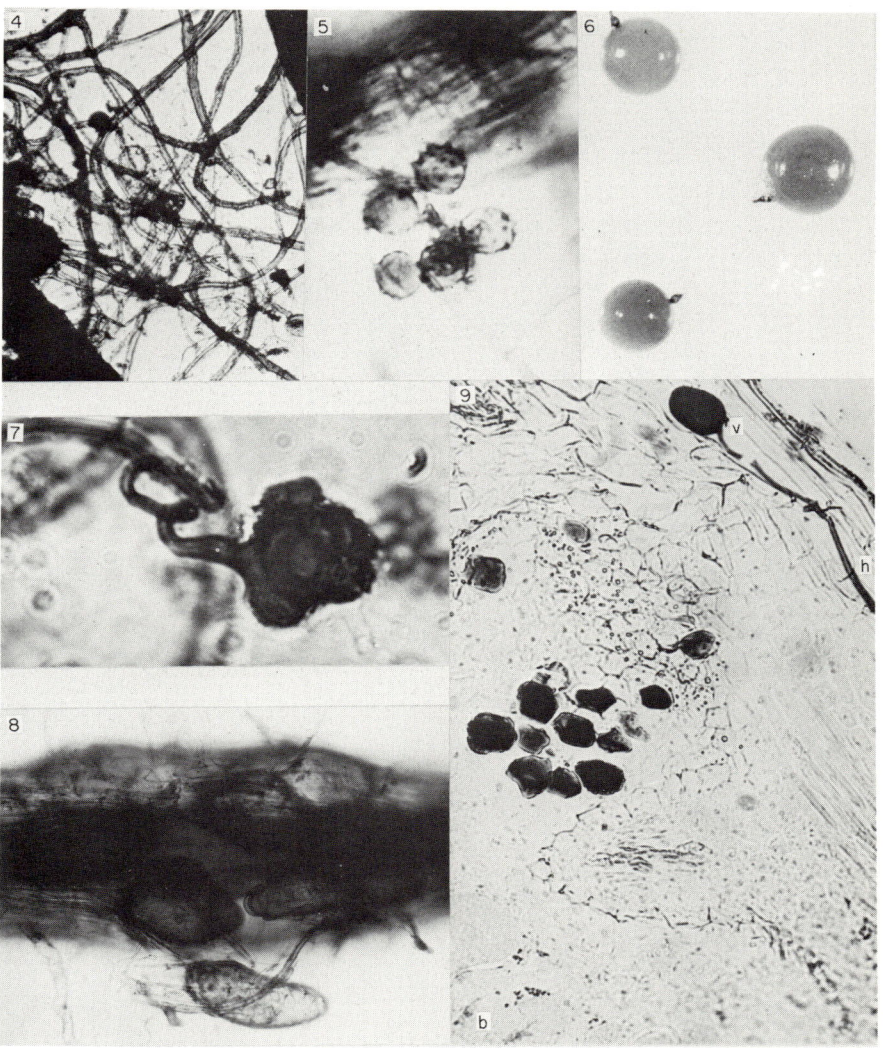

vacuoles.

Our autoradiographs show that carbon compounds deriving from host photosynthate do enter the fungal mycelium, and that within 27h from commencement of labelling label is found in the external mycelium as well as in the inter- and intracellular structures. Previous experiments of this nature have only shown transfer into spores over a period of several months (Ho & Trappe, 1973). The mean lifetime of arbuscules in the *Glomus mosseae*/onion mycorrhiza is probably around 9 days (Cox & Tinker, in preparation; Bevege & Bowen, 1975). Our autoradiographs from successive harvests show a steady decline in activity in the mycelium from 27h up to 17 days. Thus arbuscule senescence cannot be an important component of the mechanism for transfer of host-derived carbon into the mycelium as a whole, even though anatomical evidence (below) suggests that material passes from the arbuscule into the mycelium, rather than the host, when the arbuscule collapses. Our autoradiographs do not show the site of transfer of carbon, but they do show a rapid translocation of carbon throughout the mycelium. The resolution of the auto-radiographs is insufficient to show activity in individual lipid droplets, but lipid-rich intercellular hyphae show a higher activity than lipid-poor hyphae (Plate V , Fig. 9) and it therefore seems likely that carbon from host photosynthate does pass into fungal lipid within 27h.

Arbuscule senescence

When the arbuscule senesces, fine branches in an initially very localized area lose their contents, and the empty walls collapse inward (Plate VI , Figs. 12, 14). This collapse later becomes general, but the arbuscule trunk always collapses last (Cox & Sanders, 1974). This pattern means that living branches are always sealed off by the collapsed walls of dead hyphae, and perhaps also by septum formation

(Kaspari, 1973, 1975; Cox & Sanders, 1974).   It is difficult, therefore, to visualise bulk loss of fungal cytoplasm to the host taking place, nor have we ever seen any signs of such loss in electron micrographs.   It is much simpler to suppose that components of the arbuscule protoplast are resorbed by the fungal mycelium.   The presence of lipid droplets in the trunks of senescent arbuscules, seen in both light (Plate IV , Fig. 2) and electron micrographs (Fig. 14 of Cox & Sanders, 1974), hints strongly at such a process.   This can only be a recycling of components, and not a major aspect of nutrient transfer, however.

SUMMARY AND CONCLUSIONS

Granules, 0.1 - 0.2 μm in diameter are seen in the vacuoles of the fungus.   These display cytochemical properties of polyphosphate granules.   Chemical analysis suggests the presence of higher levels of polyphosphate in mycorrhizal roots than non-mycorrhizal roots, or shoots.   The granules are regularly distributed one to each vacuole, and may be in equilibrium with a vacuole sap saturated with polyphosphate.   It is suggested that these vacuoles, moved by cytoplasmic streaming, may be the means of translocation of phosphorus within the fungus.   Sometimes vacuoles are seen without granules in the fine branches of the arbuscule, but not elsewhere.   This suggests that the fine branches may be the sites where polyphosphate is converted into another form, perhaps for transfer to the host.

Carbon derived from host photosynthate is transferred to the fungus, and translocated to the external mycelium, within 27 hours.   This must involve transfer across a living interface, but the site of this interface was not established.   Lipid is abundant in the intercellular hyphae.   Its distribution, most abundant near the entry point and progressively less abundant as the growing tips are approached, suggests that it is moving either inward or outward. Since it appears to be rapidly labelled with ^{14}C from the host, this movement is probably outward. Lipid may therefore play a part in sequestration and transport of carbon derived from the host.

ACKNOWLEDGMENTS

We are most grateful to M. B. Jouffrey, M. Trinquier and the staff of CNRS Toulouse for making the 3 MeV microscope available to us, and to M. G. Dupouy for his interest and encouragement.   We are very grateful to Dr. M. J. Goringe, Dr. B. E. Juniper, and the Science Research Council for giving us time on the Oxford 1 MeV microscope.   We acknowledge gratefully a Travelling Fellowship from the Yorkshire Agriculture Society, which made the visit to Toulouse possible, and financial support from the Agricultural Research Council.

REFERENCES

ATKINSON, A. W., JOHN, P. C. L. & GUNNING, B. E. S. (1974). The growth and division of the single mitochondrion and other organelles during the cell cycle of *Chlorella*, studied by quantitative stereology and three dimensional reconstruction. *Protoplasma,* 81, 77-109.

BENNETT, J. & SCOTT, K. J. (1971). Inorganic polyphosphates in the wheat stem rust fungus and in rust-infected wheat leaves.   *Physiol. Pl. Path.,* 1, 185-198.

BEVEGE, D. I. & BOWEN, G. D. (1975). This symposium, p. 77.

COX, G. & SANDERS, F. (1974). Ultrastructure of the host-fungus interface in a vesicular-arbuscular mycorrhiza. *New Phytol.*, 73, 901-912.

EBEL, J. P., COLAS, J. & MULLER, S. (1958). Cytochemical research on inorganic polyphosphates contained in living organisms. II. Critical study of cytochemical tests specific for polyphosphates. *Expl. Cell Res.*, 15, 28-36.

GEZELIUS, K., (1974). Inorganic polyphosphates and enzymes of polyphosphate metabolism in the cellular slime mould *Dictyostelium discoideum*. *Arch. Microbiol.*, 98, 311-329.

HATTINGH, M. J., GRAY, L. E. & GERDEMANN, J. W. (1973). Uptake and translocation of ^{32}P-labelled phosphate to onion roots by endomycorrhizal fungi. *Soil Sci.*, 116, 383-387.

HAYMAN, D. S. (1974). Plant growth responses to vesicular arbuscular mycorrhiza. VI. Effect of light & temperature. *New Phytol.*, 73, 71-80.

HEPPER, C. M. & MOSSE, B. (1972). Trehalose and mannitol in vesicular arbuscular mycorrhiza. *Rep. Rothamsted Exp. Sta.*, 1972, p.81.

HO, I. & TRAPPE, J. M. (1973). Translocation of C^{14} from *Festuca* plants to their endomycorrhizal fungi. *Nature, New Biology,* 244, 30.

JENSEN, T. E. (1968). Electron microscopy of polyphosphate bodies in a blue-green alga, *Nostoc preniforme*. *Arch. Microbiol.*, 62, 144-152.

KASPARI, H. (1973). Elektronenmikroscopische Untersuchung zur Feinstruktur de endotrophen Tabakmykorrhiza. *Arch. Microbiol.* 92, 201-207.

KASPARI, H. (1975).   This symposium, p, 325.

KECK, K. & STICH, H. (1957). The widespread occurrence of polyphosphate in lower plants.   *Ann. Bot.,*
21, 611-619.

MOSSE, B. (1959). The regular germination of resting
spores and some observations on the growth requirements of an *Endogone* species causing vesicular-
arbuscular mycorrhiza.   *Trans. Br. mycol. Soc.,*
42, 273-286.

MOSSE, B. (1973). Advances in the study of vesicular-
arbuscular mycorrhiza.   *A. Rev. Phytopath.,* 11,
171-196.

PEARSON, V. & TINKER, P. B. (1975). This
symposium, p. 277.

PROTSENKO, M. A., SHEMAKHANOVA, N. M. &
METLISKIY, L. V. (1971). Nekotorye dankye ob
ultrastruktorye mikorizy gorocha *Pisum sativum* L.
Dokl. Akad. Nauk. SSSR, 199, 722-724.

SANDERS, F. E. & TINKER, P. B. (1971). Mechanism of
absorption of phosphate from soil by *Endogone*
mycorrhizas.   *Nature, London,* 233, 278-279.

SCANNERINI, S. (1972). The ultrastructure of the
endomycorrhizae of *Ornithogalum umbellatum* L. at
the early vegetative stage.   *Allionia,* 18,
129-150.

SCANNERINI, S. & BELLANDO, M. (1967). Some ultrastructural features of endotrophic mycorrhiza in
Ornithogalum umbellatum G. bot. Ital., 101, 313.

SCANNERINI, S. & BELLANDO, M. (1968). Sull'ultrastruttura delle micorrize endotrofiche di *Orni-
thgalum umbellatum* L. in attivita vegetative.

Atti Accad. Sci., Torino., <u>102</u>, 795–809.

SMITH, D., MUSCATINE, L. & LEWIS, D. (1969).
Carbohydrate movement from autotrophs to hetero-
trophs in parasitic and mutualistic symbiosis.
Biol. Rev., <u>44</u>, 17–90.

AN ULTRASTRUCTURAL MODEL FOR THE HOST-SYMBIONT INTERACTION IN THE ENDOTROPHIC MYCORRHIZAE OF ORNITHOGALUM UMBELLATUM L.

S. SCANNERINI, PAOLA F. BONFANTE, ANNA FONTANA

Centro di Studio per la Micologia del Terreno, Istituto Botanico dell'Università, Torino, Italy.

INTRODUCTION

There has been a rapid development of the study of vesicular-arbuscular (VA) mycorrhizae in recent years, particularly with respect to their effect on the nutrition of the host plant. It has been shown that they increase overall uptake of phosphorus compounds from P-deficient soils (Mosse, 1973). Less, however, is known about their influence on endophyte metabolism, though it has been made clear that the fungus is an obligate symbiont (Mosse, 1973) and that its sugar metabolism is substantially different from that of the ectomycorrhizae, as Hepper and Mosse demonstrated (Mosse, 1973).

These physiological interactions presuppose a host-endophyte exchange of substances; this, of course, must take place at their interfaces, so that the cytological picture should lead to a better understanding of interaction modalities. Strangely enough, little attention has so far been given to

FIGURE NUMBERS IN THIS PAPER REFER TO PLATES VII-XI.

this side of the matter, though much work has been done on closely related questions, such as the cytology of the host-pathogen relationship (review of the literature in Bracker & Littlefield, (1973)). The few papers dealing with the fine structure of VA mycorrhiza do not offer a uniform account of the ultrastructural dynamics of the symbiont interaction. In field specimens of *Ornithogalum umbellatum*, intracellular infection structures were observed as entire (Scannerini, 1972) or virtually digested forms (Scannerini & Bellando, 1968). These two features can be observed together at different points in the host's vegetative cycle, even in the same individual or cell (Scannerini, 1972). They are noted in a variety of VA mycorrhizae, including those of clover (Mosse, pers. comm.) A similar picture was noted by Becking (1965) in *Podocarpus* root nodules and Protsenko *et al.*, (1971) in pea endomycorrhizae. In the latter instance, ultrastructural localisation of acid β-glycerophosphatase activity pointed to the phagocytic significance of the final interaction (Protsenko, 1973). The cytological aspects of the host-VA endophyte reaction can be directly compared with those observed in the orchid (Dörr & Kollmann, 1969; Hadley *et al.*, 1971; Nieuwdorp, 1972) and, in more general terms, with those of the classic interactions between the host and obligate fungi, whose interface types have been recently revised and classified by Bracker and Littlefield (1973).

The present paper deals with the fine structure of *O. umbellatum* VA mycorrhizae, with particular reference to the structure and dynamics of their interfaces in accordance with Bracker and Littlefield's classification. A model for host-endophyte interaction applicable to this species and comparable with those given in the literature for the cytological relations of various host-fungus complexes (susceptible hosts and pathogenic fungi, ectomycorrhizae, orchid mycorrhizae, lichens) will be proposed.

MATERIALS AND METHODS

VA mycorrhizae of *Endogone* sp. in *O. umbellatum* (Peyronel, 1924) grown at the Turin University Botanical Gardens in a 6-yr period (1967-73) were employed. Material was collected in all months, except May, August and October. Since the roots of field specimens sometimes show relations with other microorganisms (particularly *Rhizoctonia*) on histological examination, care was taken to select roots infected with VA endophytes only, or at any rate areas that were free of other microorganisms. The methods used enabled optical and electron microscope features to be linked in the same cell. Individual mycorrhizae were both fragmented directly into the fixing agent and hand-sectioned into approx. 50 μm thick slices after treatment with 6% glutaraldehyde (GA) in 0.1 M cacodylate buffer (pH 6.9) or with GA and lactic acid. The sections were mounted in the same fixative and examined with a phase microscope or stained with lactic blue for histological checking. Fragments and unstained thick sections were fixed for electron microscopy according to Grove and Bracker (1970), using GA OsO_4 and infiltration with uranyl acetate, or with cold GA OsO_4 as above. Fragments were also fixed with 1% OsO_4 in a veronal acetate buffer (pH 7), with the addition of $CaCl_2$, or directly with unbuffered 2% $KMnO_4$. All the material was then dehydrated and embedded in Durcupan ACM. A closer selection of areas for study was made by the analysis of semi-fine sections obtained with an LKB pyramitome and stained with toluidine blue according to Trump *et al.,* (1961) after treatment with NaOH dissolved in methanol. Thin sections for electron microscopy were then taken from these areas with a Sorvall MT2 ultramicrotome. After contrasting with uranyl acetate and lead citrate, or lead citrate alone, the sections were observed and photographed with a Philips 300 electron microscope.

Scannerini, S. *et al.*

RESULTS

Our results (Scannerini & Bellando 1967, 1968;
Scannerini, 1972; and unpublished data) show little
change in the host-endophyte relationship during
vegetative growth of the host, and our ultrastructural model can therefore be taken as applicable to
the entire vegetative cycle. The intercellular
endophyte hyphae (Figs. 1, 2A) occur in cords that
pass through the pectic mass of the middle lamella.
The interface structures (using Bracker and Littlefield's terminology) are IT 8 (cell walls of symbiont
and host cell in direct contact and intact proto-
plasms). The hyphae of the endophyte are surrounded
by a wall of average density that is finely layered
and contains a highly organised protoplasm with a
normal complement of euplasmic organelles (Fig. 2).
In other, even adjacent, areas, this picture may be
changed due to a gradual increase in wall density,
as well as extensive cytoplasmic vacuolation
(Fig. 1). This alteration corresponds to differences in mycelium age and function, such as those
observed in cultures of Phycomycetes by Grove *et al.*,
(1970). Highly vacuolated ramifications may extend
from the mycelia into the wall and push back the
remnants of degenerated material (Fig. 3).

Penetration of the cell wall cannot be followed
through all its stages. Protuberances with organised cytoplasm and very thin walls can be seen near
the main cords: these are known as angular protuberances (Scannerini, 1972). The young haustoria
near the neck in the host cell, however, display a
distinct wall much the same as that of the cords
(Fig. 4). The haustorium pushes back the host
cytoplasm without damage and thus forms IT 24 (live,
walled host and endophyte with intervening matrix).
A poorly vacuolated, nucleated hypha with a full
complement of organelles and an intact wall is separated from the host plasma membrane, which plays the
part of a sheath membrane, by a structureless matrix

of average electron density that is rather thin and lies immediately alongside the fungus wall.   It is also separated by a large, apparently empty, clear space that took up none of the stains employed (Fig. 5).   Following its penetration, the hypha grows to form the circumvolute assembly or coils that is a familiar picture under the light microscope (Fig.   6B) and corresponds to a complex system that primarily keeps IT 24.   As the coil continues to grow, it provides an extensive network of hyphae whose cytoplasm has a normal complement of organelles.   In the older areas, vacuolation is extensive (Fig. 6).

The vacuolated areas often contain dense inclusions and bacteria-like organisms (Fig. 7).   Microorganisms are separated from the host by a type 24 interface, consisting of a large transparent space and a fungus sheath membrane, similar to that noted in the bacterial endosymbionts of higher plants (Bracker and Littlefield, 1973).   Some of these bacteria-like organisms have incipient division figures, though there are no signs of division of the sheath membrane as described by Kidby and Goodchild, (1966) for *Rhizobium.*   Progressive degeneration suggesting digestion of these organisms by fungal vacuoles playing a heterophagic role is occasionally observed.

Circumvolutions may develop to the point where a single sheath membrane encloses the entire fungal mass without following the course of its gradual ramification.   The transparent part of the sheath grows to form a single sac (Fig. 8), which is reminiscent of the so-called "vacuole" of the orchid endomycorrhizae (Dörr & Kollman, 1969).   The hyphae are not always so far from the sheath membrane, however. The size of the coil may mean it becomes contiguous with various host organelles, including the nucleus. Certain segments of the hypha may lie tangentially to the membrane and the matrix may be extremely thin (Fig. 9).   Elsewhere, on the contrary, and especially

in old coils and widely fungus-infected cells, where
there is marked endophyte vacuolation, the part of
the matrix with average electron density may become
remarkably thick (Fig. 10) and take on the appearance of the layered zone described by Scannerini and
Bellando (1968).   The end result of coil ramification is an apical transition from IT 24 to IT 25
(similar to IT 24, but symbiont dead or moribund).
This transition is continuous: the apical areas
collapse and the walls tend to fall towards each
other (Fig. 11).   The collapsed areas are in all
cases connected or connectable with a hypha, though
separated from its cytoplasm by a septum that isolates the residual protoplasm, which is nucleated and
capable of continuing apical growth, from dissolving
material (Figs. 12, 13, 14, 15).

On some occasions, lamellar structures, possibly
related to the start of other septa, (Fig. 13) are
observed alongside the septum.   This suggests that
the formation of septa is progressive and takes place
in more rearward areas as the hypha is digested.
During its differentiation, the septum has a lamellar fine structure, like the endophyte wall, but is
thinner and has a cupola shape with a basal part
running parallel to the wall (Figs. 12, 13, 14).
When differentiation is complete, its structure and
size are those of the wall, so that it in fact forms
the apical wall of the hypha and isolates it from
the lysed residues (Fig. 15).

The mass of digested material immersed in the
sheath maintains a type 25 interface (Fig. 16).   The
arbuscules seen under the light microscope are thus
the result of a mixture of fungus material, even
though their mass is initially also derived from the
sheath matrix and, in the early digestion stages,
partially from structured hyphae.   In the final
stages of digestion, the matrix becomes less dense
and less thick.   The sheath membrane forms a closer
cover around the small masses of digested material.
In some cases, the transparent space between the host

and the sheath membrane becomes of negligible size as
the area of contact between the two organisms gradually increases (Fig. 17). The endophyte acquires
relations with the host organelles, particularly the
nucleus (Figs. 17, 18), but also with the proplastids
(Fig. 17), even in the arbuscular stage.

Serious cell lesions are not observed in the host
cell.   The nuclei are hypertrophied as described by
Protsenko & Shemakanova (1971) in pea endomycorrhizae.
The plastids appear either as conventional (Fig. 17)
or atypical proplastids containing myelin membranes
according to Scannerini (1972) or dense globules
(Fig. 19).   Lastly, the smooth-surfaced (Fig. 17)
and rough-surfaced (Fig. 18) endoplasmic reticulum is
much developed, with masses of rough reticulum near
fine fungal protuberances (Fig. 18).

DISCUSSION

The host-symbiont relation in *O. umbellatum* VA
mycorrhizae can be viewed as an unequivocal series of
interface reactions, in accordance with Bracker and
Littlefield's classification (1973).   Three interface types are observed: IT 8 (live and vital host
cell and symbiont in direct contact with the cell
walls and no intervening matrix) at the intercellular
cords; IT 24 (live and vital host and symbiont with
interface formed from the host plasmalemma, matrix of
uncertain homology, endophyte wall); IT 25 (as last,
but symbiont dead or moribund) at the intracellular
hyphae, coils and arbuscules.   IT 24 structures
correspond to those described by Scannerini (1972),
while type 25 has been observed by Scannerini and
Bellando (1968) in *O. umbellatum* and Protsensko *et
al.,* (1971) in the pea.   The last type of interface
may be indicative of phagocytosis: for instance in
pea endomycorrhizae Protsenko (1973) demonstrated
β-glycerophosphatase activity around material in the
process of digestion.   The focal point of the host-
endophyte reaction can thus be seen in the transition

from IT 24 (more common in haustoria and coils) to
IT 25 (predominant in the arbuscules).   Haustoria,
coils and arbuscules can thus be considered as nothing more than three different aspects of the same
phenomenon.   The resulting interaction is directly
referable to that of the orchid endotrophic mycorrhizae described by Dörr & Kollmann (1969) and
Hadley *et al.*, (1971).   In VA mycorrhizae, in fact,
endophyte hyphae are similar to the intracellular
haustoria of obligate parasites and are enclosed in
sacs by the host plasma membrane to form an extensive
IT 24 contact area with a two-layer matrix (one
transparent, one dense) between host and fungus prior
to their digestion by the host cell.   This implies a
rather rare form of interaction wherein the host
cells survive and those of the symbiont are gradually
destroyed.   As shown by Scannerini and Bellando
(1968) and Bracker and Littlefield (1973), it is
nearer a predator-prey relationship.

The similarity with orchid mycorrhizae is also
supported by the cytological changes noted in the
host cell, though their functional meaning is not
clear.   Severe alterations to the organelles are
not observed.   Hypertrophic nuclei, however, and
accumulation of ER described for orchid cells invaded
by *Rhizoctonia* (Dörr & Kollmann, 1969; Hadley *et
al.*, 1971) can be seen in VA mycorrhizae.   Prop-
lastid polymorphosis, on the other hand, is relatively frequent in VA mycorrhizae (Scannerini & Bellando,
1968; Scannerini, 1972), but has not been reported
for the orchid.

Quite apart from the obvious distinctions arising
from the symbiont involved (*Rhizoctonia* in the
orchid; *Endogone* sp. in the Star of Bethlehem),
other more notable morphological and functional
divergences exist between the two plants in addition
to the extensive IT 8 between VA mycorrhiza cells.
It is clear, in fact, that the root-phycomycetic endophyte complex in our VA mycorrhizae is really a root-
fungus-bacterium-like organism complex.   The exist-

ence of such organisms gives indirect support to the
observation of bacteria-like organisms by Mosse
(1970) in germinating *Endogone* spores. They can
readily be demonstrated in the coils and intracellular hyphae of VA mycorrhizae. The microorganism is
separated from the fungal cytoplasm by a type 24
interface (consisting of a sheath membrane encasing
a large transparent matrix), similar to that commonly
found in the bacterial endosymbionts of the higher
plants (Bracker & Littlefield, 1973). Some endophytic bacteria are capable of multiplication and
their division, as shown by Kidby and Goodchild (1966)
for *Rhizobium,* does not result in sheath membrane
division buds; this corroborates the diagnosis of a
type 24 interface. The microorganisms often suffer
degeneration and lysis, on the other hand, leading to
transition to a type 25 interface and showing lysis
on the part of *Endogone.*

Ornithogalum VA mycorrhizae also display cytological features not yet seen in the orchid. Transition from IT 24 to IT 25 is continuous in their
individual cells, which explains the simultaneous
observation of fine structures indicative of haustoria, coils and arbuscules in the same host cell
(Mosse, 1973). Continuity is guaranteed by the
production of septa that isolate dissolution areas
from those containing vital protoplasma able to maintain apical growth. In accordance with the well-
known functional characteristics of the Phycomycetes,
which display the same isolation of older areas
(Peyronel, 1923), this process results in a gradual
transition from coils to arbuscules at the histological level. Analysis of the digested areas shows
that the arbuscules are all derived from the fungus
with a quantitatively small contribution from the
sheath matrix. Following its increase in thickness
and electron density in the initial stages of the
IT 24 - IT 25 transformation, in fact, this matrix
becomes evanescent. When digestion is complete, the
arbuscule is formed solely of fungus residues encased

Scannerini, S. *et al.*

by the sheath membrane.

Summing up: the three different types of interface, the transition and the continuity between
haustoria, coils and arbuscules, the existence of a
bacteria-like organism encased by the fungus suggest
a gradual digestion of the microorganisms (fungus and
bacterium): this gives a plausible explanation for
the continuous flow of material to the host cell, and
its replenishment with bacterial and fungal substances. This cytological relationship offers a
morphological explanation for the increased metabolite consumption displayed by the host cell and its
behaviour as a predator towards continuously renewed
endophytes following the mycorrhizal relationship.

ACKNOWLEDGMENT

This work (n. 176) is supported by a Grant of
C.N.R. (National Council of Researches).

REFERENCES

BECKING, J. H. (1965). Nitrogen fixation and mycorrhiza in *Podocarpus* root nodules. *Pl. Soil.*
23, 213-226.

BRACKER, C. E. & LITTLEFIELD, L. J. (1973). Struct-
ural concepts of host-pathogen interfaces. In:
Fungal pathogenicity and the plant's response.
R. J. W. Byde and C. V. Cutting Ed. 159-318.
Academic Press, London, N. York.

DORR, I. & KOLLMANN, R. (1969). Fine structure of
mycorrhiza in *Neottia nidus-avis* L. C. Rich.
(Orchidaceae). *Planta,* 89, 372-375.

GROVE, S. N. & BRACKER, C. E. (1970). Protoplasmic
organization of hyphal tips among fungi: Vesicles
and spitzenkörper. *J. Bact.,* 104, 989-1009.

GROVE, S. N., BRACKER, C. E. & MORRE, D. J. (1970). An ultrastructural basis for hyphal tip growth in *Pythium ultimum*. *Am. J. Bot.*, 57, 245-266.

HADLEY, G., JOHNSON, R. P. C. & JOHN, P. A. (1971). Fine structure of the host-fungus interface in orchid mycorrhiza. *Planta*, 100, 191-199.

KIDBY, D. K. & GOODCHILD, D. J. (1966). Host influence on the ultrastructure of root nodules of *Lupinus luteus* and *Ornithopus sativus*. *J. gen. Microbiol.*, 45, 147-152.

MOSSE, B. (1970). Honey-coloured, sessile *Endogone* spores: II Changes in fine structure during spore development. *Arch. Mikrobiol.*, 74, 129-145.

MOSSE, B. (1973). Advances in the study of vesicular-arbuscular mycorrhiza. *A. Rev. Phytopath.*, 11, 171-196.

NIEUWDORP, P. J. (1972). Some observations with light and electron microscope on the endotrophic mycorrhiza of Orchids. *Acta bot. neerl.*, 21, 128-144.

PEYRONEL, B. (1923). Prime ricerche sulle micorrize endotrofiche e sulla micoflora radicicola normale delle fanerogame. *Riv. Biol.*, 4, 463-485.

PEYRONEL, B. (1924). Prime ricerche sulle micorrize endotrofiche e sulla micoflora radicicola normale delle fanerogame. *Riv. Biol.*, 6, 17-53.

PROTSENKO, M. A. (1973). Elektronno-microskopisheskoe izushenie lokalizatsii kisloi fosfatazi v perevar-ivaiushih griv kletkah mikorizii goroha. *Dokl. Akad. Nauk SSSR*, 211, 213-215.

PROTSENKO, M. A. & SHEMAKANOVA, N. M. (1971).
Vliianie mikorizoovrazuiushevo griva no razteri
iadra i iadrishka v kletkah kornia *Pisum sativum*.
Mikologiya i Fitopatologiya, 5, 335-338.

PROTSENKO, M. A., SHEMAKANOVA, N. M. & METLINTSKII,
L. V. (1971). Ultrastructure of mycorrhiza of the
garden pea (*Pisum sativum* L). *Dokl. Akad. Nauk
SSSR,* 199, 120-122.

SCANNERINI, S. (1972). Ultrastruttura delle Endomicor-
rize di *Ornithogalum umbellatum* L. all'inizio
dell'attività vegetativa. *Allionia,* 18, 129-150.

SCANNERINI, S. & BELLANDO, M. (1967). Some ultrastructural features of endotrophic mycorrhiza in
Ornithogalum umbellatum. *G. Bot. ital.,* 101, 313.

SCANNERINI, S. & BELLANDO, M. (1968). Sull'ultrastruttura delle micorrize endotrofiche di *Ornithogalum umbellatum* L. in attività vegetativa.
Atti Accad. Sci., Torino, 102, 795-809.

TRUMP, B. F., SMUCKLER, E. A. & BENDIT, E. P. (1961).
A method for staining epoxy sections for light
microscope. *J. Ultrastruct. Res.,* 5, 343-348.

FINE STRUCTURE OF THE HOST-PARASITE INTERFACE IN ENDOTROPHIC MYCORRHIZA OF TOBACCO

H. KASPARI

Institut für Pflanzenkrankheiten
Bonn, W. Germany*

INTRODUCTION

Endotrophic mycorrhizas are generally considered a form of eusymbiosis (Schaede, 1962). After penetration, the fungus behaves as a parasite in a compatible host-fungus combination in which neither of the partners sustains damage. The changes in host cell metabolism induced by the fungus can be regarded as a defence reaction. They may manifest themselves in structural modifications, for example hypertrophy of host nucleus and increases in the host cytoplasm and organelles such as ER cisternae and ribosomes (Dörr & Kollman, 1969). Such host reactions are generally regarded as signs of increased metabolic activity. These visible changes probably reflect physiological changes – DNA synthesis increases in orchid mycorrhiza together with protein synthesis (Williamson, 1970; Baltruschat 1975). Changes also occur in enzyme metabolism. For instance Dehne

---

FIGURE NUMBERS IN THIS PAPER REFER TO PLATES XII–XIV.

---

* Present address: Institut für Mikrobiologie, 53 Bonn, Meckenheimer Allee 168.

Kaspari, H.

(1973) reported in tobacco mycorrhiza an increased activity in peroxidase important, among other things, in infection processes. This increase was however relatively small compared to that occurring in host-pathogen associations. The mycorrhizal endophyte thus had a smaller effect than the pathogenic parasite and induced a weaker defence reaction in the host, apparently within the limits of toleration of the endophyte. This result agrees well with the concept of a mutual toleration of both partners.

The physiological balance between attack by the parasite and defence by the host is however eventually disturbed in favour of the host. The parasite is at least partially digested during the arbuscular stage, as shown by light (review by Gerdemann, 1968) and electron microscope observations (Kaspari, 1973; Mosse, 1973). Orchid mycorrhizas show similar features (Dörr & Kollmann, 1969; Hadley et al., 1971; Nieuwdorp, 1972).

Not only biochemical but also morphological studies (especially in the field of ultrastructure) help in the understanding of interactions between two closely associated organisms. Of particular interest is the host-parasite interface, because metabolic exchanges occurring here eventually determine the character of the interaction between the partners. Therefore the individual components of this zone have, at least potentially, a function in the development of the association (Bracker & Littlefield, 1973). Numerous investigations of the ultrastructure of essentially different host-parasite associations (review by Bracker & Littlefield, 1973) have shown a remarkable similarity in the structure of the host-parasite interface. It is therefore difficult and only very partially possible to deduce a particular functional relationship between two organisms from a particular type of interface.

On the other hand the structure of the interface may change during the course of the association and this may give an indication of a changing relation-

ship during development.   This will be demonstrated
in the example of the intracellular development of
phycomycetous mycorrhiza in tobacco (var. Havanna).

OBSERVATIONS

*Stages in the intracellular development of tobacco
mycorrhiza*

After invasion of the host cell by *Endogone
mosseae* a host-parasite interface develops resembling
that in many pathogenic and non-pathogenic host-
parasite associations during their period of compat-
ibility.

Enzymes are presumably involved in the penetration process; this is suggested by the granulation
of the cell wall at the penetration point (Fig. 1)
(Akai *et al.*, 1971).   Associated with this is the
formation of an electron dense thickening (a papilla)
on the inner cell wall (Fig. 1).   Such a structure
is an additional barrier to penetration and also
occurs in many other host-parasite associations.
The papilla appears to be a swelling of the cell wall
(McKeen *et al.*, 1969) rather than additional wall
thickening by apposition (Bracker, 1968; Edwards &
Allen, 1970), because the dense zone extends into the
cell wall.

After penetration the fungal hypha (haustorium)
invaginates the plasmalemma, the cytoplasm and the
tonoplast of the host cell, thus remaining exterior
to the protoplast but within the cell (Fig. 2).   The
invaginated plasmalemma (the extra-haustorial membrane according to the terminology of Bushnell
(1972)), and the three-layered fungal wall (Fig. 3)
are separated by a zone, termed the extra-haustorial
matrix (Bushnell, 1972).   This zone, starting from
the penetrated cell wall, encloses the fungus during
the whole of its intracellular existence, similar to
the "wall-like apposition" of parasitic Oomycetes

(Hanchey & Wheeler, 1971).   It presumably contains
the remnants of the penetrated papilla, which thus
do not develop into a "collar" (Bracker, 1968) around
the penetrating hypha.

If the intracellular hypha grows through the
host cell the extra-haustorial matrix fuses with the
opposite cell wall, like the infection thread of
endo-symbiotic bacteria (Jordan et al., 1963).

The fungal cytoplasm contains normal cell components such as mitochondria, ER cisternae, nuclei
and oil bodies and is not vacuolated (Fig. 2).

The host cell reacts to invasion with an
increase in cytoplasm and an accumulation of cell
organelles such as mitochondria, dictyosomes and ER
cisternae in the vicinity of the parasite (Fig. 4).
The nucleus also is often close to the invading
hypha.   This proximity could indicate an interaction between these organelles and the hypha, in
which case they would form part of the host-parasite
interface.   The strongly developed ER system of
orchid mycorrhizas (Dörr & Kollmann, 1969) should be
remembered in this connection.

The interactions in this phase of co-existence
thus occur between an actively growing parasite and
a metabolically stimulated host cell (Fig. A).

The end of mutual parasitism is marked by the
beginning of arbuscule formation (Fig. 5), which is
however preceded by the formation of septa whose
centripetal development can be followed in the organised hyphae (Figs. 6, A).   The septa separate
parts of the arbuscule in different stages of development (Figs. 6, 8), thereby preventing a sudden
degeneration of the entire arbuscule.   Within the
same host cell seemingly healthy parts of the arbuscule co-exist with moribund and dead sections
(Figs. 5, 6, 8).

The weakening and incipient decline of the
fungus is shown by the decreasing organisation of its
protoplast.   Oil bodies and mitochondria disappear,
and the ground plasma becomes increasingly vacuolated

(Fig. 7).   However, little change occurs in the morphology of the host-parasite interface (Fig. A). Of course interactions between host and parasite have basically changed and metabolic transfer is now likely to be mainly into the host.

During the final stages of the arbuscule marked changes occur in the structure of the host-parasite interface (Fig. A).   The fungal plasmalemma disintegrates, whereupon the protoplast dies and the hyphal contents can be assimilated by the host cell (Figs. 8, 9).   The dead hyphae eventually collapse leaving behind compact masses of fungal wall (Fig. 10).   These evidently also disintegrate and no longer show a three-layered structure (Fig. 11).

The changes in the host-parasite interface (Fig. A). thus reflect the changing relationship between host and parasite although no indications are given why mycorrhizal host cells can partially digest the fungus whereas this is not possible in other host-fungus associations.

Observations on membranes

Particular attention should be given in such a closely integrated association to the membranes which, as biologically active structures, must fulfil important functions in the interactions between two cells.   Thus the extra-haustorial membrane is frequently torn or vesicular (Figs. 9, 10, A). This is certainly an artefact because the functioning of the host cell depends on the integrity of this membrane (Hirata & Kojima, 1962;   Schnepf et al., 1971).   Nevertheless this lability of the membrane could indicate changes in its nature during arbuscule development.   Changes in plasmalemma structure, presumably influencing exchange processes between parasite and host, were for instance observed by Schnepf (1972) during interactions in fungus-algae associations.   Indeed the ability of membranes to change and differentiate is considered an essential

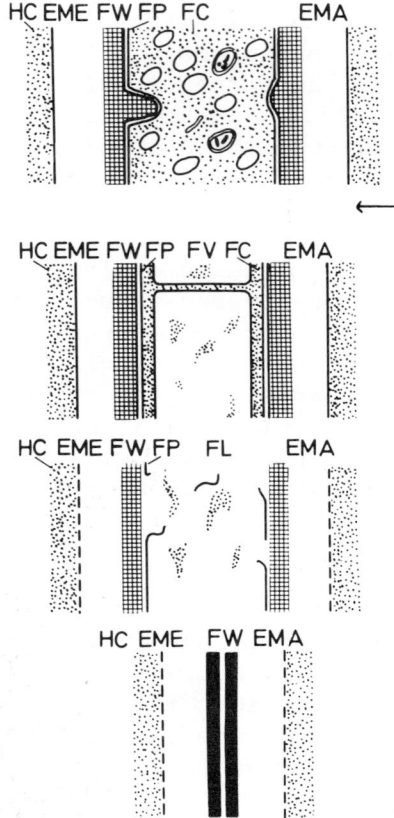

Fig. A. Diagrammatic representation of the host-fungus interface. HC, host cytoplasm; EME, extra-haustorial membrane; FW, fungal wall; FP, fungal plasmalemma; FC, fungal cytoplasm; FV, fungal vacuole; FL, lumen of dead hypha.

factor in host-parasite interactions.

The plasmalemma of organised hyphae often has invaginations pointing towards oil bodies; sometimes the two appeared to be closely associated (Fig. 3).   Hadley *et al.*, (1971) also found invaginations of the fungus plasmalemma in orchid mycorrhiza.   Attempts to interpret such structures in terms of transport mechanisms must however be treated with caution because they may, as Hadley *et al.*, (1971) also pointed out, be artefacts of preparation.

A further feature that might be connected with exchanges between host and fungus are invaginations of the host tonoplast into the vacuole which occur near the fungus during arbuscule development (Fig. 11).   These invaginations are associated with vacuole based vesicle-like structures with cytoplasm-like contents.   It is at least possible that these are evidence of a transport mechanism, similar to that described by Matile & Moor (1968), whereby cytoplasmic material might move into the vacuole in budded off vesicles, to be liberated after lysis of the vesicle membranes.

Further ultrastructural indications of possible interactions between host and parasite in the form of distinctive structures, such as protruberances of the hyphal wall as in orchids (Hadley *et al.*, 1971), tentacle-like projections of the extra-haustorial membrane (Ehrlich & Ehrlich, 1971), and connections between it and ER cisternae of the host, or even continuity between host and fungus protoplasts as in rusts (Calonge, 1969), were not found in endotrophic tobacco mycorrhiza.

ACKNOWLEDGMENTS

I am grateful to Dr. B. Mosse for her help in translating the manuscript.   This research was supported by the Deutsche Forschungsgemeinschaft.

REFERENCES

AKAI, S., HORINO, O., FUKUTOMI, M., NAKATA, A.,
KUNOH, H., & SHIRAISHI, M. (1971). Cell wall
reaction to infection and resulting change in cell
organelles. In *Morphological and biochemical
events in plant - parasite interaction* (Eds. Akai,
S., & Ouchi, S.). p329-344. The Phytopathological
Society of Japan, Tokyo.

BALTRUSCHAT, H. (1975). Untersuchungen über den Ein-
fluß der endotrophen Mycorrhiza auf das Resistenz-
verhalten von Pflanzen gegenüber pathogenen Pilzen,
insbesondere auf den Befall von *Nicotiana tabacum*
var. "Havanna" durch *Thielaviopsis basicola*. Diss.
Bonn.

BRACKER, C. E. (1968). Ultrastructure of the haustorial apparatus of *Erysiphe graminis* and its relationship to the epidermal cell of barley.
Phytopathology, 58, 12-30.

BRACKER, C. E. & LITTLEFIELD, L. J. (1973). Struct-
ural concepts of host-pathogen interfaces. In
Fungal pathogenicity and the plant's response
(Eds. Byrde, R. J. W. & Cutting, C. V.), 159-318.
Academic Press, London, New York.

BUSHNELL, W. R. (1972). Physiology of fungal haustoria. *A. Rev. Phytopath.*, 10, 151-176.

CALONGE, F. D. (1969). Ultrastructure of the haustoria or intracellular hyphae in four different
fungi. *Arch. Mikrobiol.*, 67, 209-225.

DEHNE, H. W. (1973). Die Rolle von Oxidasen in der
Pathogenese und Methoden zur Bestimmung ihrer
Aktivität. Diplomarbeit Bonn.

DÖRR, I. & KOLLMANN, R. (1969). Fine structure of

mycorrhiza in *Neottia nidus-avis* (L.) L. C. Rich. (Orchidaceae). *Planta,* <u>89</u>, 372-375.

EDWARDS, H. H. & ALLEN, P. J. (1970). A fine-structure study of the primary infection process during infection of barley by *Erysiphe graminis* f. sp. *hordei. Phytopathology,* <u>60</u>, 1504-1509.

EHRLICH, M. A. & EHRLICH, H. G. (1971). Fine structure of *Puccinia graminia* and the transfer of C[14] from uredospores to *Triticum vulgare.* In *Morphological and biochemical events in plant-parasite interaction* (Eds. Akai, S. & Ouchi, S.), 279-307. The Phytopathological Society of Japan, Tokyo.

GERDEMANN, J. W. (1968). Vesicular-arbuscular mycorrhiza and plant growth. *A. Rev. Phytopath.,* <u>6</u>, 397-418.

HADLEY, G., JOHNSON, R. P. C. & John, D. A. (1971). Fine structure of the host-fungus interface in orchid mycorrhiza. *Planta,* <u>100</u>, 191-199.

HANCHEY, P. & WHEELER, H. (1971). Pathological changes in ultrastructure: Tobacco roots infected with *Phytophthora parasitica* var. *nicotianae. Phytopathology,* <u>61</u>, 33-39.

HIRATA, K. & KOJIMA, M. (1962). On the structure and the sack of the haustorium of some powdery mildews with some considerations of the significance of the sack. *Trans. mycol. Soc. (Japan),* <u>3</u>, 43-46.

JORDAN, D. C., GRINYER, I. & COULTER, W. H. (1963). Electron microscopy of infection threads and bacteria in young root nodules of *Medicago sativa. J. Bact.,* <u>86</u>, 125-137.

KASPARI, H. (1973). Elektronenmikroskopische Untersuchung zur Feinstruktur der endotrophen Tabakmykorrhiza.   *Arch. Mikrobiol.,* <u>92</u>, 201-207.

MATILE, P. H. & MOOR, H. (1968). Vacuolation: Origin
and development of the lyosomal apparatus in root-
tip cells.   *Planta,* <u>80</u>, 159-175.

McKEEN, W. E., SMITH, R. & BHATTACHARYA, P. K. (1969)
Alterations of the host wall surrounding the in-
fection peg of powdery mildew fungi.   *Can. J.
Bot.,* <u>47</u>, 701-706.

MOSSE, B. (1973). Advances in the study of vesicular-
arbuscular mycorrhiza.   *A. Rev. Phytopath.,* <u>11</u>,
171-196.

NIEUWDORP, P. J. (1972). Some observations with light
and electron microscope on the endotrophic mycorrhiza of orchids.   *Acta bot. neerl.,* <u>21</u>, 128-144.

SCHAEDE, R. (1962). *Die pflanzlichen Symbiosen.*
G. Fischer, Stuttgart.

SCHNEPF, E., HEGEWALD, E. & SOEDER, C. J. (1971).
Elektronenmikroskopische Beobachtungen an Para-
siten aus *Scenedesmus* - Massenkulturen. 2. Über
Entwicklung und Parasit - Wirt - Kontakte von
Aphelidium und virusartige Partikel im Cytoplasma
infizierter *Scenedesmus*-Zellen.   *Arch. Mikrobiol.,*
<u>75</u>, 209-229.

SCHNEPF, E. (1972). Strukturveränderungen am Plasmalemma *Aphelidium* - infizierter *Scenedesmus*-Zellen.
Protoplasma, <u>75</u>, 155-165.

WILLIAMSON, B. (1970). Induced DNA synthesis in
orchid mycorrhiza.   *Planta,* <u>92</u>, 347-354.

ORGANIZATION AND FINE STRUCTURE OF ORCHID MYCORRHIZA

G. HADLEY

Botany Department, University of Aberdeen, Aberdeen, U. K.

INTRODUCTION

The interrelationship between orchid endosymbiotic fungi and their hosts is probably one of the most complex situations known to biologists. There is a state of balance, allowing progressive but controlled invasion by the fungus while the host achieves a partial or total appropriation of fungal metabolites. Disturbance of the balance may lead either to the elimination of the fungus from the host tissue, or to parasitism and perhaps death of the host (Hadley, 1970).

GENERAL CHARACTERISTICS OF THE SYSTEM

Germinating seeds (protocorms) of the host, lacking in food reserves, are supplied through the fungus with carbohydrate and possibly other materials. Transfer of carbohydrate from a heterotroph to an autotroph is unusual since in most symbiotic systems the reverse occurs. The fungus also infects the roots of its host, but there is at present no

---

FIGURE NOS. IN THIS PAPER REFER TO PLATES XV-XXII

direct evidence of transfer of materials from fungus
to host in this phase.   However the characteristic
features of infection, and subsequent digestion of
intracellular hyphae, are similar in roots and protocorms, whatever the strain of fungus involved.

The fungi of orchid mycorrhizas are rhizoctonias.   The genus *Rhizoctonia* is complex and includes
pathogenic species but rhizoctonias isolated from
orchids are usually benign.   They can be placed in
genera such as *Tulasnella, Sebacina* and *Ceratobasidium* (Warcup & Talbot, 1967).

Williamson and Hadley (1970), using a selection
of eight rhizoctonias including orchid symbionts and
also *Thanatephorus cucumeris (Rhizoctonia solani)*
strains known to be pathogenic to crop hosts, observed that, regardless of source, all the fungi infected
protocorms of *Dactylorhiza purpurella* by single
hyphae entering epidermal hairs, leading to a compatible intracellular infection.   Compatible infection normally results in symbiosis, evident as a stimulus to growth of the protocorm (Hadley & Williamson,
1971).   Deviations from this pattern are of two
basic sorts.   Occasionally the fungus is contained
and eliminated, or, with vigorous and aggressive
strains, the compatible phase is succeeded by an
incompatible (pathogenic) one in which the fungus
spreads into all the host cells, including the meris-
tem, and eventually kills the protocorm.   The balance between symbiosis and parasitism in culture is
related to nutrition because cellulose (Hadley, 1969)
or a controlled low concentration of glucose (Harvais
and Hadley, 1967) are conducive to symbiosis whereas
high sugar levels lead to parasitism by the fungus.

Light microscopy

Magnus (1900; see Burgeff, 1959, for descrip-
tive details) was the first worker to recognise the
presence of two types of infected cells, the 'host
cells' in which the fungus remains apparently healthy

and unaffected, and the 'digestion cells' in the
inner cortex, in which the intracellular fungus is
always transformed into a dead amorphous mass of
material.   Burgeff (1959) emphasized also that the
innermost cortical cells often remain as uninfected
'storage cells'.   But in many orchids such a clear
spatial separation has not been recorded; for instance in many tropical and some north temperate
species all the hyphae eventually become digested so
that large areas of the root, as it ages, are occupied by digested clumps of fungal material.

    In *Dactylorhiza purpurella* (T. & T. A. Steph.)
Soó the pattern of infection is similar in both germinating seeds and roots.   Intracellular hyphae in
cortical cells form coils (Fig. 1) which extend
loosely through the whole cell and then form numerous
lateral branches which anastomose into a complex
three-dimensional structure, the 'peloton' of earlier
workers (Fig. 2).   At this stage the fungus has
usually occupied most of the cell but, as will be
discussed later, there may be no actual penetration
of the plasmalemma.   When fully developed, pelotons
appear to be fairly densely packed (Fig. 3) although
thin sections usually show a space between adjacent
hyphae (Figs. 4, 5, 6).   Later the pelotons become
disorganized and collapse into a clump of material
which often persists in the cell and readily takes up
stains (Fig. 4).   Secondary infection of cells
containing such clumps is quite common in both roots
and protocorms.   Secondary infection was not frequent in most of the material examined in this work
because it was not old enough, but Fig. 5 shows part
of a cell containing both digested and healthy
hyphae.

Methods

    All the work with *Dactylorhiza purpurella* used
young protocorms infected by known fungi.   Except
where stated the fungus was a rhizoctonia, isolate T2,

obtained from roots of the same orchid and tentatively identified as a species of *Ceratobasidium*.   Sym-
biotically infected protocorms were selected using
low power magnification a few days after infection,
fixed, embedded and sectioned (Hadley *et al.*,   1971),
and stained with uranyl acetate/lead citrate.

     Where other authors' work is referred to the
names of the organisms are stated.   Throughout this
paper the orchid is referred to as the 'host' and the
term 'endophyte' refers to the fungus present in the
tissue, whether it is symbiotic or otherwise.

STRUCTURAL FEATURES OF THE SYMBIOTIC PHASE

Hyphae

     Light micrographs of thin sections of protocorms
in the early symbiotic phase (Figs. 4 & 6) showed
stages of hyphal development correlated with structural features described later.

     Intracellular hyphae at an early stage of devel-
opment were usually filled with cytoplasm and stained
heavily.   Hyphae in sub-epidermal parenchyma cells,
corresponding to the 'host-cell layer' described by
Burgeff (1959) were of this nature (Fig. 6, up).
Other hyphae, typically those in the inner cortical
parenchyma cells, showed the first stages of digestion and were angular or collapsed and flattened, and
contained less cytoplasm (Fig. 6,c).

     Under the electron microscope hyphae were seen
to be thinly enveloped by host cytoplasm (Fig. 8).
The cytoplasm of hyphae was often densely packed with
ribosomes and mitochondria, and nuclei were sometimes
seen (Fig. 8).   Vacuoles were fewer in young than in
older hyphae and the tonoplast was sometimes not very
clearly defined.   The plasmalemma sometimes invaginated to form tubes into the cytoplasm and in one
section it appeared continuous with membranes forming
a system of tubules or vesicles lying in the central

vacuole (Fig. 9).   Inclusions presumed to be reserve material were sometimes present in hyphae (Fig. 22).

Hyphal walls

The walls of intracellular hyphae usually consisted of two layers (Figs. 8, 9, 14 etc.), the inner electron dense and normally about 60 nm thick, the outer rather granular or flocculose and from 100 to 200 nm thick.   Hadley *et al.*, (1971) interpreted this outer layer as originating from the fungus alone but further examination of many samples of material suggests that it may result from the interaction between fungus and host.   Hyphae grown in culture on agar media do not possess a granular outer wall layer (Fig. 10); their walls are similar to the inner wall layer of intracellular hyphae, being 60 to 100 nm thick.   They may have an outer electron-transparent layer, not readily stainable and visible only as a gap where adjacent hyphae are in contact (Figs. 10, 11).

The variations in thickness of the granular outer wall layer (Fig. 14, and see later) and the lack of such a layer in hyphae which are parasitic (Fig. 13) also suggest that it is part of the host-fungus interface, i.e. an encasement material laid down by the host.   This is discussed later.

Intracellular features of the host

Intracellular hyphae progressively permeate the host cell, so that the host plasmalemma becomes extensively invaginated.   As the hyphae anastomose the plasmalemma must be continually reorganizing and increasing in amount due to the large area of host-fungus interface.   Much of the host cell cytoplasm is thus dispersed into a thin film around the invading hyphae (Figs. 7, 8, 9).   In some instances the host plasmalemma and tonoplast appeared to have no cytoplasm between them (Fig. 8).   Nuclei (Fig. 12)

were usually associated with larger amounts of cytoplasm containing organelles and hyphae.   Crystals
also occurred (Fig. 14).   Membrane configurations
including endoplasmic reticulum were sometimes seen
(Figs. 14, 15).

The host-fungus interface

It is important to establish what structural
features are peculiar to the interface and whether
the situation is comparable to that of any other
host-fungus interaction.   The host plasmalemma, the
encasement material around the fungal hypha and the
wall of the hypha itself are always present.   The
encasement may be of uniform thickness (Figs. 8, 14)
but may appear to be extremely thick where hyphae are
cut obliquely (Fig. 15).   If this layer is a true
encasement (see Bracker & Littlefield, 1973), it
would be most clearly seen, in median longitudinal
sections through hyphae penetrating host cells, as a
layer continuous with the host cell wall.   Such
sections were not seen in the present work but in
some instances where hyphae lay adjacent to the host
cell wall (Fig. 16) the encasement layer was continuous with the latter.

Supporting evidence that the outer wall layer is
an encasement comes from the fact that it is often
continuous between adjacent intracellular hyphae
(Figs. 17, 18).   In one instance (Fig. 18) a space
of about 1.2 μm between two hyphae was occupied by
material similar to, and continuous with, the encasement layer around the individual hyphae.   The plasmalemma of the host was usually in contact with the
encasement layer (Fig. 14, insert) and there appeared
to be nothing between them except, for example, where
small vesicles (Fig. 17) were seen, or where the
plasmalemma was separated from the encasement (Figs.
8, 15).   Whether the space between plasmalemma and
wall is occupied by electron transparent material or
whether it is an artefact is not known.   An elect-

ron-transparent zone between plasmalemma and wall in many higher plants has been termed the "espace membranaire" by Roland (1968).

Some protruberances from the encasement layer were seen, around which the plasmalemma was closely pressed (pb, Figs. 8, 14).   The proturberances may be of different composition from that of the encasement and their appearance is similar to that of the lomasomes described for many fungi and plants.   Cox and Sanders (1974) described similar structures in vesicular-arbuscular mycorrhiza as multivesicular paramural bodies.   Hadley *et al.*, (1971) showed by freeze-etching the presence of many bulges or protuberances which increased the surface area of the interface on the outside of hyphal walls.   They were assumed to be associated with, or part of, the 'outer wall layer of the fungus'.   On the evidence now available this layer and the protuberances from it probably originate from the host.

Folds or loops in the host plasmalemma occur where it separates from the encasement layer to form vesicle-like bodies extending into the cytoplasm (Fig. 19).   In some sections there appeared to be a series of membranes, connected to the plasmalemma, adjacent to healthy hyphae (Figs. 19, 20).

Where the layer of cytoplasm around a hypha is thin, the loops of plasmalemma may protrude into the host cell vacuole, carrying the tonoplast with them (Figs. 20, 21, 22).   In Fig. 20, for example, it is possible that there are cytoplasmic (ER) membranes pressed between the tonoplast and plasmalemma.   In other examples (Figs. 23, 24) only the boundary membranes appear to be present.

Nieuwdorp (1972) shows similar arrangements of membranes.   He does not show the early stages of infection but in later stages in *Corallorhiza trifida* roots, he describes an 'undulating cytoplasmic membrane' (the plasmalemma?) which eventually forms groups of pinocytic vesicles around hyphae which are in an advanced stage of disorganization and collapse.

Similar undulations of the plasmalemma around hyphae are evident in micrographs by Dörr and Kollman (1969) although not in detail.

STRUCTURAL FEATURES ASSOCIATED WITH DEGENERATION OF HYPHAE

Protocorms sectioned at a stage when digestion was commencing showed disorganized fungal material in some of their central cells and collapsed, flattened hyphae in others (Figs. 4, 6).

Hyphae in the first stages of digestion appear in light microscope sections as angular shapes with fewer contents, and are less densely stained, than healthy hyphae in outer cortical cells (Fig. 6). Under the electron microscope they are seen as bizarre shapes, with vacuoles in the cytoplasm (Fig. 25). Their walls may become very electron-dense and thicken, while their contents may start disintegrating (Fig. 26).

As the hyphae lose turgor they presumably collapse, lose their soluble cell contents and become completely flattened (Fig. 27). At this stage the encasement layer often appears to thicken and become more granular or flocculose (Figs. 14, 27). The host plasmalemma may draw away leaving a space between it and the encasement (Fig. 27). However, this may be of no significance since in other instances the plasmalemma remains in close contact with the encasement layer (Fig. 28).

Collapsed hyphae are often surrounded by portions of host cytoplasm containing mitochondria and complex ER (Fig. 28). The hyphal wall may become irregularly folded (Fig. 29), sometimes enclosing a matrix of material the origin of which is unclear.

The encasement layer around degenerating hyphae may take on a reticulate appearance and appears to thicken, especially at the margin of the clump

(Fig. 28 and inset).   Whether the encasement material helps to coagulate hyphal remains together is
difficult to determine, but sections taken near the
edge of a clump show the continuity of the encasement (Fig. 30).   What happens to the host plasmalemma is equally difficult to detect.   It appears to
become re-organized around degenerating hyphae as
they condense into larger masses which form the
'digested peloton' seen under the light microscope
(Fig. 4).

Elaborate configurations of membranes occurred
in the digestion phase, often adjacent to the remains of hyphae (Fig. 31), with very little cytoplasm
evident between adjacent membrane layers.   In other
areas the space between adjacent membranes was occupied by cytoplasmic material (Fig. 32), including
ribosomes and mitochondria.   Associated with the
tonoplast were series of electron dense hemi-ellip-
soidal lipid bodies (Figs. 31, 32).

Dörr and Kollman (1969) showed very complex
arrangements of layers of membranes running through
the host cell cytoplasm in *Neottia*.   They regarded
these as a well developed ER system although it was
not certain that the membranes were associated with
endophytic hyphae because they occurred throughout
most of the cytoplasm.

Non-infected protocorms of *D. purpurella* were
not examined in this study and it is not possible to
say whether complex ER is a normal feature of host
cell cytoplasm.   It was not usually found around
healthy intracellular hyphae, but occurred near
hyphae which were collapsing (Fig. 14) as well as
around disorganized hyphal remains.

In the ultimate phase of degeneration individual
hyphae become indistinguishable as they congeal into
a mass (Fig. 34).   These masses are variously referred to as clots (Nieuwdorp, 1972), clusters (Dörr &
Kollmann, 1969), 'exkretklumpen' (Borriss, et al.,
1971) or digested pelotons.   The electron dense
strands (Fig. 34) are the remains of walls of hyphae

and they are interspersed with less dense zones of
material originating, presumably, from the encasement
layers.   Portions of double membrane, the origin and
function of which is unclear although they are possibly the remains of ER described above, occur in
fragments among the hyphal remains (Fig. 33).   In
infections by *Rhizoctonia solani* isolate RS 16 (but
not with isolate T) digested pelotons often enclosed
some pockets of material apparently containing membranes, vesicles and perhaps other organelles of
unknown origin (Fig. 35).

Digested pelotons may be surrounded by the complex membranes already described (Fig. 31).   A layer
of material referred to by Nieuwdorp (1972) as a
cellulose slime layer which is 'continuous with the
original plant cell wall', analogous to the reticulate encasement material seen in the present work, may
form the boundary of the mass.   But in the present
samples of *D. purpurella* the process of digestion has
probably been incomplete and the peripheral layer of
the peloton has been in a state of flux (Fig. 30).
The material described by Nieuwdorp may have been
completely digested.

STRUCTURAL FEATURES ASSOCIATED WITH
PARASITISM OF THE HOST

Protocorms of *D. purpurella* were infected with a
Rhizoctonia solani isolate (Rs 16) which was originally obtained from garden soil.   In symbiosis tests
(Hadley, 1970) this isolate was found to establish a
compatible infection but it subsequently became parasitic.   Material selected about three weeks after
infection showed intracellular hyphae contained in
host cells lacking an orderly structure (Fig. 36).
Some of the hyphae were surrounded by a thin encasement layer characteristic of the compatible system
but others were not encased and were robust, fully
turgid and full of cytoplasm.   Detail of an intra-

cellular hypha alongside the wall of a disorganized
cell, with the host plasmalemma in an adjacent
healthy cell, is shown in Fig. 37.   In one instance
where a hypha was growing in an intercellular space
(Fig. 38) no encasement layer was present.

DISCUSSION

In their exhaustive review Bracker and Littlefield (1973) emphasize that intracellular hyphae are
often surrounded by a layer of mucilaginous or other
material.   Where host cells are disrupted, as with
pathogens that kill the cells during or in advance of
penetration, the layer surrounding the hyphae consists of either a sheath-like matrix of fungal origin
or a mixture of disrupted host cytoplasm and vacuolar
breakdown products.   In the case of obligate parasites and intracellular mycorrhizal fungi, host cells
are not disrupted and there is normally a highly
organized interface to which both host and fungus
may contribute components.

With obligate parasites the intracellular hypha
is usually a specialised haustorium, enveloped in a
structure consisting of the haustorial sheath or
extra-haustorial matrix (Bushnell, 1972).   This is a
flexible encapsulating layer, differing in texture
from the haustorial wall and different from the host
cell wall or any wall appositions which may be deposited by the host.   The haustorial sheath is usually
only a fraction of a micron in thickness (Bracker &
Littlefield, 1973).   As it ages it often becomes
thicker, although artefacts may lead to false interpretations (Bushnell, 1971).

It is difficult to determine whether in the
orchid-fungus system there is any layer analogous to
the sheath although there are similarities to several
of the examples shown by Bracker & Littlefield (1973).
They regard orchid mycorrhizal hyphae in compatible
host cells as falling into their interface type IT24

in which a matrix or sheath, but no wall apposition,
separates the endophyte from the host cell plasma
membrane.   This interpretation may have been based
on the work of Hadley *et al.*, (1971).   However, if
as is now suggested the host is responsible for
depositing what the latter authors termed the 'outer
wall layer' of the intracellular hyphae, then this
deposit is an apposition and the interface is type
IT21 of Bracker and Littlefield, and is an example of
encased intracellular hyphae.   This view is supported by other evidence, for example where hyphae are
adjacent to the host cell wall and the encasement
layer appears to be identical to, and continuous with,
the wall material.   Also the encasement layer was
often continuous between two adjacent hyphae (see
Figs. 8, 16, 17).

Dörr & Kollman (1969), describing the endophyte
of *Neottia*, stated only that 'the hyphal wall under-
goes various changes in thickness and fine structure".
But Nieuwdorp (1972) describes an enveloping 'cellulose slime layer' which surrounds the hyphae and is
'continuous with the original plant cell wall'.
He suggests that such a layer is a feature of a late
stage of development when the hyphae are older, and
suggests that a cellulose layer is rarely found
around 'young hyphae with an active protoplast',
hypothesizing that the fungus prevents deposition of
cellulosic materials by the host.   But the observations reported here do not support such a hypothesis.   It is important to note that other workers have
looked only at material of roots collected from the
field with infections of unknown age, whereas in the
present work protocorms were infected in the laboratory by a known symbiont.   The layered wall structure described was always found in intracellular
hyphae in the symbiotic phase, regardless of age.

In some recent work Strullu and Gourret (1975)
have described and illustrated intracellular hyphae
in the roots of *Dactylorchis maculata* (L.) Verm.
(Dactylorhiza maculata (L.) Soó) which clearly

possess a two-layered wall structure correlating very closely with that described here for *Dactylorhiza purpurella*.   A granular outer layer, of variable thickness, surrounded all intracellular hyphae.

Evidence from studies of vesicular-arbuscular mycorrhizas suggests that encasement material of host origin occurs around their intracellular hyphae and arbuscule branches.   For example, Scannerini and Bellando (1968), working with *Ornithogalum umbellatum*, illustrate a 'stratified zone' of variable thickness and similar in appearance to the host cell wall, between hyphal wall and host plasmalemma.   In *Allium cepa*, Cox and Sanders (1974) describe the deposition of encasement material around arbuscules of the endophyte.   They suggest that the encasement is a feature of senescent arbuscules and that young arbuscules are surrounded only by an 'espace membranaire' between the hyphal wall and the host plasmalemma.

In another mycorrhizal system, that of *Monotropa uniflora* which is partially endotrophic, Lutz and Sjolund (1973) describe a fairly complex encapsulation, which they regard as a transfer region consisting of components of host origin, around invading intracellular haustoria.

Orchid mycorrhizas may therefore have some features in common with other endotrophic forms, and with some obligate parasites, by the host depositing material around the intracellular hyphae as an encasement.   The encasement, which may be largely cellulosic material (Nieuwdorp, 1972), becomes more amorphous as the hyphae become disorganized and it probably remains as a component of, and a layer surrounding, the residual digested hyphae.

There is little evidence from fine structure as to whether or how metabolites are transferred from fungus to host across a living interface, but changes in the metabolism of the host such as the channelling of carbohydrate from production of storage materials into growth processes (Purves and Hadley, this Symposium) and the endoreduplication of host nuclear DNA

(Williamson, 1970) are probably initiated in this
way.   Nor is there at present, any clear evidence as
to whether the degeneration of hyphae is actually a
digestion, i.e. initiated by enzymes of host origin,
or whether it is a form of lysis (autolysis?) in
which enzymes originating from the hyphae themselves
initiate self-digestion.   The appearance of complex
membranes with associated ribosomes and other components around degenerating hyphae suggests that these
may be the sites of vigorous enzyme activity.
Williamson (1973) found that there was an increase in
the activity of two acid phosphatase enzymes (which
are regarded as a marker for lysosome activity)
shortly before hyphae lysed but it was not possible
to conclude whether this increase was due to enzymes
synthesized by the fungus or the host.   Protsenko
(1973), working with mycorrhizal pea roots, also
found that acid phosphatase activity was enhanced
during the digestion phase.   He hypothesized a phagocytic function originating from lysosome-like bodies
in the host cell.   There is no evidence from the
present work, or from other studies of orchid mycorrhiza, as to whether or not lysosomes are involved in
the digestion sequence.
     The most striking feature of the orchid-fungus
system is the similarity to vesicular-arbuscular
mycorrhizas.

ACKNOWLEDGMENTS

     I am grateful to Dr. G. C. Cox and Dr. F. E. T.
Sanders, and to MM. D. G. Strullu and J. P. Gourret,
for providing information which was unpublished at
the time of writing; and to Dr. B. Williamson for
Figure 5.   It is a pleasure to thank Dr. R. P. C.
Johnson and Mr. D. A. John who did much of the elec-
tron microscopy, and also Mrs. Linda Johnston and
Mr. E. Middleton for photographic work.

REFERENCES

BORRIS, H., JESCHKE, E. M. & BARTSCH, G. (1971).
Elektronen-mikroskopische Untersuchungen zur Ultra-
struktur der Orchideen - Mykorrhiza. *Biologische
Rundschau*, 9, 177-180.

BRACKER, C. E., & LITTLEFIELD, L. J. (1973). Structural concepts of host-pathogen interfaces. In:
Fungal Pathogenicity and the Plant's Response
(eds. Byrde, R. J. W. & Cutting, C. V.), 159-318.
Academic Press, London & New York.

BURGEFF, H. (1959). Mycorrhiza of orchids. In:
The Orchids; a Scientific Survey (ed. Withner, C.),
361-395. Ronald Press, New York.

BUSHNELL, W. R. (1971). The haustorium of *Erysiphe
graminis:* An experimental study of light microscopy. In: *Morphological and Biochemical Events
in Plant-Parasite Interaction* (eds. Akai, S. &
Ouchi, S.), 229-254. Mochizuki Publishing Co.,
Japan.

BUSHNELL, W. R. (1972). Physiology of fungal haustoria. *A. Rev. Phytopathol.*, 10, 151-176.

COX, G. & SANDERS, F. (1974). Ultrastructure of the
host-fungus interface in a vesicular-arbuscular
mycorrhiza. *New Phytol.*, 73, 901-912.

DÖRR, I. and KOLLMAN, R. (1969). Fine structure of
mycorrhiza in *Neottia nidus-avis* (L) L.C. Rich
(Orchidaceae). *Planta*, 89, 372-375.

HADLEY, G. (1969). Cellulose as a carbon source for
orchid mycorrhiza. *New Phytol.*, 68, 933-939.

HADLEY, G. (1970). Non-specificity of symbiotic

infection in orchid mycorrhiza. *New Phytol.*, 69, 1015-1023.

HADLEY, G., JOHNSON, R. P. C. & JOHN, D. A. (1971). Fine structure of the host-fungus interface in orchid mycorrhiza. *Planta,* 100, 191-199.

HADLEY, G. & WILLIAMSON, B. (1971). Analysis of the post-infection growth stimulus in orchid mycorrhiza. *New Phytol.,* 70, 445-455.

HARVAIS, G. & HADLEY, G. (1967). The development of *Orchis purpurella* in symbiotic and inoculated cultures. *New Phytol.,* 66, 217-230.

LUTZ, R. W., & SJOLUND, R. D. (1973). *Monotropa uniflora:* ultrastructural details of its mycorrhizal habit. *Am. J. Bot.,* 60, 339-345.

NIEUWDORP, P. J. (1972). Some observations with light and electron microscope on the endotrophic mycorrhiza of orchids. *Acta bot. neerl.* 21, 128-144.

PROTSENKO, M. A. (1973). Electron-microscope study of the acid phosphatase localization in fungus-digesting cells of pea mycorrhiza. *Dokl. Akad. Nauk. SSSR,* 211, 213-215.

PURVES, S. & HADLEY, G., (1975). This symposium,p.175.

ROLAND, J. C., (1968). Recherches sur l'infrastructure de l'espace membranaire des cellules végétales. *C. r. Acad. Sc., Paris,* 267, 712-715.

SCANNERINI, S. & BELLANDO, M. (1968). Sull'ultrastruttura delle micorrize endotrofiche di *Ornithogalum umbellatum* L. in attivita vegetativa. *Atti Accad. Sci., Torino,* 102, 795-809.

STRULLU, D. G. & GOURRET, J. P. (1975). Ultrastruct-

ure et evolution du champignon symbiotique des
racines de *Dactylorchis maculata* (L.) Verm.
J. de Microscopie, 20, 285-294.

WARCUP, J. H. & TALBOT, P. H. B. (1967). Perfect
states of rhizoctonias associated with orchids.
New Phytol., 66, 631-641.

WILLIAMSON, B. (1970). Induced DNA synthesis in
orchid mycorrhiza. *Planta,* 92, 347-354.

WILLIAMSON, B. (1973). Acid phosphatase and esterase
activity in orchid mycorrhiza. *Planta,* 112,
149-158.

WILLIAMSON, B. & HADLEY, G. (1970). Penetration and
infection of orchid protocorms by *Thanatephorus
cucumeris* and other *Rhizoctonia* isolates.
Phytopathology, 60, 1092-1096.

SOIL CHEMISTRY OF PHOSPHORUS AND MYCORRHIZAL EFFECTS ON PLANT GROWTH

P. B. TINKER

Department of Plant Sciences,
University of Leeds, U.K.

INTRODUCTION

It is by now largely agreed that infection with arbuscular-vesicular mycorrhizal fungi can cause growth increases in higher plants, and that the main mechanism whereby this is caused is an increase in the supply of phosphorus (Mosse, 1973; Tinker, 1975). In nearly all cases this phosphorus comes from the soil, and this paper discusses the various processes which have been proposed to account for the increased supply.

It is a surprising fact that, some 130 years after the invention of superphosphate fertilizer, there are still wide gaps in our understanding of the chemical forms of phosphorus in the soil, and of its relationships with the mineral and biological surfaces present there. A special difficulty lies in the heterogeneity of soil phosphorus. Its low mobility causes local depletions and concentrations to disappear very slowly, and a natural soil must be a mosaic of differing phosphate compounds, properties and concentrations. At present we have no practical or theoretical tools for dealing with such variability in any useful way, and we must therefore consider average properties only.

Any supply mechanism must utilize some form of soil phosphorus. The most logical way of classifying these sources seems to be, (1) phosphate ions in solution or held on surfaces in such a way that it is in rapid labile equilibrium with the phosphate in the soil solution, (2) mineral phosphate held in crystal lattices, or otherwise such that it is not in rapid equilibrium with the soil solution, (3) phosphorus in organic compounds.

LABILE POOL PHOSPHATE

The 'labile pool' is defined as phosphate which is in isotopic equilibrium with the soil solution over short periods of time, but the real situation is by no means as simple as this operational definition implies. It is quite essential to ignore many of these complications if any progress is to be made at all, but we should be aware of what these simplifications are.

There are three essential points concerning the labile pool: its size, the phosphate concentration which it maintains in the soil solution, and the form of the isotherm which relates concentration in solution to amount adsorbed. The size of the pool is determined by isotopic dilution of ^{32}P; the labelled pool can be sampled either physically via the soil solution (in which case the result is called an 'E value') or biologically by growing a plant in the soil (in which case the result is an 'L value') (Larsen, 1967). A comparison of the E and L values gives an insight into the origin of the phosphorus in the plant: if they are closely similar, then the plant is absorbing only from the labile pool, and it is likely that it is absorbing directly from the soil solution.

These methods have been applied to mycorrhizal plants by Sanders and Tinker (1971) and Hayman and Mosse (1972). The detailed results of the former authors are in Table 1; L values for both mycorrhizal and non-mycorrhizal onions were essentially identical at all harvests, though both increased with time

Table 1. L and E values for a sandy loam soil at different times after ^{32}P addition; value obtained with onions ($\mu g\ g^{-1}$).

Time, days from planting	-32	-15	+23	26	37	42	44	54	60
L values, infected	-	-	111	-	127	-	137	140	-
L value, uninfected	-	-	112	-	125	-	134	140	-
E value	116	133	-	152	-	152	-	-	158

Tinker, P. B.

as expected, since equilibration continues slowly over a long period following addition of ^{32}P. The E value of similar uncropped soil was always higher (since the measuring technique tends to promote equilibrium), but the values are clearly converging, and it is inferred that all plants absorb from the same source, which is probably the soil solution. The results of Hayman and Mosse (1972) agreed with this conclusion, though Benians and Barber (1972) found a small non-significant difference in L values in their work. However, such studies should certainly be repeated on other soils and with other crops, particularly where the soil solution is extremely dilute or where a very large fraction of the total phosphate is in organic form.

The second important factor is the soil solution concentration. At concentrations above 10^{-6} M chemical analysis is possible, but at lower concentrations, such as are found in very deficient (and hence interesting) soils, it becomes very difficult. In most of our own work so far, we have used a soil (Sanders & Tinker, 1971; 1973) with a soil solution just within the measurable range, which we estimated could be determined chemically with a reproducibility of about 10%.

The third important factor is the sorption isotherm, which connects solution and sorbed phosphate, and thereby defines phosphate mobility in the soil (Drew & Nye, 1970). The sorption is usually defined satisfactorily by a Langmuir or similar isotherm (Olsen & Watanabe, 1957; Gunary, 1970), but a much more complete treatment, which takes account of the simultaneous sorption of protons and hydroxyl ions from the soil solution has been given by Bowden, et al., (1973), who use it to explain the variation of the Langmuir maxima with soil pH. There are serious difficulties connected with the use of phosphate sorption isotherms, however. They are affected by variations in pH, electrolyte concentration, time and temperature. They also show strongly

marked hysteresis (see Tinker, 1971) so that adsorption and desorption curves can differ greatly, and
all isotherms are curved so that no single value for
the slope can be given.   The technical problems in
measuring the isotherms are also quite formidable,
and it is therefore difficult to be precise in either
the practical measurement or the theoretical interpretation and use of phosphate sorption isotherms.
Since the mobility of the phosphate ion in a given
soil depends upon the isotherm, this could form a
very serious source of uncertainty in the application
of diffusion theory to availability of phosphate
ions, and this problem is discussed below.

SUPPLY OF LABILE POOL PHOSPHATE TO PLANTS

We take it as axiomatic that phosphate ions are
supplied to absorbing surfaces in the soil by diffusion, and that this may be the rate-limiting step in
the uptake process (Nye 1969;   Bhat & Nye, 1974).
However, application of diffusion theory to nutrient
uptake requires a knowledge of the 'boundary condition' at the root or hyphal surface, and this is
still very uncertain.   In the absence of specific
information, we adopt the simplest 'boundary condition':   that the concentration is constant.   As a
special case of particular interest, we choose a surface concentration of zero, which gives the maximum
which can be brought to the root.

This approach was developed, for cations, by
Brewster and Tinker (1970), and applied to phosphate
uptake by mycorrhizal onions by Sanders and Tinker
(1971).   They used the modified equation:

$$I = 2\pi D_p \ (C_{li} - C_{la}) \ \gamma$$

Here I is the mean uptake rate per unit length of
root per second, which may be derived from successive
measurements of nutrient uptake and root length
(Brewster & Tinker, 1972), D_p is the porous system
diffusion coefficient, which depends upon soil geom-

etry and water content, C_{li} the initial bulk soil
solution concentration, C_{la} the solution concentration at the root surface. γ is a complex term that
depends upon Dt/a^2, where D is the diffusion coefficient, t time and a the root radius.   Since
$D = D_p$/slope of isotherm, the isotherm slope affects
the uptake rate via this term. γ varies rather
slowly with Dt/a^2 when the latter is above 1, (Table
2) (Brewster, 1971), and it is obvious that for
hyphae, γ becomes fairly constant after very short
times.   It changes more for roots, but even when D
changes by a factor of 10, γ only changes by a factor
of 3 or less for moderate times.   The important
value is therefore that of D_p, whilst D has much less
effect.   Consequently the isotherm slope value is
not critical, and approximate values are acceptable.
     From this argument, we conclude that the mean
inflow I for a root will only exceed that of a
hyphae, at the same time and with the same value of
C_{la}, by a factor which will rarely be greater than
10, and this explains the great efficiency of a
hyphal network.   It is obviously speculative to
apply diffusion theory for a homogeneous medium to
hyphae, which are often smaller than the individual
particles of the soil, but the conclusions should be
approximately correct.

Table 2.   Values of the dimensionless
parameter γ, for two times and diffusion
coefficients, and for roots and hyphae,
of radius 0.025 cm and 4µm respectively.

D cm^2 s^{-1}	1 day		10 Days	
	Root	Hyphae	Root	Hyphae
10^{-8}	0.84	0.21	0.47	0.15
10^{-9}	2.16	0.27	0.84	0.21
10^{-10}	5.32	0.38	2.16	0.27

Table 3.   Summarised inflow value for roots
and hyphae (from Sanders and Tinker, 1973).

Mean inflow into uninfected onions measured	4×10^{-14} mol $cm^{-1}s^{-1}$
Mean inflow into infected onions measured	17×10^{-14} mol $cm^{-1}s^{-1}$
Mean maximum inflow into uninfected onions calculated	4×10^{-14} mol $cm^{-1}s^{-1}$
Mean inflow into hyphae, from measured whole plant inflow and hyphal length	2×10^{-15} mol $cm^{-1}s^{-1}$
Mean maximum inflow into hyphae, calculated approx.	8×10^{-15} mol $cm^{-1}s^{-1}$

This approach was used by Sanders & Tinker
(1971; 1973) to calculate the maximum inflow of
phosphate (with $C_{1a} = 0$) to a root system of onions,
using a time-weighted average value of γ.   They
concluded that the inflow to the roots of uninfected
plants was near the maximum possible, and that the
inflow to infected plants, which is 4 times greater,
could only be explained by uptake by, and transport
along, hyphae (Table 3).   The present hypothesis is
therefore that the mycorrhizal effect arises quite
simply by uptake of phosphate from the soil solution
by the hyphae, which then translocate it to the host
root.   I do not, however, feel that this must
necessarily be correct for all hosts and all soils,
and it is essential for it to be more widely tested.
The rest of this paper therefore considers some of
the implications of these ideas, and also some of the
alternative supply mechanisms and sources which must
be kept in mind.

Root hairs and mycorrhizal hyphae in uptake from labile pool P

Professor Baylis' work (Baylis, 1975) has focussed attention on the relative efficiency of root hairs and hyphae. Their diameters are very similar, so many of the arguments given above apply to both. It is as yet uncertain whether their uptake efficiency (Nye, 1966) is the same, i.e. whether they absorb at similar rates from the same solution concentration. If it is, the plant could in theory use either mechanism, and it is interesting that the figure of 80 cm of hyphae per cm of onion root (Sanders & Tinker, 1973) would not be far from the total length of root hairs with 1.5 mm length and 50 hairs mm^{-1}, which is a dense growth, but by no means unusual. In some other soils there is less mycelium, and the total length of root hairs in a piliferous species could well exceed it. However, root hairs often have short lives, and some of the root hair length may be less effective, because of inter-hair competition (Fig. 1). The individual depletion zones set up around each hair will overlap quickly near the root surface. Theory indicates (Baldwin, Tinker & Nye, 1972) that cylindrical zero-sink absorbing organs begin to compete seriously (uptake reduced by more than 10%) when the distance between them is less than $\sqrt{Dt}$. This corresponds to 0.29 mm for $D = 10^{-8}$ cm^{2}s^{-1} and $t = 1$ day. At 50 hairs/mm, this suggests that hairs up to a distance of 0.67 mm from the centre of the root are competing with each other more or less seriously by this time. Hyphae are more dispersed, and will suffer less competition.

Root exudates

These arguments in favour of simple uptake by hyphae do not exclude one other possible mechanism for increasing phosphate uptake: if the adsorption

HYPHAL AND ROOT HAIR UPTAKE

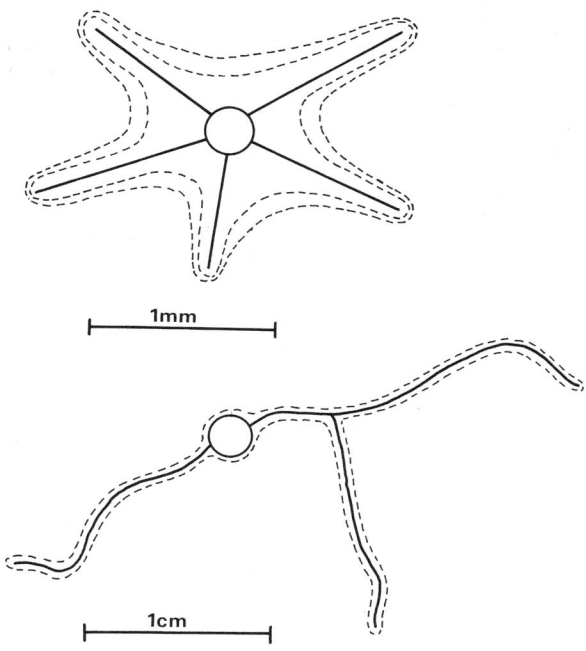

Figure 1.   Schematic diagram to illustrate
uptake competition between root hairs.
Dotted lines indicate limits of depletion
zones.

isotherms were altered locally around the root, the
soil solution concentration would rise, and a higher
uptake rate of phosphate would become possible,
though the source would still be the labile pool.
It has often been suggested in the past that root
exudates, e.g. citrates, can displace phosphate ions
from the soil and thereby alter the adsorption isotherms.   The calculations of Tinker and Sanders
(1975) however indicate that the total amount exuded
is unlikely to be large enough to have a major effect,
there is no direct evidence that the displacement

occurs, and there is no reason at present to believe
that exudation is greater in mycorrhizal than in non-
mycorrhizal plants.   However, if this process does
occur the presence of either root hairs or hyphae
around the root might make the whole system very much
more efficient, since they would serve to trap displaced phosphate ions, which otherwise could diffuse
out into the soil and be re-adsorbed there.

ORGANIC SOIL PHOSPHORUS

It has been suggested that mycorrhizal roots
possess special powers of utilizing organic forms of
phosphorus in the soil (Bowen, 1973), and it is known
that roots and ectomycorrhizas can hydrolyse organic
phosphates by surface and soluble phosphatases
(Harley, 1969; Bartlett & Lewis, 1973).   It is
inferred from this that soil organic phosphates are
hydrolysed on the root surface.   However, inositol
hexaphosphate, and probably other soil phosphate
esters, are strongly adsorbed by the soil surfaces,
in a very similar way to phosphate ions themselves
(Anderson, Williams & Moir, 1974), though sorption
decreases rapidly with pH.   Wild & Oke (1966) showed
that axenically cultured plants could utilize soluble
organic phosphates, and detected up to 1 µg g^{-1} in
soil extracts, but they had first to pretreat their
soils at 40°C to produce such amounts in solution.
Even if amounts are found in the soil solution which
are comparable to those of inorganic phosphate, it is
unlikely that the organic phosphorus compounds are
buffered to the same extent.   It thus seems unlikely
that root surface phosphatases could be effective,
since little organic phosphorus would reach them.
Reference is sometimes made to organic phosphorus
'in contact' with root surfaces, but the amount for
which this is literally true must be very small
indeed.   From the results of Sanders & Tinker (1973)
it seems that, for onions in one soil, the total

hyphal surface in mycorrhizal plants was only about
0.5 times that of the true root, which is not a
sufficient addition to change the above conclusion.
Alternatively, soluble phosphatases could diffuse out
into the soil, and free phosphate ions be produced
there.   The latter will be adsorbed by the soil in
the same way as other phosphate ions, and will form
part of the labile pool.   It is thus possible that
a small part of the rhizosphere will get an enlarged
labile phosphate pool, but this is hardly an efficient way of providing more phosphate to the plant.
Tinker & Sanders (1975) have suggested that bacterial
debris, of colloidal size, containing organic phosphates could be carried in the transpiration stream
to the root surface and there hydrolysed by the surface phosphatases.   However, such a mechanism, if it
exists, seems more appropriate to the true root than
to mycorrhizal hyphae, and can probably be neglected
for our purposes.

NON-LABILE INORGANIC PHOSPHORUS
AND ROCK PHOSPHATE

A considerable part of the phosphorus in most
soils is held in mineral lattices, below the crystal
surface, so that it is not immediately exchangeable
with ions in the soil solution, mainly as hydroxy- or
fluor apatite or aluminium or iron phosphates of
varying purity and composition, depending upon the
pH.   The fertilizer known as 'rock phosphate'
(Rogers, Pearson & Ensminger, 1953), can be considered in the same category.   Daft and Nicolson (1967)
have found larger responses with mycorrhizal than
non-mycorrhizal plants to added rock phosphate. This
suggests 'solubilization' of phosphates by mycorrhizal fungi, either alone or in association with rhizosphere bacteria (Barea et al. 1975).

There is in fact, little evidence that bacterial
solubilization can improve P supply to plants (Rovira
& Davey, 1974; Tinker & Sanders, 1975).   Suitable

bacteria or fungi in pure agar culture can certainly
dissolve various poorly soluble phosphates: the
question is whether this occurs in soil and, if so,
whether this phosphate can be transferred to higher
plants.   This last point is of particular importance
for our topic, since transfer to a higher plant is
difficult to explain in relation to bacterial solubilization, but there is no such difficulty with a
mycorrhizal fungus.   It is therefore worth looking
critically at the possible solubilization processes.
The formula of apatite is $Ca_5(PO4)_3(OH,F)$, fluorapat-
ite being the least soluble, with a solubility
product of the order of 10^{-56} (Fassbender, *et al.*,
1965).   Many authors have published solubility data
on hydroxyapatites and rock phosphates, and most give
values for total phosphate ion concentration of 10^{-5}
- 10^{-6} M in the equilibrium solution at near neutral
pH, though Jacob and Hill (1953) give 10^{-6} - 10^{-7} M.
It may be as low as 10^{-8} M in a neutral soil.
     The dissolution of hydroxyapatite may be caused
by reducing the solution pH (thereby reducing OH and
PO_4^{3-} ion activities), reducing the cation (Ca, Al
or Fe) activity, or reducing the total phosphate ion
(P_i) activity.   Significant change in Ca ion activity locally at the root surface is unlikely, and the
pH seems as likely to increase as to decrease.
The most general way of dissolving rock phosphate by
root action is thus by reducing the phosphate ion
concentration, i.e. by root uptake of phosphate.
     We can visualise three major possibilities:
a) hyphae simply act as well-distributed absorbing
surfaces for P_i in the soil; b) hyphae proliferate
preferentially near to, on the surface of or (for
porous materials) inside phosphate sources of low
solubility; c) hyphae have some power of dissolving
the mineral phosphate directly.
     If roots, hyphae and rock phosphate particles
are in a non-sorbing medium, such as sand culture,
then all depends upon the distance between an absorbing surface and a rock phosphate surface, since this

determines the concentration gradient between them. For example, the distance between individual grains of rock phosphate, of diameter 0.1 mm, with 1 ton ha^{-1} mixed to 10 cm, could be of the order of 0.15 cm. The flux of phosphate between two surfaces half this distance apart, one (the source) being 10^{-6} M and the other (the sink) 10^{-7} M in equilibrium concentration, would only be about 1 x 10^{-14} moles cm^{-2} s^{-1}. If hyphae are distributed at random, their advantage will lie entirely in reducing the mean distance between each rock phosphate grain and the nearest absorbing surface. Exactly the same result should be gained by using more finely ground rock phosphate, which is known to increase responses when it is used as a fertilizer. This effect seems sufficient to account for the responses reported so far.

If the plants are in a sorbing medium such as soil, a rather different situation will occur. If the equilibrium concentration of the phosphate source is the same as or less than the soil solution concentration, nothing will happen unless a grain falls within the depletion zone of a root or hyphae, in which case it increases soil buffering power. If the equilibrium concentration is higher than the soil solution concentration, then the rock phosphate dissolves to give enriched zones of slowly increasing radius. The presence of hyphae allows these to be exploited in the way described earlier in this paper, and if the original soil is very deficient, an increased mycorrhizal effect may be produced. If the zones are sufficiently numerous and concentrated, roots alone may absorb sufficient for the plant, and the mycorrhizal effect may be eliminated by rock phosphate additions.

If hyphae tend to proliferate near a source of phosphate, the situation is rather different, in that fineness of division and evenness of distribution would no longer be so important. The mean distance between source and sink would then also be much reduced, and considerably higher rates of uptake could be expected. It is well known that roots tend to develop in soil zones enriched with phosphorus (and nitrogen), and it seems important to determine whether hyphae have an analogous property.

Finally, there is the possibility that hyphae have some special property of dissolving rock phosphate surfaces, beyond that caused by simple uptake of phosphate ions. This might arise from changes in pH in or cation concentration in the very small volume of solution between the phosphate mineral surface and an adpressed hyphae. For such an effect to be of importance, it seems essential that the hyphae should proliferate near phosphate sources, since random distribution would hardly produce a sufficient degree of contact.

This gives a range of interesting possibilities, and at first it seemed possible to me that studies on mycorrhizas would explain most differences between crops in utilization of rock phosphate. However, it has been accepted for a long time that 'efficient' users of rock phosphate include lupins, buckwheat, clover, mustard, Swiss chard, rape and cabbage (Rogers, *et al.* 1954; Dean & Fried, 1953), whereas cotton, cowpeas, and most cereals are inefficient. Several non-mycorrhizal species are thus users, whereas the heavily infected cereals are not. I suspect, therefore, that mycorrhizas alone cannot explain the use or non-use of rock phosphate, and that some other factors are effective.

Russell, Russell and Marais (1958) tested L values in various soils with barley, rye and cabbage (there is no information on mycorrhiza development of their plants, but it seems likely the cereals were infected). They found that the L values for rye and barley were closely similar in three soils, but differed by a factor of 2 in a strongly sorbing basaltic soil. The L values of barley and cabbage were the same in unfertilized soil, but the latter became relatively greater with increasing phosphate

additions.   It is possible that relative differences
in either absorbing power (see Russell *et al.* 1958),
or in frequency of root hairs can explain these
effects, but some further process may be at work.
In our discussions of the merits of mycorrhizal
hyphae in phosphate uptake, I therefore believe that
we should pay some attention to the Cruciferae, who
seem to manage so very well without them.

ACKNOWLEDGMENT

I thank Dr. F. E. Sanders for useful discussions
during the development of these ideas.

REFERENCES

ANDERSON, G., WILLIAMS, E. G. & MOIR, J. O. (1974).
A comparison of the sorption of inorganic orthophosphate and inositol hexaphosphate by six acid
soils.   *J. Soil Sci.*, 25, 51-62.

BALDWIN, J. P., TINKER, P. B. H. & NYE, P. H. (1972).
Uptake of solutes by multiple root systems from
soil.   II.   The theoretical effects of rooting
density and pattern of uptake of nutrients from
soil.   *Pl. Soil,* 36, 693-708.

BARTLETT, E. M. & LEWIS, D. H. (1973). Surface phosphatase activity of mycorrhizal roots of beech.
Soil Biol. Biochem., 5, 249-257.

BAYLIS, G. T. S. (1975).   This symposium, p, 373.

BENIANS, G. J. and BARBER, D. A. (1972). *A. Rep.*
ARC Letcombe Laboratory, 1971, 7-9.

BAREA, J. M., AZCON, R. & HAYMAN, D. S. (1975).
This symposium., p.

BHAT, K. K. S. & NYE, P. H. (1974). Diffusion of phosphate to plant roots in soil.   Depletion around onion roots without root hairs.   *Pl. Soil,* 41, 383-394.

BOWDEN, J. W., BOLLAND, M. D. A., POSNER, A. M. & QUIRK, J. P. (1973). Generalised model for anion and cation adsorption at oxide surfaces.   *Nature, London,* 245, 81.

BOWEN, G. D. (1973). Mineral nutrition of ecto-mycorrhizae.   In *Ectomycorrhizae,* eds. Marks, G. C. & Kozlowski, T. T.   Academic Press, New York, pp. 151-206.

BREWSTER, J. L. (1971). D. Phil. thesis, Oxford University.

BREWSTER, J. L. & TINKER, P. B. H. (1970). Nutrient cation flows in soil around plant roots. *Proc. Soil Sci. Soc. Amer.,* 34, 421-426.

BREWSTER, J. L. & TINKER, P. B. H. (1972). Nutrient flow rates into roots.   *Soils Ferts.,* 35, 355-359.

DAFT, M. J. and NICOLSON, T. H. (1966). Effect of *Endogone* mycorrhiza on plant growth.   *New Phytol.,* 65, 343-350.

DEAN, L. A. & FRIED, M. (1953). Soil plant relation-ships in the phosphorus nutrition of plants.   In *Soil and Fertiliser Phosphorus,* eds. Pierre, W. H. & Norman, A. G., Academic Press, New York. pp. 43-58.

DREW, M. C. & NYE, P. H. (1970). The supply of nut-rient ions by diffusion to plant roots in soil. III. Uptake of phosphate by roots of onion, leek and rye-grass.   *Pl. Soil,* 33, 545-563.

FASSBENDER, H. W., LIN, H. C. and ULRICH, B. (1966). Solubility and solubility product of hydroxylapatite and rock phosphates. *Z. Pflanzenernaerh. Dung. Bodenk.*, 112, 101-113.

GUNARY, D. (1970). A new adsorption isotherm for phosphate in soil *J. Soil Sci.*, 21, 72-77.

HARLEY, J. L. (1969). *The biology of mycorrhiza.* Leonard Hill, London.

HAYMAN, D. S. & MOSSE, B. (1972). Plant growth responses to vesicular-arbuscular mycorrhizas. III. Increased uptake of labile P from soil. *New Phytol.*, 71, 41-47.

JACOB, K. D. & HILL, W. L. (1953). Laboratory evaluation of phosphate fertilisers. In *Soil and Fertiliser Phosphorus,* eds. Pierre, W. H. & Norman, A. G.   pp. 299-346.

LARSEN, S. (1967). Soil phosphorus. *Adv. Agron.*, 19, 151-210.

MOSSE, B. (1973). Advances in the study of vesicular-arbuscular mycorrhizas. *A. Rev. Phytopath.*, 11, 171-195.

NYE, P. H. (1966). The effect of nutrient intensity and buffering power of a soil and the absorbing power and size of root hairs on absorption, *Pl. Soil,* 25, 81-106.

NYE, P. H. (1969). The soil model and its application to plant nutrition. In *Ecological aspects of the mineral nutrition of plants,* ed. Rorison, I. H., *Symp. Br. ecol. Soc.*, 9, 201-213.

OLSEN, S. R. & WATARNABE, F. S. (1957). A method to determine a phosphorus adsorption maximum of soils as measured by the Langmuir isotherm. *Pro. Soil Sci. Soc. Amer.*, 21, 144–149.

ROGERS, H. T., PEARSON, R. W. & ENSMINGER, L. E. (1953). In *Soil and Fertiliser Phosphorus,* eds. Pierre, W. H. and Norman, A. G., Academic Press, New York.

ROVIRA, A. D. & DAVEY, C. B. (1974). Biology of the rhizosphere. In *The plant root and its environment,* ed. Carson, E. W., University Press of Virginia, Charlottesville pp. 153–204.

RUSSELL, R. S., RUSSELL, E. W. & MARAIS, P. G. (1958). Factors affecting the ability of plants to absorb phosphate from soil. II. A comparison of the ability of different species to absorb labile soil phosphate. *J. Soil Sci.*, 9, 101–108.

SANDERS, F. E. T. & TINKER, P. B. H. (1971). Mechanism of absorption of phosphate from soil by *Endogone* mycorrhizas. *Nature, London,* 233, 278–279.

SANDERS, F. E. T. & TINKER, P. B. H. (1973). Phosphate flow into mycorrhizal roots. *Pestic. Sci.,* 4, 385–395.

TINKER, P. B. H. (1971). Some problems in the diffusion of ions in soils. In *Sorption and Transport processes in Soils,* S.C.I. Monograph No. 37, pp. 120–134.

TINKER, P. B. (1975). Effects of vesicular-arbuscular mycorrhizas on higher plants. *Symp. Soc. exp. Biol.,* 29 (in press).

TINKER, P. B. & SANDERS, F. E. T. (1975). Rhizosphere microorganisms and plant nutrition. *Soil Sci.,* (in press).

WILD, A. & OKE, O. L. (1966). Organic phosphate compounds in calcium chloride extracts of soils: identification and availability to plants. *J. Soil Sci.*, 17, 356-371.

THE MAGNOLIOID MYCORRHIZA AND MYCOTROPHY[1] IN ROOT SYSTEMS DERIVED FROM IT

G. T. S. BAYLIS

Botany Department, Otago University
Dunedin, New Zealand

INTRODUCTION

In modern attempts to assemble the living orders of flowering plants in phyletic sequence the woody Magnoliales are always chosen as the initial, most primitive group, ancestral to the other dicotyledons and to the monocotyledons. Surveys of the Magnoliales and other orders for features which seem to preserve the pattern of their Mesozoic ancestors have always neglected their roots (e.g. Sporne, 1973). I have recently suggested that the mycorrhizal phycomycetes have exercised a controlling influence on the evolution of roots since their origin from rhizomes in the Devonian (Baylis, 1972, 1974). This paper draws attention to the form they have apparently induced in the roots of the first flowering plants, in so far as the existing Magnoliales may be taken to preserve it, and discusses evolutionary trends thereafter.

THE MAGNOLIOID ROOT AND MYCORRHIZA

The present distribution of the Magnoliales, in

---

[1] Mycotrophy is used in the restricted sense of deriving a growth stimulus from mycorrhizas.

the broader usage adopted by Cronquist (1968), is
centred in the tropics and subtropics where the first
adaptive radiation of the flowering plants apparently
occurred.　This places the majority beyond reach,
but I have been able to examine the following genera[1]
- *Annona, Apollonias, Chimonanthus, Cryptocarya,
Cinnamomum, Beilschmiedia, Drimys, Hedycarya, Laur-
elia, Laurus, Liriodendron, Magnolia, Manglietia,
Michelia, Monodora, Persea, Pseudowintera* and *Talauma*
- with such consistent results that they seem an
adequate basis for defining the morphology of the
magnolioid root.　It is coarsely branched so that
the ultimate roots are rarely less than 0.5 mm in
diameter (Fig. 1).　The most finely divided sample
came from *Annona cherimolia* with a minimum of 0.35 mm
and a mean of 0.56 mm, the coarsest from *Drimys
winteri* with a mean of 1.5 mm.　These roots have a

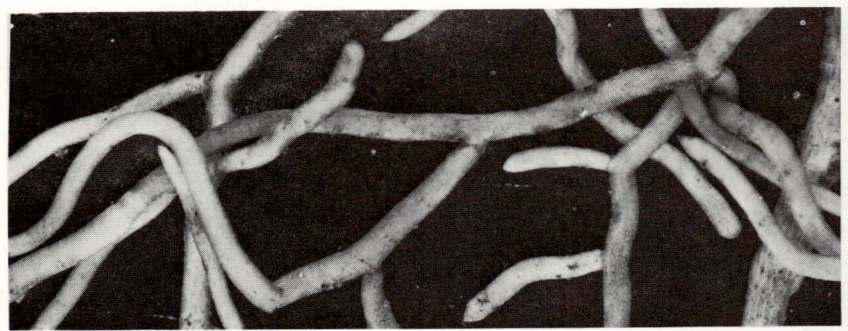

Figure 1.　Magnolioid roots, *Magnolia
stellata.*

[1] I am indebted to the Director, Royal Botanic Gardens, Kew and to Mr. Alan Esler, Botany Division,
D.S.I.R., Auckland for help in obtaining material.

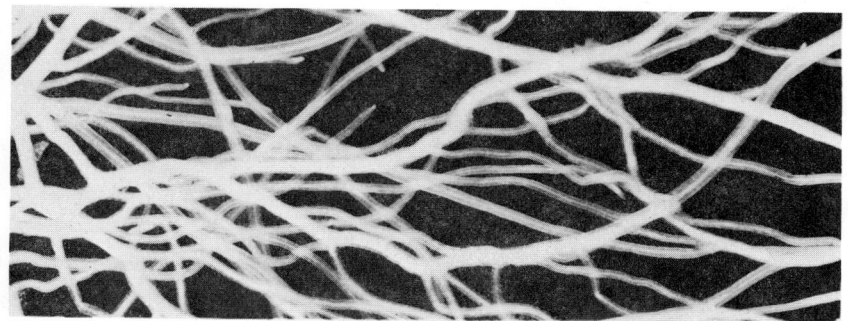

Figure 2. An intermediate root system,
Solanum nigrum.

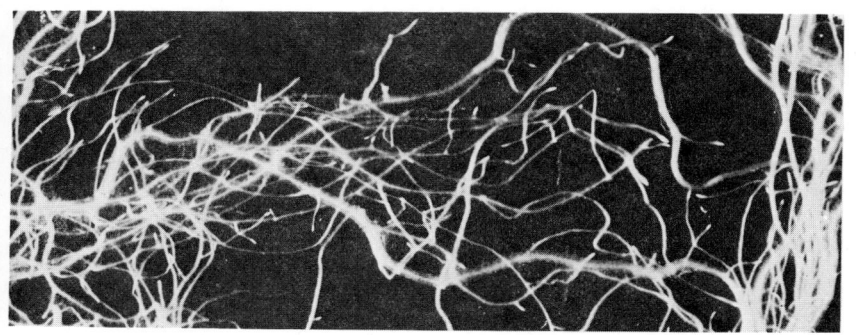

Figure 3. Graminoid roots, *Lolium perenne.*

compact stele with 2 - 4 protoxylem groups so that
their cross section consists mainly of cortex.   This,
even in well manured shrubberies, almost invariably
contains a zone of cells densely packed with the
mycelium of phycomycetous endophytes.   There are
normally no root hairs: I have seen none in *Magnolia,* but they may occur singly or in small groups in
Pseudowintera colorata, Liriodendron tulipifera and
Annona cherimolia. These hairs are always wide and
stumpy - 15-20 µm in diameter and usually much less
than 100 µm in length.

Soft thick roots with at most a sparse and erratic scatter of short, stout and presumably vestigial
hairs are by no means confined to the Magnoliales.
As mycorrhizas they have been studied in *Liriodendron
tulipifera* (Clark, 1963) and in *Griselinia littoralis*
(Baylis, 1959, 1967).   This is the type of root most
dependant upon conversion to a mycorrhiza for phosphorus uptake from the low levels of available P
normally found in mature forest soils.   For an uninfected plant to equal the growth rate of one that
has mycorrhizas, sufficient soluble P has to be added
to the soil to raise the Truog value 5-10 fold.

THE GRAMINOID ROOT

Flowering plant evolution has diverged towards
several climaxes.   The rushes, sedges and grasses
are sometimes treated as the group furthest removed
from the ancestral stock (e.g. Hutchinson, 1973).
Certainly their roots contrast sharply with the
magnolioid type.   The ultimate branches may be less
than 0.1 mm in diameter and they are consistently and
densely covered with slender root hairs that are
often 1-2 mm in length (Fig. 3).

Grasses with truly graminoid roots are mycotrophic only when the availability of P is extremely
low — lower perhaps than in some natural grassland
soils (Mosse, 1972;   Mosse *et al.,* 1973;   Crush,

1973 a, b). But phosphorus uptake by a non-mycorrhizal graminoid root system reaches its greatest
efficiency not in the grasses but in the sedges and
rushes. In these families mycorrhizal infection is
relatively slight in the field, and seems impossible
to establish in pot experiments. Uninfected rushes
and sedges grow steadily where uninfected grasses do
not (Fig. 4).

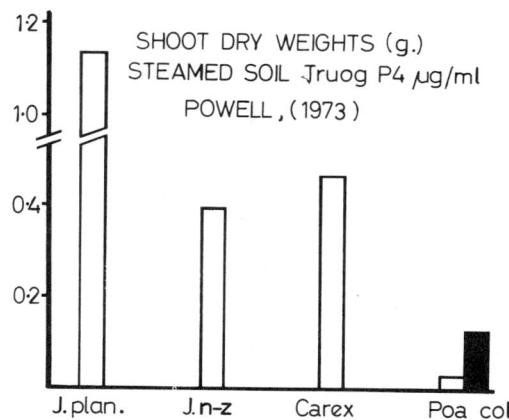

Figure 4. Shoot growth of non-mycorrhizal
plants of *Juncus planifolius*, *J. novae-zeland-
iae*, *Carex coriacea* and *Poa colensoi* and of
mycorrhizal plants of *Poa colensoi* (black column) after four months in soil very low in
available P. (Powell, 1975).

The branching of rush and sedge roots may be no finer
than that of grasses but their average root hair
length can exceed 2 mm.

INTERMEDIATE TYPES

There appears to be little published on the
branching of roots and the abundance and length of
their hairs. Metsävainio (1931) in a study of moorland plants found that the species without root

hairs were the most constantly mycorrhizal and credits Höveler with suggesting this in 1892.   We agree with Dittmer (1949) in finding branching and root hair characteristics fairly constant in a species and Table 1, which is not limited to flowering plants, illustrates that most root systems are neither magnolioid nor graminoid but somewhere between the two (Fig. 2).   This table presents the species in a series ranging from those which have shown a growth response to mycorrhizas in relatively fertile soils to those that have not proved mycotrophic even in the poorest soils.   All the values could be refined by better sampling methods but the order of the species should be secure, as it rests on growth responses in a few soils with graded additions of phosphorus and supplies of other nutrients sufficient to ensure that no other element was limiting growth.

In each species a characteristic balance has been struck between dependence on phycomycetous symbionts to secure phosphorus and formation of a root system that is itself adequate.   This balance brings independence at a point that seems capable of precise chemical definition.   The data suggest that every level of compromise between root hair development and mycotrophy will be found viable for some species or other.

THE BASIS OF NON-MYCOTROPHIC GROWTH

When a plant is able to grow without mycorrhizas in a soil low in available phosphorus there are two simple alternatives (a) it is unusually economical in using phosphorus (b) it obtains more phosphorus than the species which fail to grow.

(a) A reduction in the percentage of phosphorus in the shoot dry matter usually occurs when a seedling remains uninfected in a poor soil.   In the more mycotrophic species like *Griselinia littoralis* and *Coprosma robusta* the stem apex becomes dormant and

Table 1. Species in a descending series from those mycotrophic in soils relatively rich in available P to those that are never mycotrophic; no other factor limiting growth.

Truog P (µg/ml)	species	habit	growth rate	rootlet[b] dia. (mm)	root hairs[b] mn lgth (mm)	frequency	% P in[c] shoot	reference
> 32	Lycopersicum esculentum	crop plant	fast	0.1-0.2	0.3	constant	.25	Hall, 1975
	Podocarpus totara	conifer	slow	> 1.0	0.1	rare	.08	Baylis, 1967, 1972
	Griselinia littoralis	hardwood	slow	> 1.0	0.1	rare	.10	Johnson, 1973
15-32	Weinmannia racemosa	hardwood	slow	0.2-0.3	0.07	rare	.04	Hall, 1975
	Coprosma robusta	shrub	mod.	0.2-0.3	0.07	rare	.05	Hall, 1975; Crush, 1973a
	Solanum laciniatum	soft shrub	fast	0.2-0.3	0.8	constant	.07	Cooper, 1973
	Leptospermum scoparium	shrub	mod.	0.1-0.2	0.3	inconst.	.03	Cooper, 1973
8-15	Fuchsia excorticata	shrub	mod.	0.1-0.2	0.2	constant	.10	Johnson, 1973
	Histiopteris incisa	fern	mod.	0.3-0.4	1.5	constant	.07	Cooper, 1973
	Solanum nigrum	herb	mod.	0.2-0.3	0.7	constant	.08	Cooper, 1973
	Metrosideros umbellata	hardwood	slow	0.1-0.2	0.3	inconst.	.04	Hall, 1975
	Chionochloa rigida	grass	v.sl.	0.1-0.2	0.5	inconst.	.04	Crush, 1973a
	Pteridium esculentum	fern	mod.	0.2-0.3	1.6	constant	.10	Cooper, 1973
4-8	Lolium perenne	grass	fast	0.1-0.2	1.3	constant	.05	Crush, 1973b
	Poa colensoi	grass	slow	0.1-0.2	0.7	constant	.11	Crush, 1973a; Powell, 1975
	Anthoxanthum odoratum	grass	fast	0.1-0.2	1.8	constant	.04	Crush, 1973b
NM[a]	Chionochloa macra	grass	v.sl.	0.1-0.2	0.8	constant	.05	Crush, 1973a
	Carex coriacea	sedge	mod.	0.1-0.2	1.0	constant	.10	Powell, 1975
	Juncus planifolius	rush	fast	0.2-0.3	1.1	constant	.07	Powell, 1975
	Asplenium bulbiferum	fern	v.sl.	0.2-0.3	2.2	constant	.12	Cooper, 1973

a = Never mycotrophic; some mycorrhizal infection may occur.   b = Figures adjusted for differences in sampling method.

c = The lowest values recorded in seedlings making growth but still strongly responsive to added P. Plants mycorrhizal excepting the NM series.

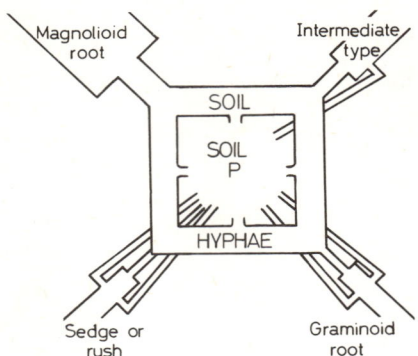

Figure 5. Diagrammatic illustration of the uptake of P from soil via Endogonaceous hyphae and/or root hairs.

the phosphorus content falls to .01–06%.   The values given in Table 1 are not these values, but the lowest P concentrations recorded in shoots of actively growing mycorrhizal seedlings that are still strongly responsive to added phosphorus.   They range from .03% to .11% (excluding *Lycopersicum*) and no pattern is discernible in their distribution.   Still more important, the P concentrations of species that are never mycotrophic are very similar, i.e. .05% to .12%.   Plants with a rigid economy in the use of phosphorus exist (Beadle, 1968) but obviously this is not the basis for the general seriation of Table 1. The % P in shoots of plants growing on poor soils without mycorrhizas was similar to that of partially or strongly mycotrophic species.

(b) The texture of the root system is indicated in Table 1 by the diameter of the ultimate rootlets and the mean length and frequency of the root hairs. No plant with a truly magnolioid root, i.e. >0.5 mm diameter would be expected to come below *Griselinia littoralis* in the table.   The rootlet diameter of species below *Griselinia* ranges from 0.1 - 0.4 mm, so that fineness of branching has no overriding effect upon the degree of mycotrophy.   But the last 8 species have their finer roots continuously covered in root hairs longer than the grand average (0.42 mm)

of Dittmer's (1949) sample.   The short root hairs
(< 1 mm) of *Chionochloa macra* and *Poa colensoi* are
offset by a low growth rate.   Correspondingly, roots
that constantly produce hairs *c.* 1 mm long stand
higher in the table if the roots bearing them are
unusually coarse, as in *Histiopteris incisa,* or the
plant is capable of very rapid growth like *Solanum
laciniatum.*   The short hairs of *Fuchsia excorticata*
are effective because of their abundance throughout a
finely divided root system.   The separation of the
two Myrtaceous species *Leptospermum scoparium* and
Metrosideros umbellata, so similar in root morphology
and shoot phosphorus levels, depends on differences
in potential growth rate.   *Asplenium bulbiferum,*
probably New Zealand's only non-mycotrophic ground
fern, combines long hairs with slow growth and a
natural restriction to young soils.   It is richer in
phosphorus than other ferns even when planted in a
poor soil, but growth is then, of course, slower
still.

The length and frequency of the root hairs is
clearly the best single index of a plant's capacity
for non-mycotrophic growth.   Calculation of Spear-
man's rank coefficient between the two gave the value
0.761 significant at 1% level.*

EVOLUTIONARY TRENDS IN THE ENDOPHYTES

As soils mature phosphorus becomes less and less
available (Walker, 1965) while combined nitrogen
becomes more plentiful.   The mycorrhizal fungi in a
particular habitat are thus under pressure to increase their efficiency, unlike the nitrogen fixing

---

* I am grateful to Dr. J. B. Wilson for suggesting
  this method which is preferable to the usual
  correlation coefficient because the population
  includes different plants and different soils.

symbionts which can become inactive. Strains of
phycomycetous symbionts which are ineffective in
phosphorus uptake should not persist, and all tendencies for the fungi to be harmful should be eliminated. The demand that they make upon their hosts
must be mainly for carbohydrate, and even for shaded
seedlings their assistance seems to be worth its
cost. Thus in *Leptospermum scoparium* which is only
moderately mycotrophic, infection remained remarkably
constant and seedlings still responded strongly to it
even when their growth rate was reduced to one-third
by shading (Fig. 6). Nevertheless inoculation with

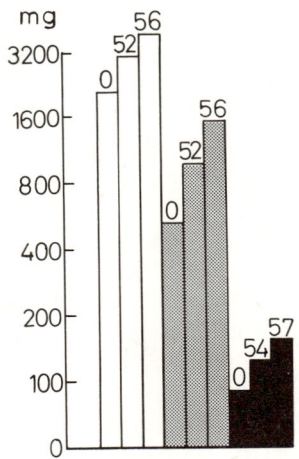

Figure 6. Dry weight of
Leptospermum scoparium in
full sunlight (clear col-
umns), light shade (stip-
pled) or heavy shade
(black); non-mycorrhizal
(0) infected with *Acaulospora laevis* (A) or
Rhizophagus tenuis (R).
Superscript is the %
infected roots. Mean of
8 replicates Johnson
(1973).

mycorrhizal fungi has sometimes retarded growth,
particularly during the early establishment phase of
infection (Crush, 1973 a, b; Furlan & Fortin, 1973;
Cooper, 1973). However, Cooper's observations on
Solanum nigrum suggest that this is peculiar to
partially sterilized soils.

    Within a natural plant community there is no
evidence of host specificity. Crush (1973 a) found
that all his grasses selected the same fungus from a
mixed inoculum and Johnson (1973) studying a forest
community found that inoculum from any part of it had
similar effects, no matter what species dominated the

site of collection (Table 2)。 Any benefits from
developing a special relationship with a particular

Table 2. Mean dry weight (mg) of inoculated
seedlings after 4 months in steamed soil,
Truog P 7ppm. (Johnson, 1973).

| Seedling | Source of inoculum (soil from) | | |
	Melicytus	Griselinia	Pittosporum
Melicytus	735	856	810
Griselinia	535	587	398
Pittosporum	1049	1392	1165

No significant differences within a species

---

host must have been continually offset by the advantage in dispersal of using every root as a staging
post through the soil.

But there is good evidence of soil specificity
from crop plants (Mosse, 1973) and from *Coprosma
robusta* (Table 3). Since the Endogonaceae are not

Table 3. Mean dry weight (mg) of *Coprosma
robusta* seedlings after 4 months growth in
two steamed soils. (Johnson, 1973).

| Growth medium | Inoculum | |
	Soil A	Soil B
Soil A	2513	228
Soil B	143	277

Soil A, Truog P 11ppm; endophyte *Glomus* sp.
Soil B, Truog P 5ppm; endophyte *Gigaspora* sp.

All differences are significant.

readily dispersed and soils can be greatly changed on
conversion to agriculture, much farm land may now
lack the endophytes most appropriate to it.   Inoculated seedlings will thus grow better than those
which derive what symbionts they can from the soil
(Mosse & Hayman, 1971;   Khan, 1972).

Higher fungi, especially Basidiomycetes, have
replaced the Phycomycetes as symbionts in some families of woody Angiosperms centred in the North Temperate Zone, in orchids, and in heaths and epacrids.
The copious airborne spores of the Basidiomycetes
remove the barrier of immobility which has obstructed
the evolution of host specificity in the Endogonaceae, and the mycorrhizal Basidiomycetes are relatively, indeed sometimes absolutely, host specific
(Trappe, 1962).   A host that develops pure stands
might be greatly assisted towards dominance by an
exclusive fungus partner in the soil: this, rather
than any fundamental physiological advantage, may be
the basis of preference for Basidiomycetous symbionts.   There are indeed indications that there is
no physiological advantage in that ferns, though
capable of forming ectomycorrhizas, do so only in pure
stands of beech or pine where Endogonaceous inoculum
is presumably lacking.   Moreover seedlings of
Leptospermum scoparium, often ectomycorrhizal with
Boletaceae in pure stands (McNabb, 1968), established
better in pot experiments with an inoculum of endomycorrhizal fungi (Cooper, 1973).

EVOLUTIONARY TRENDS IN HOSTS

The ferns and gymnosperms have root hairs, but
in gymnosperms they are usually poorly developed, or
displaced (Hatch, 1937) by an ectomycorrhizal mantle.
The hairless state of the magnolioid root could well
have been inherited from a gymnospermous ancestor.
It would have evolved along with the transfer of
functions from root hairs to symbiotic mycelium.
This seems most likely in an environment in which

uptake of phosphorus was the most exacting of these
functions. For this, mycelium might prove superior
to hairs because of the more economical generation of
interface with the soil, because of the possible
absence of any threshold value for uptake of soil
phosphate by hyphae (Mosse et al., 1973) and because
of better access to less soluble forms of phosphate
(Hall, 1975; Mosse, 1973).

The Magnoliaceae retain the typical magnolioid
root even when they have developed the intermittent
growth associated with the deciduous habit.   There
are tropical and subtropical trees that have advanced (Young & Watson, 1970) sympetalous flowers yet
retain this ancestral root pattern, e.g. Elingamita
(Myrsinaceae).   But in most advanced families root
evolution has been plastic enough to provide a range
of types.   Thus the Rubiaceae includes Coprosma
robusta with rare, short root hairs and Coffea arab-
ica in which they are highly developed (Kramer, 1969).
Even in the Gramineae Setaria viridis has limited
branching and short hairs (Dittmer, 1949).

The circumstances that favour development of
root hairs rather than symbiotic mycelium are obviously to be sought in grasslands and sedge and rush
habitats.   The varying degrees of mycotrophy among
the species of mixed New Zealand forests suggest that
in mesic environments there is no over-riding advantage either in hairs or in mycelium for generation of
the root/soil interface.   Absolute immunity to endomycorrhizal infection seems to be rare - even the
Cyperaceae and Juncaceae are not immune (Powell,
1975).

Crop plants have been selected in soils fertilized with soluble phosphate which depresses mycorrhizal infection.   As the tomato illustrates
(Table 1) crops that are not showing actual phosphorus deficiency symptoms have about double the phosphorus concentration of comparable wild species
(Bingham, 1966).   The world's supply of phosphate is

perhaps its most limited resource (Kucera, 1973). It seems time to begin reselecting varieties under conditions that exploit the mycorrhizal endophytes fully.   If they do sometimes give access to sources of phosphorus that root hairs do not exploit, this seems a better approach than seeking hormones that will give tomatoes the root hairs of a sedge.

REFERENCES

BAYLIS, G. T. S. (1959). Effect of vesicular arbuscular mycorrhizas on growth of *Griselinia littoralis* (Cornaceae).   *New Phytol.*, 58, 274-280.

BAYLIS, G. T. S. (1967). Experiments on the ecological significance of phycomycetous mycorrhizas. *New Phytol.*, 66, 231-243.

BAYLIS, G. T. S. (1972). Fungi, phosphorus, and the evolution of root systems.   *Search*, 3, 257-258.

BAYLIS, G. T. S. (1974). The evolutionary significance of phycomycetous mycorrhizas.   In "Mechanisms of regulation of plant growth", eds.   11, Bieleski, R. L., Cresswell, M., Ferguson, A. R. *Roy. Soc. N.Z. Bull.* 12, 191-193.

BEADLE, N. C. W. (1968). Some aspects of the ecology and physiology of Australian xeromorphic plants. *Aust. J. Sci.*, 30, 348-355.

BINGHAM, F. T. (1966). Phosphorus.   *Diagnostic criteria for plants and soils* (ed. Chapman, H. D. University of California, Riverside, p324-361.

CLARK, F. B. (1963). Endotrophic mycorrhizae influence yellow poplar seedling growth.   *Science, N.Y.*, 140, 1220-1221.

COOPER, K. M. (1973). Mycorrhizal associations of New Zealand ferns. Thesis, Otago University.

CRONQUIST, A. (1968). *The evolution and classification of flowering plants.* Nelson, London. 396pp.

CRUSH, J. R. (1973a). Significance of endomycorrhizas in tussock grassland in Otago, New Zealand. *N.Z. J. Bot.,* 11, 645-660.

CRUSH, J. R. (1973b). The effect of *Rhizophagus tenuis* mycorrhizas on ryegrass, cocksfoot and sweet vernal. *New Phytol.,* 72, 965-973.

DITTMER, H. J. (1949). Root hair variations in plant species. *Am. J. Bot.,* 36, 152-155.

FURLAN, V. & FORTIN, J. A. (1973). Formation of endomycorrhizae by *Endogone calospora* on *Allium cepa* under three temperature régimes. *Naturaliste can.,* 100, 467-477.

HALL, I. R. (1975). Endomycorrhizas of *Metrosideros umbellata* and *Weinmannia racemosa.* N.Z.J. Bot., (in press).

HATCH, A. B. (1937). The physical basis of mycotrophy in the genus *Pinus. Black Rock For. Bull.,* 6, 1-168.

HUTCHINSON, J. (1973). *The families of flowering plants.* Clarendon, Oxford. 968pp.

JOHNSON, P. N. (1973). Mycorrhizas of coniferous-broadleaved forest. Thesis, Otago University.

KHAN, A. G. (1972). The effect of vesicular-arbuscular mycorrhizal associations on growth of cereals. I. Effects on maize growth. *New Phytol.,* 71, 613-619.

KRAMER, P. J. (1969). *Plant and soil-water relationships: a modern synthesis.* McGraw-Hill, New York. 482pp.

KUCERA, C. L. (1973). *The challenge of ecology,* Mosby, Saint Louis, 226pp.

McNABB, R. F. R. (1968). The Boletaceae of New Zealand. *N.Z. J. Bot.,* 6, 469-483.

METSÄVAINIO, K. (1931). Untersuchungen über das Wurzelsystem der Moorpflanzen. *Ann. Soc. Zool.-Bot. Fenn. Vanamo,* 1, 1-417.

MOSSE, B. (1972). Effects of different *Endogone* strains on the growth of *Paspalum notatum. Nature, London,* 239, 221-223.

MOSSE, B. (1973). Advances in the study of vesicular-arbuscular mycorrhizas. *A. Rev. Pl. Path.,* 11, 171-196.

MOSSE, B. & HAYMAN, D. S. (1971). Plant growth responses to vesicular-arbuscular mycorrhiza. II. In unsterilized field soils. *New Phytol.,* 70, 29-34.

MOSSE, B., HAYMAN, D. S. & ARNOLD, D. J. (1973). Plant growth responses to vesicular-arbuscular mycorrhiza. V. Phosphate uptake by three plant species from P-deficient soils labelled with ^{32}P. *New Phytol.,* 72, 809-815.

POWELL, C. L. (1975). Rushes and sedges are non-mycotrophic. Plant and soil (in press).

SPORNE, K. R. (1973). The survival of archaic dicotyledons in tropical rain forests. *New Phytol.,* 72, 1175-1184.

TRAPPE, J. M. (1962). Fungus associates of ectotrophic mycorrhizae.   *Bot. Rev.,* 28, 538-606.

WALKER, T. W. (1965). The significance of phosphorus
in pedogenesis.   In *Experimental Pedology*
eds. Hallsworth, E. G. & Crawford, D. V.,
Butterworths, London. p295-316.

YOUNG, D. J. & WATSON, L. (1970). The classification
of dicotyledons: a study of the upper levels of
the hierarchy.   *Aust. J. Bot.,* 18, 387-433.

GROWTH RESPONSES TO THE FORMATION OF ENDOTROPHIC MYCORRHIZAS IN SOLANUM, LEPTOSPERMUM, AND NEW ZEALAND FERNS

KAREN M. COOPER

Botany Department, University of Otago, Dunedin, New Zealand*

INTRODUCTION

Phycomycetous mycorrhizas are common in northern hemisphere ferns (Boullard, 1957; Hepden, 1960). An extensive survey of the leptosporangiate ferns of New Zealand has established that, with few exceptions, they are constantly mycorrhizal in the field (Cooper, 1973).

The uptake by roots of plant nutrients, from soil and applied fertilisers, is dependent on the morphology of the root system (Barley, 1970). The importance of root hairs in the uptake of P has been stressed by many workers (Nye, 1966; Drew & Nye, 1970; Lewis & Quirk, 1967a,b,c; Newman & Andrews, 1973; Farr & Vaidyanathan, 1972; Bhat & Nye, 1973). It has been suggested (Baylis, 1970; 1972) that genera deficient in root hairs have greater dependence on mycorrhizas or added P for growth in P-deficient soils than plants with finely branched root systems and copious root hairs. Ferns, in general, have long persistent root hairs and well developed root systems (Baylis, 1970; Cooper, 1973).

---

* Present address: Department of Forestry, University of Oxford.

However, sporophytes of *Pteridium aquilinum, Histio-pteris incisa* and *Dryopteris filix-mas* are myco-trophic[1] in soils sufficiently low in available P (Cooper, 1973).

     Solanum nigrum, S. laciniatum and *Lolium perenne* are fine hairy rooted plants, while the roots of *Leptospermum scoparium* are coarser and less constant-ly hairy (Baylis, 1970; Cooper, 1973). In a soil containing 8 ppm available P[2], Baylis (1971) observed a growth response to inoculation with an endogonac-eous endophyte in *Leptospermum* but not in either *Solanum nigrum* or *Lolium*. However, Crush (1973) established that *Lolium*, when inoculated with *Rhizophagus tenuis* in a soil very low in available P (4 µg ml^{-1}), became strongly mycotrophic. All four species have subsequently been shown to be dependent on mycorrhizas for growth in soils containing <8 µg ml^{-1} available P (Cooper, 1973).

     The intention of this paper is to define the conditions, particularly those of P availability, under which selected ferns, and *Solanum, Leptospermum* and *Lolium* species may become mycotrophic in pot culture.

MATERIALS AND METHODS

     The mycorrhizal response was tested using two rhizomatous ferns *(Pteridium aquilinum* and *Histio-pteris incisa), Leptospermum scoparium* (Myrtaceae), *Lolium perenne* cv. 'Grasslands Ruanui' (ryegrass), *Solanum nigrum* (black nightshade), and the shrubby

---

[1] The term mycotrophic is used of a plant which is benefited by mycorrhizal infection

[2] Unless otherwise stated, soil P levels given are by the Truog test and expressed as µg P ml^{-1} soil.

New Zealand poroporo, *Solanum laciniatum*.

Fern sporophytes were raised from spores sown on trays of steamed soil (8 μg ml^{-1} P), and seed of *Solanum, Leptospermum* and *Lolium* was germinated in pots of steamed soil (4 μg ml^{-1} P).

Mycorrhizal and non-mycorrhizal plants were compared in four soils (Table 1) sterilized by steaming between electrodes. There were two forms of inoculum: a) 50 mg wet weight of chopped, heavily infected roots from either ferns or dicotyledons were placed in the planting holes and 20 ml of a sieved soil suspension (< 250 μm) added. Controls received corresponding quantities of autoclaved choppings and soil suspension passed through a 53 μm sieve. b) As a) but the chopped roots were omitted.

Nutrient treatments were based on Hewitt's (1952) solution. At intervals throughout the experiment, phosphorus (P) or complete minus phosphorus (C-P) nutrient was added at full strength or one third dilution in doses of 20 ml per 200 ml soil. Pots were arranged in randomised blocks and sampled 2 - 4 times.

Table 1. Chemical analyses of experimental soils. Soil set nomenclature follows the N.Z. Soil Bureau (1968).

Soil Set	Horizon	Community	Altitude m	pH	Truog P μg ml^{-1}	Olsen P μg ml^{-1}
Leith 1	A	Forest	289	5.6	12	32
Leith 2	B	Forest	289	4.7	10	19
Leith 3	B	Forest	289	4.5	8	14
Te Anau	A	Bracken	210	5.2	4	6

Root systems were cleared and stained using the method of Phillips and Hayman (1970). Infection was recorded over the entire root system (Cooper, 1973).

The P content of the leaf or frond dry matter was estimated using the molybdate blue method (Jackson, 1962) preceded by wet digestion with perchloric acid (Johnson & Ulrich, 1959).

RESULTS

Pteridium, Histiopteris, and *Leptospermum* (Table 2).

The patterns of growth response to P and C-P fertilisers were established early in growth and changed little between samplings 1 & 2. Infection was highest where the natural P level was lowest and always decreased with the addition of P to the soil. In general, infection increased between harvests.

Increased growth accompanied mycorrhizal infection if available soil P was sufficiently low. In *Pteridium,* this condition occurred in the poorest soil. With the addition of C-P it also occurred in the intermediate soil following a build-up of infection between harvests. In the most fertile soil, *Pteridium* grew well without mycorrhizas and infection was scant and ineffective.

Histiopteris was strongly mycotrophic in soils with <8 µg/ml P. However, where a treatment supported good non-mycotrophic growth (in the P treatment in the intermediate soil and in all treatments in the most fertile soil), a persistent growth depression was associated with mycorrhizal infection.

By the second harvest of *Leptospermum,* infection had benefited all plants in the poorest soil and in the intermediate soil when P had not been added. In the soil with the highest natural P, the initial growth depression associated with mycorrhizas did not persist and by the second sampling mycorrhizal and non-mycorrhizal plants grew similarly.

Table 2. Growth and infection of mycorrhizal (M) and non-mycorrhizal (NM) plants of Pteridium (10 replicates), Histiopteris (10 replicates) and Leptospermum (6 replicates) in three steamed soils. Inoculum as in method (a). 20 ml of one-third strength P or C-P nutrient were applied 2, 4, 8 and 12 weeks after transplanting. Plants were harvested in two equal random samplings 12 - 15 weeks (sample 1) and 21 - 24 weeks (sample 2) after transplanting.
* M and NM plants within each treatment differ significantly (P <0.05).

Species	Soil Nutrient P µg ml^{-1}	SAMPLE 1 Shoot dry weight (g) NM	M	% mycorrhizal infection	SAMPLE 2 Shoot dry weight (g) NM	M	% mycorrhizal infection
Pteridium	12 P	0.76	0.70	2	1.73	1.70	3
	NIL	0.64	0.60	1	1.63	1.60	4
	C-P	0.59	0.56	3	1.68	1.65	3
	8 P	0.40	0.45	3	1.16	1.10	4
	NIL	0.11	0.08	29	0.41	0.44	49
	C-P	0.12	0.10	35	0.36	0.54*	56
	4 P	0.08	0.10	4	0.26	0.30	32
	NIL	0.01	0.04*	27	0.08	0.18*	48
	C-P	0.01	0.04*	35	0.06	0.27*	56
Histiopteris	12 P	0.48	0.25*	13	1.26	0.96*	16
	NIL	0.45	0.23*	36	1.21	0.86*	30
	C-P	0.41	0.22*	35	1.26	1.03*	33

Table 2. (Contd.)

Species	Soil Nutrient P μg ml⁻¹		SAMPLE 1			SAMPLE 2		
			Shoot dry weight (g)		% mycorrhizal infection	Shoot dry weight (g)		% mycorrhizal infection
	Soil P	Nutrient	NM	M		NM	M	
Histiopteris (*contd.*)	8	P	0.32	0.22*	14	0.96	0.70*	28
		NIL	0.01	0.06*	29	0.36	0.44	60
		C-P	0.01	0.05*	43	0.30	0.53*	66
	4	P	0.12	0.07	47	0.26	0.23	51
		NIL	0.01	0.06*	78	0.03	0.19*	78
		C-P	0.01	0.05*	85	0.05	0.25*	84
Leptospermum	12	P	0.95	0.72*	14	1.68	1.76	18
		NIL	0.60	0.48*	22	0.80	0.88	40
		C-P	0.42	0.38*	24	0.54	0.62	48
	8	P	0.73	0.45*	18	0.70	0.73	57
		NIL	0.006	0.17*	27	0.008	0.39*	76
		C-P	0.004	0.13*	30	0.007	0.30*	78
	4	P	0.11	0.10	43	0.16	0.31*	43
		NIL	0.004	0.07*	45	0.005	0.23*	73
		C-P	0.003	0.06*	49	0.005	0.16*	69

Table 3.   Growth and infection of mycorrhizal (M)
and non-mycorrhizal (NM) seedlings of *Lolium
perenne, Solanum nigrum* and *S. laciniatum* in
four steamed soils.
Soil - 12 µg ml^{-1} P: Inoculum (b); 10 replicates
   per treatment harvested after 6, 10, 15 and
   26 weeks; full strength C-P fertiliser
   applied after 10, 15 and 20 weeks.
Soil - 10 µg ml^{-1} P: Inoculum (a); 10 replicates
   in sample 1 (10 weeks) and 20 in sample 2
   (24 weeks); one-third strength C-P fertiliser applied 2 and 12 weeks after transplanting (†).
Soil - 8 µg ml^{-1} P: Inoculum (a); 10 replicates
   in each of two harvests after 10 (sample 1)
   and 20 (sample 2) weeks.
Soil - 4 µg ml^{-1} P: Inoculum (b); 10 replicates
   of 15 replicates after 10 weeks (sample 1)
   and 24 weeks (sample 2).
* M and NM plants within each treatment differ
   significantly (P<0.05).

Species	Soil P µg ml^{-1}	Sample time weeks	Shoot dry weight (g) NM	M	% mycor- rhizal infection
Lolium	12	6	0.36	0.39	0
		10	1.08	1.04	1
		15	2.38	2.36	3
		26	3.48	3.52	4
	10	10	0.62	0.34*	17
			†0.53	0.32*	22
		20	1.21	0.93*	72
			†0.90	1.18*	74
	8	10	0.11	0.28*	50
		20	0.38	0.69*	91
	4	10	0.003	0.06*	31
		24	0.004	0.14*	80

Table 3. (Contd.)

Species	Soil P µg/ml	Sample time weeks	Shoot dry weight (g)		% mycorrhizal infection
			NM	M	
Solanum nigrum	12	6	0.11	0.03*	0
		10	0.41	0.33*	9
		15	1.28	1.09	41
		26	2.92	3.67*	70
	10	10	0.08	0.18*	23
			†0.06	0.16*	20
		20	0.17	0.43*	59
			†0.10	0.62*	64
	8	10	0.003	0.07*	33
		20	0.11	0.37*	70
	4	10	0.002	0.03*	24
		24	0.003	0.12*	96
S.laciniatum	12	6	0.24	0.11*	0
		10	0.84	0.79	9
		15	0.98	2.44*	73
	10	10	0.17	0.23*	24
			†0.08	0.20*	30
		20	0.45	0.63*	73
			†0.29	0.81*	75
	8	10	0.08	0.17*	49
		20	0.11	0.53*	87
	4	10	0.003	0.04*	25
		24	0.004	0.15*	90

Solanum laciniatum, *S. nigrum* and *Lolium* (Table 3)

The pattern of growth and the spread of infection were similar to those in the ferns. Growth of all species was improved by mycorrhizas at soil P levels $\leqslant 8\mu g$ ml^{-1}. In the soil containing 10 µg ml^{-1} P, both species of *Solanum* were mycotrophic throughout,

the C-P fertiliser intensifying this effect by the
second harvest.   In *Lolium,* mycorrhizas depressed
growth but the addition of C-P permitted a mycotrophic response in this treatment by the second
harvest.

At 12 µg ml^{-1} soil P, all species grew vigorously.   Initially, growth was depressed by the mycorrhizal inoculum.   There was little or no infection at
this stage but mycelium was plentiful in the rhizosphere and there were some penetration points. By the
second sampling, the mycorrhizal and non-mycorrhizal
treatments were less dissimilar, and by 15 weeks,
Solanum laciniatum was strongly mycotrophic.
S. nigrum reached a similar endpoint but took longer
to do so.   The growth of *Lolium* was never affected
by the inoculum and indeed, root infection, as in
Pteridium, failed to reach more than 4%.

DISCUSSION

The four soils used in these experiments differed in available P content, but as they are different
soils, the effect of other soil properties should not
be ignored.   Nevertheless, there appears to be some
critical soil available P level for each species,
above which it grows well without mycorrhizas and
below which it depends increasingly on the symbiosis
(Table 4).   This critical P level for mycotrophy
appears sharply defined and it is probable that roots
and mycorrhizas respond to the amounts of P actually
reaching them.   These amounts are determined by
soil P content and diffusion rate in the particular
soil.   Soil chemical analysis does not necessarily
take into account the difference in diffusion rates.
The dependence of a plant on mycorrhizas, particularly in soils near the threshold value for mycotrophy,
can be increased by adding nutrients other than P,
(ie C-P).   This increased mycorrhizal response to
C-P may be caused in part by the enhanced growth of

Table 4. Critical levels of soil available P for the development of mycotrophy.
M mycotrophic   NM non-mycotrophic
* Growth depression
( ) Infection <10% after 22 - 26 weeks
[ ] Became mycotrophic when C-P added
(Data from present paper and Cooper (1973)).

	Soil P (μg ml^{-1})			
	12	10	8	4
Pteridium	(NM)	NM	[NM]	M
Lolium	(NM)	[*]	M	M
Histiopteris	*	*	M	M
Solanum nigrum	[*]	M	M	M
S. laciniatum	[*]	M	M	M
Leptospermum	[*]	[*]	M	M

the plant producing a greater demand for P. However, the addition of C-P apparently lowers the soil available P (Table 5) providing therefore, a greater P stress.

Where available P was not sufficiently limiting for a species to be mycotrophic, the mycorrhizal inoculum often depressed growth (Table 4). This effect was usually associated with a wide range of infection levels (mean 0 - 50%). However, individual plants with 70 - 90% infection were often relatively small (Cooper, 1973). When there was little or no infection, endogonaceous mycelium was often plentiful in the rhizosphere and there were some penetration points. In general, the depression phase was transitory, preceding heavier infection and an acceleration of plant growth. However, there is

Table 5. Final chemical analyses of experimental steamed soils given 20 ml full
strength (*) or one-third dilution C-P
fertiliser 2, 4, 8 and 12 weeks after the
start of the experiment and incubated under
glasshouse conditions for 4 months during
spring and summer.

Soil Set	Nutrient	pH	Truog P $\mu g\ ml^{-1}$	Olsen P $\mu g\ ml^{-1}$
Leith 1	NIL	5.5	12	32
	C-P	5.4	12	30
	C-P*	5.5	9	24
Leith 2	NIL	4.7	10	19
	C-P	4.7	8	16
Leith 3	NIL	4.5	8	14
	C-P	4.4	8	12
Te Anau	NIL	5.2	4	6
	C-P	5.0	2	5

evidence that in *Histiopteris,* at least, the growth
depression may be long-lasting.

Previous records of consistent growth depression in inoculated plants are scarce.   Baylis (1967)
noted a depression of growth by mycorrhizas in
Coprosma when high levels of P were artificially
maintained.   He interpreted the transient growth
depression in *Leptospermum* as a pathogenic phase in
the establishment of the endophyte (Baylis, 1971).
The depression of growth in inoculated onions at low
temperatures has been attributed by Furlan and
Fortin (1973) to the parasitic effect of the inoculum.   Rhizosphere organisms are well known to

compete with plants for nutrients, especially P
(Barber & Loughman, 1967; Barber, 1968; Bowen &
Rovira, 1966).   Soil organisms added with a crushed
spore mycorrhizal inoculum have been shown to temporarily depress the growth of *Solanum* and *Leptospermum* (Cooper, 1973).   That the endomycorrhizal fungus,
Rhizophagus tenuis can also act as a typical rhizosphere organism and compete with the root for P has
been suggested by Crush (1973).   Some support for
this theory from the present data comes from the
shoot P content of plants analysed according to plant
size and infection level (Cooper, 1973).   Where
growth depression occurred in *Histiopteris,* heavily
infected (>60%) small plants had substantially lower
percentage P by grouped analysis (0.084%) than their
poorly infected (<10%) counterparts (0.173%).   These
shared similar values with the non-mycorrhizal plants.
It is doubtful whether P toxicity as suggested by
Mosse (1973) would occur at these relatively low
levels of internal P.

Sanders and Tinker (pers. comm.) have found that
the quantity of external mycelium extracted from the
soil is small and it is doubtful whether this external mycelium could take up sufficient soil P to compete with the root in any way.   The paucity of
internal mycelium and penetration points sometimes
associated with growth depression, makes it difficult
to hypothesise much carbohydrate drain from the host
to account for the stunted plants.   That the
depression was caused by the presence of other pathogenic organisms introduced with the inoculum of
chopped roots seems unlikely as a similar effect was
obtained in *Solanum* and *Lolium* in the absence of
chopped roots.   Roots of the inoculated plants were
healthy when harvested and there was no sign of
attack by pathogens.   The effect of growth depression has only been observed in steamed soils and could
therefore possibly be an effect of the partially
sterilised medium.

In summary, it seems that the level of P

available to each species is crucial in determining
the pattern that mycorrhizal infection will take
(Fig. 1).   *Solanum laciniatum* and *Lolium perenne*

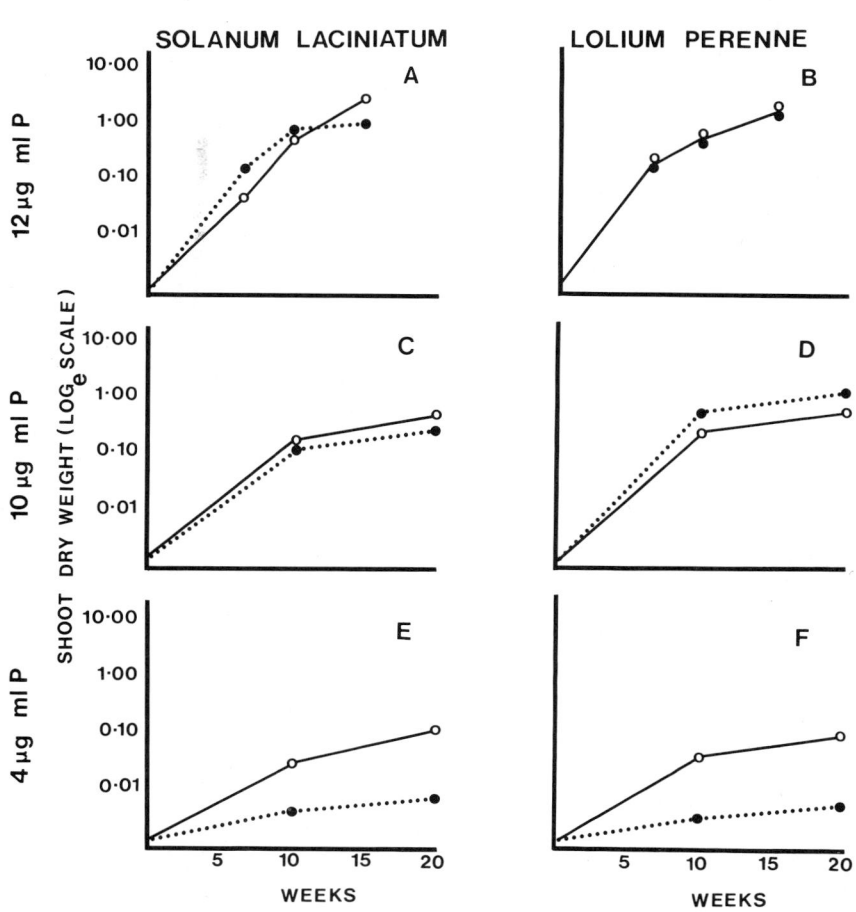

Figure 1.   Relative growth rates of *Solanum
laciniatum* and *Lolium perenne* in soils of
different fertilities (from Tables 2 and 3)
o————o   Mycorrhizal   ●-----● Non-mycorrhizal

illustrate this point but the other species studied
fit various combinations of these patterns.   When
the P available is well above the threshold value for
mycotrophy, plants can grow similarly with or without
mycorrhizas (B) and indeed, colonisation of the root
by the endophyte may be negligible.   At the other
extreme, when the available P levels severely retard
non-mycorrhizal growth, the symbiosis exerts apparently immediate beneficial effects (C,E,F).   At the
intermediate P levels, there appears to be two
phases in the establishment of the mycorrhizal
symbiosis.   Phase 1 is characterised by a slow build
up in infection levels accompanied by growth depression (A,D).   This is followed at some critical time
by phase 2 when the symbiosis becomes beneficial to
the host (A).   It is probable that the increased
demand of the host for P at this stage is caused by
a drop in soil P during the experiment.   The change
from growth depression to growth stimulation by the
mycorrhizas is a response to the lower soil P levels.

ACKNOWLEDGMENTS

        I thank Professor G.T.S. Baylis for assistance
and advice;   Dr. J. Wilson for assistance with the
statistical analyses;   Dr. J. R. Crush for helpful
discussion;   Mr. J. R. Grigg and staff at Invermay
Research Station for soil analyses.

REFERENCES

BARBER, D. A. (1968). Micro-organisms and the
    inorganic nutrition of higher plants. *A. Rev. Pl.
    Physiol.*, 19, 71-88.

BARBER, D. A. & LOUGHMAN, B. C. (1967). The effects
    of micro-organisms on the absorption of inorganic
    nutrients by intact plants.   II. Uptake and

utilisation of phosphate by barley plants grown under sterile and non-sterile conditions. *J. exp. Bot.*, 18, 170-176.

BARLEY, K. P. (1970). The configuration of the root system in relation to nutrient uptake. *Adv. Agron.*, 22, 159-201.

BAYLIS, G. T. S. (1967). Experiments on the ecological significance of phycomycetous mycorrhizas. *New Phytol.*, 66, 231-243.

BAYLIS, G. T. S. (1970). Root hairs and phycomycetous mycorrhizas in phosphorus-deficient soil. *Pl. Soil*, 33, 713-716.

BAYLIS, G. T. S. (1971). Endogonaceous mycorrhizas synthesised in *Leptospermum* (Myrtaceae). *N.Z. J. Bot.* 9, 293-296.

BAYLIS, G. T. S. (1972). Minimum levels of available phosphorus for non-mycorrhizal plants. *Pl. Soil*, 36, 233-234.

BHAT, K. K. S. & NYE, P. H. (1973). Diffusion of phosphate to plant roots in soil. I. Quantitative autoradiography of the depletion zone. *Pl. Soil*, 38, 161-175.

BOULLARD, B. (1957). La mycotrophie chez les ptéridophytes. Sa fréquence, ses charactères, sa signification. *Botaniste*, 45, 1-185.

BOWEN, G. D. & ROVIRA, A. D. (1966). Microbial factors in short term phosphate uptake studies with plant roots. *Nature London*, 211, 665-668.

COOPER, K. M. (1973). Mycorrhizal associations of New Zealand ferns. Ph.D. Thesis, University of Otago.

CRUSH, J. R. (1973). The effect of *Rhizophagus tenuis* mycorrhizas on ryegrass, cocksfoot and sweet vernal. *New Phytol.*, 72, 965-973.

DREW, M. C. & NYE, P. H. (1970). The supply of nutrient ions by diffusion to plant roots in soil. III. Uptake of phosphate by roots of onion, leek and ryegrass. *Pl. Scil,* 33, 545-563.

FARR, E. & VAIDYANATHAN, L. V. (1972). The supply of nutrient ions by diffusion to plant roots in soil. IV. Direct measurement of changes in labile phosphate content in soil near absorbing roots. *Pl. Soil,* 36, 609-616.

FURLAN, V. & FORTIN, J. A. (1973). Formation of endomycorrhizas by *Endogone calospora* on *Allium cepa* under three temperature regimes. *Naturaliste can.*, 100, 467-477.

HEPDEN, P. M. (1960). Studies in vesicular - arbuscular endophytes. II. Endophytes in the Pteridophyta with special reference to leptosporangiate ferns. *Trans. Br. mycol. Soc.,* 43, 559-570.

HEWITT, E. J. (1952). Sand and water culture methods used in the study of plant nutrition. *Commonw. Bur. Hort. Plantation Crops Tech. Comm.*, 22, p.189.

JACKSON, M. L. (1962). Soil Chemical Analysis. Constable, London. 498 pp.

JOHNSON, C. M. & ULRICH, A. (1959). Analytical methods for use in plant analysis. *Californian Agric. Exp. Stat. Bull.,* 766, 78 pp.

LEWIS, D. G. & QUIRK, J. P. (1967a). Phosphate diffusion in soil and uptake by plants. I. Self diffusion of phosphate in soils. *Pl. Soil,* 26, 99-118.

LEWIS, D. G. & QUIRK, J. P. (1967b). Phosphate diffusion in soil and uptake by plants. III. P^{31} movement and uptake by plants as indicated by P^{32} autoradiography. *Pl. Soil*, 26, 445-453.

LEWIS, D. G. & QUIRK, J. P. (1967c). Phosphate diffusion in soil and uptake by plants. IV. Computed uptake by model roots as a result of diffusive flow. *Pl. Soil*, 26, 454-468.

MOSSE, B. (1973). Plant growth responses to vesicular-arbuscular mycorrhiza. IV. In soil given additional phosphate. *New Phytol*, 72, 127- 136.

NEWMAN, E. I. & ANDREWS, R. E. (1973). Uptake of phosphorus and potassium in relation to root growth and root density. *Pl. Soil*, 38, 49-69.

NEW ZEALAND SOIL BUREAU. (1968). General survey of the soils of the South Island, New Zealand. *N. Z. Soil Bur. Bull.*, 27, 404 pp.

NYE, P. H. (1966). The effect of nutrient intensity and buffering power of a soil, and the absorbing power, size and root hairs of a root, on nutrient absorption by diffusion. *Pl. Soil*, 25, 81-105.

PHILLIPS, J. M. & HAYMAN, D. S. (1970). Improved procedures for clearing roots and staining parasitic and vesicular-arbuscular mycorrhizal fungi for rapid assessment of infection. *Trans. Br. mycol. Soc.*, 55, 158-161.

POSSIBLE SYNERGISTIC INTERACTIONS BETWEEN ENDOGONE AND PHOSPHATE-SOLUBILIZING BACTERIA IN LOW-PHOSPHATE SOILS

J. M. BAREA[1], R. AZCÓN[1]
AND D. S. HAYMAN[2]

[1] *Estacion Experimental del Zaidin, Granada, Spain*

[2] *Rothamsted Experimental Station, Harpenden, Herts., U.K.*

INTRODUCTION

There are many recent reports that plants inoculated with vesicular-arbuscular (VA) mycorrhizal fungi often take up more phosphate and grow better than non-mycorrhizal plants (see Mosse, 1973). In addition, bacteria able to dissolve insoluble phosphates have been used as seed inoculants with a view to improving plant growth (see Barea, 1969; Brown, 1974). Unlike the mycorrhizal fungi, which can become well established inside the plant root, the phosphate-solubilizing bacteria must establish themselves in the rhizosphere in competition with many other microorganisms. Bacteria introduced into the rhizosphere as seed inoculum cannot generally maintain a high population and their numbers drop to sub-optimal levels.

The possibility of synergistic interactions between VA mycorrhizal fungi (*Endogone* spp.) and phosphate-solubilizing bacteria seemed a logical sequence to these earlier studies. Two aspects

merited most attention: (i) the influence of
Endogone on the establishment of phosphate-solubilizing bacteria in the root zone and, conversely, of
the bacteria on root infection by *Endogone;* and (ii)
the combined activities of these fungi and bacteria
in terms of effects on plant growth.   The results of
such a study, in soils deficient in plant-available
phosphate, are presented here.

MATERIALS AND METHODS

Two crop species, maize (*Zea mays* L.) and
lavender (*Lavandula spica* var. vera L.), were grown
in each of two phosphate-deficient soils in pots in
the glasshouse.   The soils were collected from
Granada Province, Spain.   Soil number 3 was a "Red
Mediterranean" type, pH 7.5, containing 4.0 mg P per
kg soil soluble in 0.5M $NaHCO_3$ (Olsen *et al*, 1954)
and 0.68 µmoles per litre soluble in 0.01M $CaCl_2$
(Aslyng, 1954).   Soil number 7 was a "Grey Merid-
ional" type, pH 7.6, containing 8.2 mg P per kg soil
soluble in 0.5M $NaHCO_3$, and 0.26 µmoles per litre
soluble in 0.01M $CaCl_2$.   Both soils were sterilized
by steaming twice before use.
There were four inoculation treatments:
C = control (autoclaved bacteria); B = + bacteria;
E = + *Endogone;* E + B = + both bacteria and
Endogone.   Two bacterial species, *Pseudomonas* sp.
and *Agrobacterium* sp., were selected as efficient
solubilizers of rock phosphate in pure culture.
Inoculum was prepared by growing the bacteria in the
medium of Brown (1972) in shake culture for ten days
at 28°C.   At the time of inoculation the two species
were combined.   The maize was inoculated by dipping
the seeds in the bacterial suspension before sowing,
and the lavender by dipping the roots of four-week-
old seedlings in the suspension before transplanting.
The *Endogone* inoculum consisted of a mixture of
spores, hyphae and infected root fragments obtained

by wet-sieving rhizosphere soil from a plant infected with the E_3 spore type (Mosse, 1972).   A small pad of this inoculum was placed in each planting hole.

During the experiment samples of rhizosphere soil were collected at 15-day intervals and the numbers of phosphate-solubilizing bacteria counted on a medium containing 0.02% rock phosphate.   Solubilization was detected by halo formation around colonies.   Bacterial numbers were related to 1 g of dry rhizosphere soil.

The plants were grown for twelve weeks, during which time they were fed with nutrient solution (Hewitt, 1958) lacking phosphate.   At harvest a small portion of the root system of each plant was cleared and stained (Phillips and Hayman, 1970) and the proportion of root tissue infected with *Endogone* was assessed microscopically.   Dry weights of roots and shoots were recorded, and tissues were analysed for   phosphorus by the molybdenum blue method after dissolving in H_2SO_4 and decolourizing with H_2O_2 (Lachica *et al.*, 1965).

RESULTS

Table 1 shows that the population of phosphate-solubilizing bacteria, introduced in the inoculum, declined in the rhizosphere during the twelve weeks of plant growth.   This is usual.   However, during the first 30 days in soil 3 and 45 days in soil 7, the numbers of bacteria remained significantly higher in the rhizospheres of mycorrhizal than of non-mycorrhizal plants.   Thus inoculation with *Endogone* favoured the early maintenance of phosphate-solubilizing bacteria introduced around plant roots.

The phosphate-solubilizing bacteria did not appear to affect greatly the development of VA infection which was 40-60% of the root system in both the E and E + B treatments.   Infection was not very dense and no qualitative differences in infection were

Table 1. Numbers of phosphate-solubilizing bacteria in the rhizospheres of maize and lavender.

Plant	Soil number	Inoculation treatment	Numbers of phosphate-solubilizing bacteria (in millions) per g dry rhizosphere soil				
			Days after inoculation				
			15	30	45	60	75
Lavender	3	B	48.5	10	2.5	1.5	2
		E+B	70*	25**	2	1.5	2.2
	7	B	50	5	4	1.5	2
		E+B	90*	25**	8**	1	1
Maize	3	B	85	18	6	5	4
		E+B	87	30*	5	4.5	3
	7	B	140	57	25	6	6
		E+B	120	95*	45*	15**	7

B = bacteria, E = *Endogone*   * and ** indicate that, when comparing E + B with B in a particular soil-plant combination, mean values were significantly different at 5% and 1% respectively.

observed between *Endogone*-inoculated plants with and
without bacteria.   Nevertheless, maize in soil 7
inoculated with *Endogone* plus bacteria was slightly
more mycorrhizal than plants in the other treatments.
        Table 2 shows that plants with either mycorrhiza
or bacteria usually took up more P and grew better
than the control plants with neither, though some of
these benefits from inoculation were too small to be
statistically significant.   Plants inoculated with
both *Endogone* and bacteria took up most P and grew
best in three cases out of four, but this effect only
reached statistical significance with maize in soil
7.   The amount of P taken up by inoculated maize in
soil 7 above that taken up by the uninoculated
controls was greater for the plants inoculated with
Endogone and bacteria together (5.22 mg) than the sum
of the extra amounts taken up by plants inoculated
with bacteria or *Endogone* separately (1.33 + 3.25
mg P, respectively).

DISCUSSION

        Because root growth and metabolism are affected
by infection with *Endogone,* it is likely that root
exudates and hence the rhizosphere population will
also be affected.   In the conditions of the present
experiment the mycorrhizal infection enabled the
phosphate-solubilizing bacteria introduced into the
rhizosphere to maintain high numbers for longer
periods than in the absence of mycorrhiza.   This
effect should prolong the metabolic activities of
these bacteria in the rhizosphere, thus increasing
their opportunities to influence plant growth,
which could result in a synergistic interaction
between the mycorrhiza and the bacteria.
        Another factor contributing to a possible syn-
ergistic effect would be the uptake by *Endogone*
hyphae and translocation into the plant of any soluble phosphate ions that the bacteria released from

Barea, J. M. *et al.*

Table 2.   Dry weights and P contents of maize and lavender grown with and without *Endogone* and/or phosphate-solubilizing bacteria.

Inoculation treatment		Shoots		Roots		Shoots + Roots
		Dry weight (mg)	% P	Dry weight (mg)	% P	Total P taken up (mg)
Lavender						
	C	340	0.34	162	0.30	1.63
Soil 3	B	355	0.35	148	0.34	1.74
	E	360	0.42	166	0.41	2.19
	E+B	372	0.46	170	0.40	2.39
LSD at 5%		*29*	*0.11*	*21*	*0.10*	*0.30*
	C	236	0.22	170	0.28	0.98
Soil 7	B	290	0.27	197	0.31	1.39
	E	290	0.36	180	0.49	1.92
	E+B	290	0.35	188	0.48	1.91
LSD at 5%		*28*	*0.06*	*12*	*0.11*	*0.41*
Maize						
	C	3120	0.21	1025	0.29	9.52
Soil 3	B	3240	0.23	1005	0.36	11.07
	E	3670	0.28	1340	0.40	15.63
	E+B	3690	0.29	1525	0.41	16.95
LSD at 5%		*295*	*0.03*	*254*	*0.04*	*3.15*
	C	1920	0.17	625	0.29	5.07
Soil 7	B	2050	0.20	721	0.32	6.40
	E	2220	0.20	810	0.48	8.32
	E+B	2690	0.23	875	0.47	10.29
LSD at 5%		*280*	*0.02*	*109*	*0.08*	*1.90*

C = Control, B = Bacteria, E = *Endogone*
The figures are the means for 5 replicate pots, with 1 plant per pot for maize and 3 plants per pot for lavender.

insoluble soil phosphate and did not immediately absorb themselves.   Any such fungal uptake would also
prevent these ions from accumulating sufficiently to
inhibit the phosphatase enzyme activity of the bacteria.   However, the fact that the combined inoculum
of *Endogone* plus bacteria significantly increased
plant uptake of phosphate above that achieved with
either separately only in one case, and the similarity in % P in plants in the E and E + B treatments,
suggest that the quantities of soluble phosphate
released by the bacteria were small.   The lack of
any inhibition of VA infection, which can occur when
P concentrations are high, also suggests limited solubilization.   It might well be, therefore, that the
effects of the bacteria on plant growth are more the
result of the plant growth substances that they produce (see Barea *et al.*, 1974) than of soluble phosphate ions released.   In this context it would be of
interest to determine the effect of VA infection on
the establishment in the rhizosphere of other bacteria known to produce plant growth substances in sufficient quantity to make synergistic effects likely.

The enhancement of mycorrhizal infection in
maize by the bacteria in soil 7 may have been a pH
effect because the E_3 spore type of *Endogone* is more
suited to acid than to alkaline soils.   The smaller
stimulation of plant growth by E_3 in these soils
compared to other soils (see Mosse, 1972) may also be
accounted for by too high a pH.   Indeed, the case
where infection was enhanced by the bacteria (maize
in soil 7) was the only one where the E + B treatment
increased plant growth significantly above the other
three treatments, and this may have been due in part
to acid conditions produced by the bacteria.

Other experiments (Azcón *et al.*, 1975) suggested
that both the phosphate-solubilizing bacteria and
Endogone could use P more effectively in the form of
insoluble rock phosphate added to soil than in the
insoluble forms of phosphate that occur naturally in
soil.

416 Barea, J. M. *et al.*

ACKNOWLEDGMENTS

We thank the Chemistry Department at Rothamsted for the soil P analysis and Dr. B. Mosse at Rothamsted for the E_3 inoculum. One of us (D.S.H.) is indebted to the Royal Society, London, for a travel grant which helped to make this collaborative study possible.

REFERENCES

ASLYNG, H. C. (1954). The lime and phosphate potentials of soils; the solubility and availability of phosphates. *Yb. R. vet. agric. Coll., Copenhagen,* pp. 1-50.

AZCÓN, R., BAREA, J. M. & HAYMAN, D. S. (1975). Utilization of rock phosphate added to neutral-alkaline P-deficient soils by plants inoculated with mycorrhizal fungi and phosphate-solubilizing bacteria (in preparation).

BAREA, J. M. (1969). Estudio sobre germenes del suelo capaces de mineralizar los fosfatos organicos. I. Introducción y revisión bibliografica. *Ars Pharmaceutica,* 10, 117-128.

BAREA, J. M., NAVARRO, E. & MONTOYA, E. (1974). Plant growth substances produced by phosphate-solubilizing bacteria (in preparation).

BROWN, M. E. (1972). Plant growth substances produced by microorganisms of soil and rhizosphere. *J. appl. Bact.,* 35, 443-451.

BROWN, M. E. (1974). Seed and root bacterization. *A. Rev. Phytopath.,* 12, 181-197.

HEWITT, E. J. (1958). *Sand and water culture methods*

used in the study of plant nutrition.   Commonwealth Agricultural Bureau Tech. Comm. 22.

LACHICA, M., RECALDE, L. & ESTEBAN, E. (1965).
Analisis foliar.   Metodos analiticos utilizados
en la Estación Experimental del Zaidin.   *An.
Edafol. Agrobiol.,* 24, 589-610.

MOSSE, B. (1972). The influence of soil type and
Endogone strain on the growth of mycorrhizal
plants in phosphate-deficient soils.   *Rev. Ecol.
Biol. Sol,* 9, 529-537.

MOSSE, B. (1973). Advances in the study of vesicular-
arbuscular mycorrhiza.   *A. Rev. Phytopath.,*
11, 171-196.

OLSEN, S. R., COLE, C. V., WATANABE, F. S. &
DEAN, L. A. (1954).   Estimation of available
phosphorus in soils by extraction with sodium bicarbonate.   *Circ. U.S.,* Dep. Agric. No. 939.

PHILLIPS, J. M. & HAYMAN, D. S. (1970). Improved
procedures for clearing roots and staining parasitic and vesicular-arbuscular mycorrhizal fungi for
rapid assessment of infection.   *Trans. Br. mycol.
Soc.,* 55, 158-161.

GROWTH EFFECTS OF VA MYCORRHIZA ON CROPS IN THE FIELD

A. G. KHAN

Department of Biology, University of Islamabad, Pakistan.

INTRODUCTION

Much recent work has shown that plant growth may be improved by infection with vesicular-arbuscular mycorrhizal fungi. The growth response is normally associated with an improved supply of phosphorus from the soil (Mosse, 1973). Most of the studies reported have used sterilized media in pots, so that little is known about the effects of VA infection on plant growth in nonsterile field soils. Gerdemann (1964) reported growth responses of maize plants in unsterilized soil to inoculation with *Endogone* sporocarps, although these were smaller in non-sterile than in sterile soils. Mosse *et al.*, (1969) also found responses of onions in non-sterile soil to inoculation with VA mycorrhizal fungi. Mosse & Hayman (1971) and Hayman & Mosse (1971), also demonstrated increased growth of various plants by transplanting them, after pre-inoculation with a VA endophyte, to unsterilized soil in pots. Ross (1971) obtained similar results with soybeans grown in large bins filled with sandy loam sterilized by chloropicrin and methyl bromide and infected with soil containing *Endogone* spores. Kleinschmidt & Gerdemann (1972) found that the dry weights of citrus plants in a fumigated field plot inoculated with *E. mosseae* were significantly greater than those for non-inoculated control plants.   Jackson, Franklin and Miller (1972), using lyophilized, ground VA mycorrhizal roots as inoculum, obtained a 50% increase in the corn yield over the increases due to indigenous mycorrhizal fungi.

Although experiments in pots, bins or cans are unnatural, they have established that plant yield on an unsterilized P-deficient soil can be increased by inoculation with VA mycorrhizal fungi.   These observations were extended by Khan (1972a, 1974b) who planted pre-inoculated maize and wheat in a field with no form of soil sterilisation.   The experiments were designed to study the following points, in relation to the inoculation of field crops with VA mycorrhizal fungi.
(a) the effects of pre-inoculation with VA mycorrhizal fungi on phosphorus uptake and growth of crop plants under natural field conditions.
(b) the effects of added P-fertilizer on infection, phosphate uptake and growth of mycorrhizal and non-mycorrhizal plants in the field.

MATERIALS AND METHODS

Field data

The field selected was a non-cultivated fallow, pH 7.5-8, containing an indigenous population of *Endogone* spores (550-600/kg soil) and known to be low in phosphorus.

Plants

Mycorrhizal seedlings were obtained by sowing surface-sterilized wheat (*Triticum aestivum* L. var. "Maxi-Pak") and maize (Zea mays L.) seeds in sand which was steam-sterilized and inoculated with *Endogone mosseae* spores.   Controls were supplied

with *Endogone*-free filtrate from the soil to include
only the contaminating microorganisms.   After 4
weeks the seedlings were transplanted to the selected
field treated as below.

Field soil treatment

    After weeding the field, it was ploughed and
divided into four 18m x 15m plots.   Four week old
non-mycorrhizal (control) seedlings were transplanted
into two plots and mycorrhizal (inoculated) seedlings
into the other two.   The plants were 10 cm apart in
rows 30 cm apart.
    One each of the control and inoculated plots was
supplied with triple superphosphate at 280 kg/ha.
An application of half the fertilizer was made by
broadcasting before transplanting the seedlings and
the remainder was broadcast 25 days after transplanting.   In the case of wheat, all the plots were
fertilized with ammonium sulphate at 23 kg/ha and
potassium sulphate at 45kg/ha before transplanting.
Another 23kg/ha of ammonium sulphate was applied in
March before heading.
    The experiments were repeated the following year
in a similar field, and a further wheat experiment
was performed with a randomized block design.
    An analysis of variance was performed on the
bulked data.   The effects were similar in the replicates and the results presented here are the averages
for the experiments.

Measurements

    Total soil P and soluble P contents (Olsen *et
al.*, 1954) of the field soil was estimated before
transplanting and after harvest.   Phosphate contents
of shoots and roots at each harvest of maize were
determined by a molybdenum-blue method (Khan, 1972a).
The extent of mycorrhizal infection was measured by
recording the length of infected tissue in KOH-

cleared and trypan blue-stained root segments
(Phillips & Hayman, 1970).   Numbers of *Endogone*
spores/kg soil were determined by a wet sieving and
decanting technique (Khan, 1971).   The growth of
wheat plants was measured monthly by uprooting 100
plants at random from each plot for five months, and
that of maize fortnightly by uprooting 18 plants at
random from each plot for 3 months.

RESULTS AND DISCUSSION

Soil

The *Endogone* spore population increased most in
the plot with inoculated(M) plants in soil with no
added phosphate.   The number of external spores was
less with inoculated wheat plants(MP) than with
non-inoculated wheat(NMP) when both were fertilized
with triple-superphosphate   (Fig. 1), presumably
because plants with MP treatment grew better than
with NMP treatment.   Large applications of P greatly reduce spore formation in the field as found by
Holevas (1966).   Proportionately the spore number
increased more after five months with wheat than
after three months with maize;   in both cases, however, spore number increased most after the period of
maximum root growth, a conclusion already drawn by
previous workers (Mason, 1964;   Sutton and Barron,
1972;   Saif and Khan, 1974).   There appears to be a
close relation between the number of external spores
and the degree of root infection, as found by Daft
and Nicolson (1972) and Hayman (1970).
The wheat field soil was initially alkaline
(pH 8) but it became neutral, and the neutral maize-
field soil became slightly acidic (pH 6.1) after
addition of superphosphate.   A soil test after the
plants had been harvested indicated that the mycorrhizal plants (M) removed more phosphorus from soil
than non-mycorrhizal (NM) ones, and the percentage of

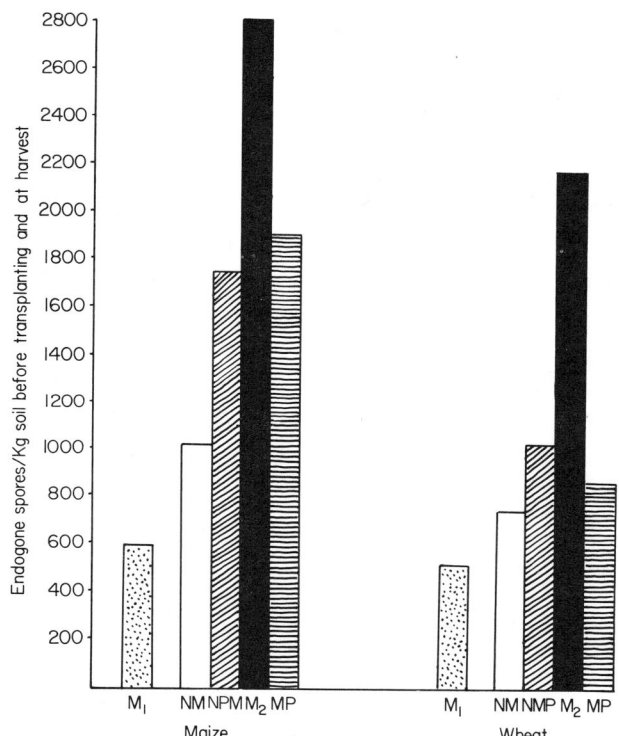

Figure 1. Number of *Endogone* spores/kg soil before transplanting and at harvest of maize and wheat. M_1, original population in field; M_2, final population. ▮ , inoculated without added phosphate (M_2); ▤ , inoculated with added phosphate (MP); ▨ , non-mycorrhizal with added phosphate (NMP); ▢ , non-mycorrhizal without added phosphate (NM).

phosphorus was significantly higher in mycorrhizal plants than in non-mycorrhizal. The total soil phosphorus (988 mg/kg soil) did not change during the growth period but the Na(HCO$_3$) soluble P decreased from 14.2 mg/kg soil to 12.6 mg/kg in the wheat soil.

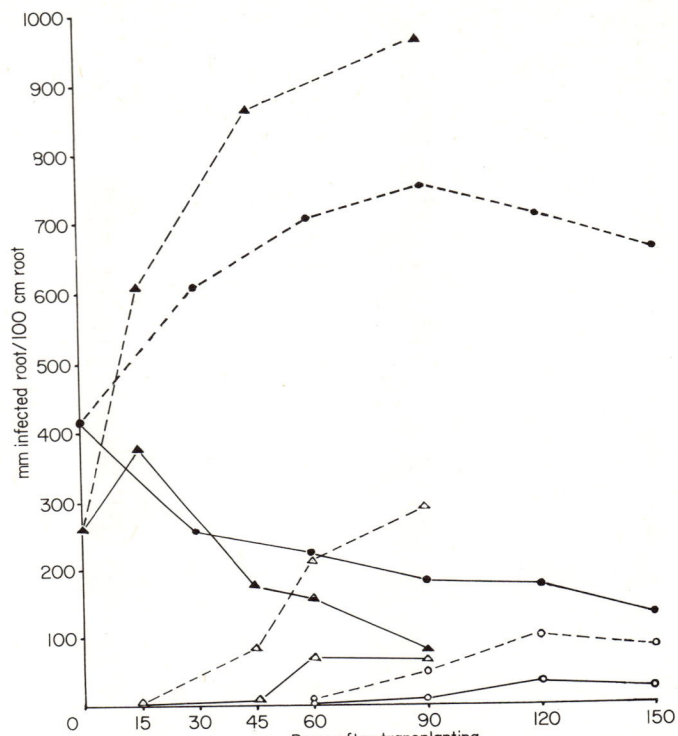

Figure 2.   Amount of infection in inoculated
and control maize and wheat seedlings after
transplanting into an unsterilized field soil.
    ● , inoculated wheat; ▲ , inoculated maize;
    O , non-inoculated wheat; Δ , non-inoculated
maize; ———— , without added phosphate;
———— , with added phosphate.

Effects on growth of shoots

    Although there was extensive mycorrhizal infect-
ion (Fig. 2), mycorrhizal and non-mycorrhizal plants
grew similarly in the short period before transplanting.   After 45 days in maize and 30 days in wheat,
a greater dry weight was noted in mycorrhizal (M)
than in controls (NM) (Fig. 3).   The uninoculated

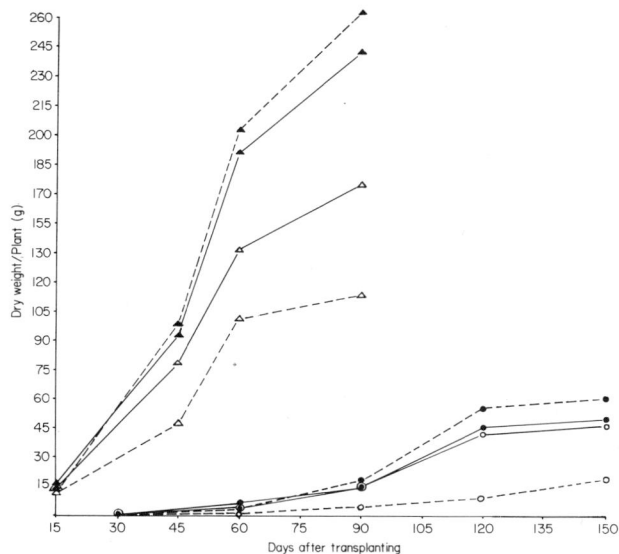

Figure 3. Amount of dry weight per plant in inoculated and control maize and wheat seedlings after transplanting into an unsterilized field soil. (Same notation as Fig. 2).

(NM) plants had symptoms of P-deficiency: stunting, with lower leaves small, desiccated and greenish or reddish brown. The rate of increase of leaf area of maize was also stimulated by mycorrhizal infection in the initial stages of growth. After 90 days however the leaf areas of inoculated and control plants were similar. Daft and Nicolson (1966) also noted an initial stimulation of leaf area in highly infected tobacco plants. Although uninoculated plants (NM) had far less VA infection from the indigenous *Endogone* population, their growth was also improved (Fig. 2), and the relative improvement of wheat from the 4th to the 5th month was greater than for the inoculated (M) treatment (Fig. 4). The symptoms of P deficiency were also observed in the non-inoculated

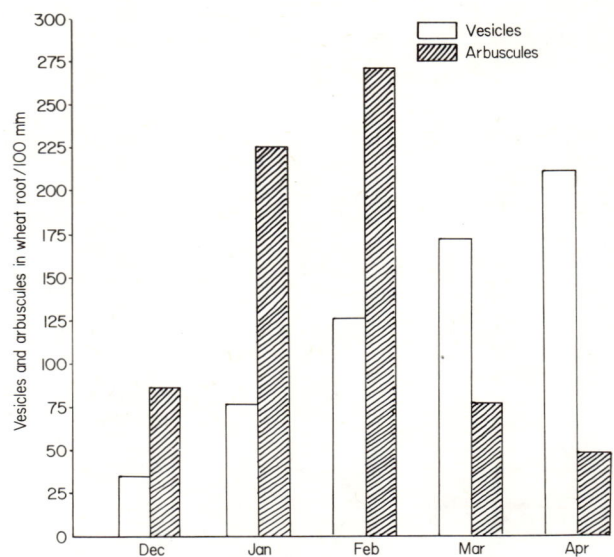

<u>Figure 4</u>.   Proportion of formation of ves-
icles and arbuscules in inoculated wheat
seedlings after transplanting in an unsterilized field soil.

wheat (NM).   P deficiency in the (NM) treatments
may be due to the slow development of mycorrhizal
infection in the roots, caused by the small population of indigenous *Endogone* spores in the selected
field and/or to the competition of normal soil
microflora in the field soil with the spore population.   The non-mycorrhizal non-phosphate (NM) plants
had shorter peduncles and smaller epidermal cells
than mycorrhizal wheat plants (M).   There was no
great difference in dry weight between mycorrhizal
(M) and non-mycorrhizal (NM) plants with phosphate
fertilizer (Fig. 3).   Mosse (1973a) observed that in
light or sandy soils, where the added P remains
available, plants with mycorrhiza grew worse when
much P was added to the soil.   The soil in the

present study was also a sandy loam, where most of
the added P remained available, and thus the inoculated plants with P-fertilizer (MP) grew slightly
worse than those without (M) (Fig. 3), indicating
that beneficial effects of VA mycorrhiza on the host
plant may be inversely related to the amount of
available phosphate.

Effects on growth of roots

The growth of the inoculated (M) maize roots
was reduced in the initial periods of growth but
infection was greater in them than those from control
(NM) plot.   The reduction in the root growth was
transient because, later, mycorrhizal (M) roots grew
better, had greater dry weight and were more branched
than non-mycorrhizal controls (NM).   The root/shoot
fresh weight ratios of non-mycorrhizal (NM) control
maize were smaller (0.29) than those of inoculated
(M) plants (0.40).   In wheat, however, the root/
shoot ratio of non-mycorrhizal controls (NM) was
actually greater (0.29) than in mycorrhizal (M)
(0.21) as found by Hayman and Mosse (1971), but the
results of Hayman & Mosse (1972) indicate that the
root/shoot ratio responses to mycorrhizal infection
differ between soils.

Effects on incidence of mycorrhizal infection

(M) Plants had 65% of root length infected at
transplanting.   During the period of vegetative
growth of wheat infection was mostly arbuscular,
rising to a peak in February followed by a sharp fall
in March and April, at the time of flowering and
ripening (Fig. 4).   Formation of vesicles increased
steadily to a peak in April, following the period of
maximum root growth, as found by earlier workers
(Neil, 1944;   Mosse, 1963;   Saif and Khan, 1974).
With the addition of P, many fewer vesicles and
arbuscules were formed, and many parallel hyphae

ramified in the root cortex.   In the mycorrhizal
plus phosphate (MP) treatment, the external mycelium
attached to the wheat roots also tended to extend
less outwards into the adjacent soil and grew mostly
along the root surface, as compared to that in the
mycorrhizal non-phosphate (M) plants.

The extent of root infection was greatest in
mycorrhizal non-phosphate (M) wheat and maize plants
(Fig. 2).   The incidence of mycorrhizal infection
in inoculated wheat seedlings after transplanting
into the field was not as high as for the maize.
The infection declined with the addition of phosphate
(MP) (Fig. 2), as reported by other workers (Holevas,
1966;   Baylis, 1967;   Daft and Nicolson, 1966, 1969;
Mosse, 1973a;   Ross, 1971).   The reduction of
infection and growth of mycorrhizal plants by adding
soluble P may be attributed to P-toxicity, as suggested by Mosse (1973a).

Infection did not produce any modification in
the external morphology or colour of the wheat roots.
Freshly dug mycorrhizal maize roots were, however,
more yellow than non-mycorrhizal, but this colour
disappeared soon after their exposure to light.   The
intensity depended upon the extent of infection
(see Gerdemann, 1961, 1964;   Barrett, 1961;
Krushcheva, 1961).   The invasion by the endophyte
did not affect the size of the host cortical cells.
The cytoplasmic contents, however, appeared denser,
largely due to the disintegration of fungal structures.   It was also observed that the infected cortical cells of wheat were usually multinucleate.   No
mitotic divisions were observed and it is not known
whether the increased number of nuclei resulted from
fragmentation or some form of equational division.

Effects on reproduction

Harvest data (Table 1) show that the plants
with the addition of phosphorus and fungus (MP) or
fungus alone (M) grew well reproductively as well as

Table 1.   Effects of VA mycorrhiza on reproduction of maize and wheat.

Host	Character	Treatments*			
		NM	NMP	M	MP
Maize	Length of ear (cm)	14.0	19.5	22.0	20.7
	No. of grains/ear	31	279	354	321
	1000-grain weight (g)	2.4	19.8	23.7	20.9
	Embryos	Partly aborted	Fully matured	Fully matured	Fully matured
Wheat	Spike bearing culms/m^2	20	26	32	28
	Spikes/plant	3	5	8	4
	Fertile spikelets/spike	11	20	21	19
	No. of grains/spike	36	69	72	70
	1000-grain weight (g)	38.50	39.00	38.00	37.50
	Calculated yield (g/m^2)	85.80	215.28	275.51	235.20

* NM, non-mycorrhizal without added phosphate; NMP, non-mycorrhizal with added phosphate; M, mycorrhizal without added phosphate; MP, mycorrhizal with added phosphate.

vegetatively.   In the control plots of maize (NM),
however, only fifty nine plants survived at the time
of final harvest.   Of these, only twenty seven
developed male inflorescences, while nine or ten
plants developed small and poorly filled ears with
seeds possessing partly aborted embryos.   The number
of ear bearing tillers per wheat plant was also greatest in mycorrhizal (M) plants as compared to that in
other treatments.   Inoculated wheat plants with
fertilizer (MP), however, had fewer spikes per plant,
contained fewer fertile spikelets per plant and
grains per spike, and yielded lesser than those without (M) (Table 1).   The number of grains per ear was
almost doubled in wheat and increased twelve times in
maize by the fungus.   The favourable effects on the
grain production in treatments M, MP and NMP, as
compared to those in treatment NM, can be attributed
to the improved uptake of phosphorus.   Grain weight
in maize was also increased almost twelve times by
the fungus.   Although weight/1000 grains in mycorrhizal (M) wheat was slightly less than in non-mycorrhizal (NM) wheat (Table 1), weight distribution of
grains was not affected by the fungus or added phosphorus.   The greater number of seeds per wheat spike
in the M treatment is probably determined quite early
in the growth period and it may account for slightly
smaller weight/1000 grains.   Ross and Harper (1970)
obtained a 34-40% increase in the yield of soybean
plants infected with pure inoculum of *Endogone* spores
although seed weight was not affected in their
studies.   Ross (1971) also found that seeds from all
mycorrhizal treatments were heavier than those from
non-mycorrhizal, as with maize in the present study.
Jackson *et al.* (1972) obtained a 50% increased corn
yield by inoculation with VA fungi over the increases
due to indigenous mycorrhizal fungi, but found no
significant differences in P concentration among any
of the treatments or controls in contrast to the
present work (Fig. 2).

CONCLUSIONS

These results were obtained under field conditions, using maize and wheat, both economically important crops, of which the latter is well supplied with root hairs. The positive responses of wheat and maize plants to mycorrhiza, confirms the previous green-house findings.

A significant feature arising from the present study is the practical implication that mycorrhizal inoculation of crop seed may be beneficial. It would be of interest to investigate subsequent development of seeds soaked in *Endogone* spore suspensions prior to planting, a method comparable to that commonly employed for inoculation of legumes with *Rhizobium*. This technique could be of great practical value in increasing grain production in those areas of the world where phosphorus limits plant growth and phosphorus fertilization is not economical. There is but one possible limitation to this technique i.e. *Endogone,* unlike *Rhizobium,* cannot be grown in pure culture and, therefore, large quantities of inoculum may be difficult to obtain. Also, *Endogone* spores are about 100 times as big as *Rhizobium* cells, so that techniques for seed-pelleting would require modifications.

The feasibility of inoculation on a field scale with chopped mycorrhizal wheat roots from the present experiment (M), is under investigation. Similarly Pakistani soils rich in certain types of *Endogone* spores are also being used to inoculate P-deficient experimental field plots with fewer indigenous spores. Kleinschmidt and Gerdemann (1972) suggested raising inoculum in quantity in small fumigated field plots and then incorporating such inoculum into the row at the time of planting in the field. The successful use of freeze dried ground mycorrhizal root material as storable inoculum by Jackson *et al*. (1972) has provided an alternative method of large scale field inoculation. The difficulties in obtaining pure cultures of VA endophytes should not be an obstacle to further progress in inoculation on a field scale.

REFERENCES

BARRETT, J. T. (1961). Isolation, culture and host relation of the phycomycetoid vesicular-arbuscular endophyte *Rhizophagus*. *Rec. Adv. Bot.*, 2, 1725-1727. Univ. Toronto Press.

BAYLIS, G. T. S. (1967). Experiments on the ecological significance of phycomycetous mycorrhizas. *New Phytol.*, 66, 231-243.

DAFT, M. J. & NICOLSON, T. H. (1966). Effect of *Endogone* mycorrhiza on plant growth. *New Phytol.*, 65, 343-350.

DAFT, M. J. & NICOLSON, T. H. (1969). Effect of *Endogone* mycorrhiza on plant growth. II. Influence of soluble phosphate on endophyte and host in maize. *New Phytol.*, 68, 945-952.

DAFT, M. J. & NICOLSON, T. H. (1972). Effect of *Endogone* mycorrhiza on plant growth. IV. Quantitative relationships between the growth of the host and the development of the endophyte in tomato and maize. *New Phytol.*, 71, 287-295.

GERDEMANN, J. W. (1961). A species of *Endogone* from corn causing vesicular-arbuscular mycorrhiza. *Mycologia,* 53, 254-261.

GERDEMANN, J. W. (1964). The effect of mycorrhiza on the growth of maize. *Mycologia,* 56, 342-349.

HAYMAN, D. S. (1970). *Endogone* spore numbers in soil and vesicular-arbuscular mycorrhiza in wheat as influenced by season and soil treatment. *Trans. Br.*

mycol. Soc., 54, 53-63.

HAYMAN, D. S. & MOSSE, B. (1971). Plant growth
responses to vesicular-arbuscular mycorrhiza.
I. Growth of *Endogone*-inoculated plants in phosphate-deficient soils.   *New Phytol.*, 70, 19-27.

HAYMAN, D. S. & MOSSE,B. (1972). Plant growth responses to vesicular-arbuscular mycorrhiza. III. Increased uptake of labile P from soil.   *New Phytol.*,
71, 41-47.

HOLEVAS, C. D. (1966). The effect of a vesicular-
arbuscular mycorrhiza on the uptake of soil phosphorus by strawberry (*Fragaria* sp. var. Cambridge
Favourite). *J. hort. Sci.*, 41, 57-64.

JACKSON, N. E., FRANKLIN, R. E. & MILLER, R. H.
(1972). Effects of VA mycorrhizae on growth and
phosphorus content of three agronomic crops.
Proc. Soil Sci. Soc. Am., 36, 64-67.

KHAN, A. G. (1971). Occurrence of *Endogone* spores in
West Pakistan soils.   *Trans. Br. mycol. Soc.*,
56, 217-224.

KHAN, A. G. (1972a). The effect of vesicular-arbuscular mycorrhizal associations on growth of cereals.
I. Effects on maize growth.   *New Phytol.*, 71,
613-619.

KHAN, A. G. (1974b). The effect of vesicular-arbuscular mycorrhizal associations on growth of
cereals. II. Effects on wheat growth.   *J. appl.
Biol.*   In press.

KRUSHCHEVA, E. P. (1961). On some morphological properties of maize mycorrhiza (Russian).   *Agrobiol-
ogy*, 4, 595-598.

KLEINSCHMIDT, G. D. & GERDEMANN, J. W. (1972).
Stunting of citrus seedlings in fumigated nursery
soils related to the absence of endomycorrhiza.
Phytopathology, 62, 1447-1453.

MASON, D. T. (1964). A survey of numbers of *Endogone*
spores in soil cropped with barley, raspberry and
strawberry.   *Hort. Res.,* 4, 98-103.

MOSSE, B. (1963). Vesicular-arbuscular mycorrhiza
an extreme form of fungal adaptation.   In *Symbiotic Associations,* eds. Nutman, P. S., Mosse, B,
Symp. Soc. gen. Microbiol., 13, 146-170.

MOSSE, B. (1973a). Plant growth responses to vesicular-arbuscular mycorrhiza. IV. In soil given
additional phosphate.   *New Phytol.,* 72, 127-136.

MOSSE, B. (1973b). Advances in the study of vesicular
-arbuscular mycorrhiza.   *A. Rev. Phytopath.,*
11, 171-196.

MOSSE, B. & HAYMAN, D. S. (1971). Plant growth
responses to vesicular-arbuscular mycorrhiza.
II. In unsterilised field soils.   *New Phytol.,*
70, 29-34.

MOSSE, B., HAYMAN, D. S. & IDE, G. J. (1969). Growth
responses of plants in unsterilized soil to inoculation with vesicular-arbuscular mycorrhiza.
Nature, London, 224, 1031-1032.

NEILL, J. C. (1944). *Rhizophagus* in citrus.   *N.Z.Jl.
Sci. Technol.,* 25, 191-201.

OLSEN, S. R., COLE, C. V., WATANABE, F. S. &
DEAN, L. A. (1954). Estimation of available phosphorus in soil by extraction with sodium bicarbonate.   *Circular U.S.D.A.,* 939.

PHILLIPS, J. M. & HAYMAN, D. S. (1970). Improved procedures for clearing roots and staining parasitic and vesicular-arbuscular mycorrhizal fungi for rapid assessment of infection. *Trans. Br. mycol. Soc.*, 55, 158-161.

ROSS, J. P. (1971). Effect of phosphate fertilization on yield of mycorrhizal and non-mycorrhizal soybeans. *Phytopathology*, 61, 1400-1403.

ROSS, J. P. & HARPER, J. A. (1970). Effect of *Endogone* mycorrhiza on soybean yields. *Phytopathology*, 60, 1552-1556.

SAIF, S. R. & KHAN, A. G. (1974). The influence of season and stage of development of plant on *Endogone* mycorrhiza of field grown wheat. *Can. J. Microbiol.*, In Press.

SUTTON, J. C. & BARRON, G. L. (1972). Population dynamics of *Endogone* spores in soil. *Can. J. Bot.*, 50, 1909-1914.

EFFECTS OF VESICULAR-ARBUSCULAR MYCORRHIZAE ON LOWLAND TROPICAL RAINFOREST TREES[1]

D. P. JANOS [2]

Department of Botany, University of Michigan, Ann Arbor, Michigan, U.S.A.

INTRODUCTION

VA mycorrhizae may affect natural plant community composition through effects on the growth of component species.   Such ecological significance cannot however be inferred from growth effects alone.   In order for mycorrhizae to have synecological effects two conditions must obtain: 1) mycorrhizae must affect species' fitnesses, and 2) mycorrhiza formation must be temporally or spatially heterogeneous.   Fitness has two components, reproduction and survival, with growth effects perhaps affecting survival in a competitive milieu or with herbivore damage.   Baylis (1959), Gerdemann (1965), and Kleinschmidt and Gerdemann (1972) reported cessation of growth of uninfected seedlings but did not indic-

[1]This paper submitted in partial fulfilment of the requirements for the degree of Doctor of Philosophy in the Horace H. Rackham School of Graduate Studies at The University of Michigan, Ann Arbor.

[2]Present address: Organizacíon Estudios Tropicales, Universidad de Costa Rica, Ciudad Universitaria, Costa Rica, Central America.

cate how long these seedlings survived.   Little is
also known of mycorrhizal effects on reproduction.
Yield increases have been recorded for mycorrhizal
corn (Khan, 1972) and soybeans (Ross, 1971, Ross &
Harper, 1970).   Environmental heterogeneity in mycorrhiza formation may result from differences in inoculum availability due to death of mycelium in root
fragments or production or introduction of infective
root fragments, spore death due to predation and
parasitism or sporulation *in situ,* and spore dispersal into or out of the soil volume.   Redhead (1968)
found infection lacking only in newly cleared and
burnt nurseries in Nigeria, but Baylis (1967)
observed patchiness in seedling infection under a
parent tree.   Thus the synecological effect of VA
mycorrhizae is a very open question.   This paper
records an effect of mycorrhizae on fitness, and
discusses sporulation of endomycorrhizal fungi in
lowland tropical rainforest.

METHODS

This study was carried out in an area of lowland
tropical rainforest located in Heredia Province in
northeastern Costa Rica at a latitude of about $10^\circ 25'$
N. Mean annual precipitation is about 4000 mm., and
the monthly mean temperature is approximately 24°C.
A "dry" season of somewhat decreased rainfall extends
from January through April.   The study was begun
during the wet season and carried on through the
beginning of the dry.

The four species used were trees and come from
different habitats.   *Inga oerstediana* (Mimosaceae)
is most commonly found in second-growth areas on
older alluvial soils of moderate fertility.   (See
Bourgeois, *et al.,* 1972 for a description of the
soils discussed here).   *Sickingia maxonii* (Rubiaceae) is common along river edges on recent alluvial
soils of high fertility.   *Vitex cooperi* (Verben-
aceae) is a rare primary forest species on either old

alluvial or residual soils developed from basaltic parent materials.   The very rare undescribed Euphorbiaceous species is also a primary forest species.

The soil used for the experiments was an old alluvial soil of moderate fertility sterilized at a rate roughly equivalent to 1780 pounds per acre of methy-bromide + chloropicrin gas.   Large tin cans were filled with loose, screened soil to give good gas penetration, placed in plastic garbage bags, and gas was sprayed into the bag just to inflate it. This method of sterilization has proved very practical in the field where autoclaving, sterilization between electrodes, or irradiation are impossible.

Seeds were collected, cleaned, and surface sterilized in aqueous sodium hypochlorite before sowing in germination trays of sterilized soil. Seedlings were transplanted to cans of three treatment groups.   Mycorrhizal plants (treatment 3) were obtained by adding approximately 15 gm. fresh weight of finely chopped field collected cacao roots (observed to contain endomycorrhizal infection) to the cans in a layer about 2-3 cm below the soil surface. Control cans (treatment 2) were prepared by adding 15 gm of sterilized chopped cacao roots plus 200 ml of a suspension made by soaking the treatment 3 inoculum overnight in water, and filtering it through a 74 micron soil sieve which excluded fungi. Treatment 1 received no pretreatment other than sterilization.

Cans of all treatments of each species were randomized and spaced a minimum of 15 cm apart on tables set outside in the light shade of a pejiballe palm plantation.   The plants were watered only by the rain which was quite sufficient.   Larvae found eating plants were removed, but no other attempt was made to discourage herbivorous insects.   The environment was similar to that in an area with a canopy opening in the forest.

Plants were measured bi-weekly.   The height from the ground surface to the tallest shoot apex, length of the largest leaf or leaflet, the number of leaves, and where appropriate, cotyledon retention were recorded for each plant. Measurement was continued for about six months, at which time plants of all species except *S. maxonii* were measured at monthly intervals for an additional three months. At the close of the experiments plant roots were sampled and examined to confirm that their mycorrhizal condition conformed to treatment.

RESULTS

The data collected demonstrate that at the close of the experiments mycorrhizal plants were significantly taller (Table 1), and had more and larger leaves (Table 2) than plants in either of the control treatments. For those two species having fleshy persistent photosynthetic cotyledons, *S. maxonii* and the Euphorbiaceous species, more mycorrhizal plants had retained both cotyledons until the final week of measurement (Table 2). Differences between mycorrhizal and control plants were obvious for *I. oerstediana, V. cooperi,* and the Euphorbiaceous species. The slight response to infection shown by *S. maxonii* is surprising. This difference took three to four months longer to appear than in the other three species.

Two interesting responses to herbivorous insect damage were observed. The Euphorbiaceous seedlings are hollow stemmed, and because mycorrhizal plants had larger stems, they were preferentially inhabited by the larvae of a Gyromitid which destroyed the shoot apices. Mycorrhizal plants responded dramatically to shoot apex loss by release of one or more axillary buds which developed into vigorous new leaders; half of the larger uninfected plants so damaged died. *V. cooperi* also suffered intense herbivore damage by both beetles and lepidopteran larvae. Plants in all three treatment groups

Table 1. Growth comparisons of three treatments for *Inga oerstediana, Sickingia maxonii, Vitex cooperi,* and an unidentified Euphorbiaceous species. Treatments are: 1) sterilized soil, 2) sterilized soil + microbial filtrate + sterilized diced cacao roots, and 3) sterilized soil + diced cacao roots. Treatment 3 plants were mycorrhizal while plants of the control treatments 1 and 2 were uninfected.

Inga oerstediana

Weeks after inoculation	Mean height (cm) Treatment		
	1	2	3
1	6.83	6.44	6.28
2	7.39	6.97	6.70
3	8.53	8.40	8.21
4	9.11	8.97	9.04
5	9.60	9.57	9.53
7	10.62	10.61	10.59
9	10.83	10.48	10.89
11	11.03	10.76	11.35
13	11.28	10.93	11.96
15	11.61	11.14	12.25
17	11.73	11.08	13.39
19	11.89	11.20	14.59
21	11.98	11.22	16.90
23	12.14	11.42	18.32
25	12.33	11.52	19.42
	(16)	(16)	(16)

Sickingia maxonii

Weeks after inoculation	Mean height (cm) Treatment		
	1	2	3
1	0.69	1.11	1.28
2	1.29	1.19	1.24
3	1.61	1.67	1.49
4	2.82	2.86	2.90
5	3.74	3.77	4.32
6	4.60	5.01	5.30
8	5.92	6.46	6.56
10	6.92	7.23	7.30
12	8.07	8.28	8.32
14	8.72	9.08	8.90
16	9.26	9.96	10.42
18	10.10	10.75	11.09
20	10.91	11.47	11.72
22	11.30	11.81	12.00
24	11.95	12.42	12.71
28	12.22	12.72	13.31
32	12.86	13.16	14.21
36	13.20	13.39	14.57
	(32)	(32)	(30)

Vitex cooperi

Weeks after inoculation	Mean height (cm) Treatment		
	1	2	3
1	1.20	1.40	1.19
2	1.56	1.79	1.66
3	2.36	2.58	2.58
4	3.20	3.54	3.46
5	3.68	4.03	4.10
6	4.42	4.77	4.78
8	5.42	5.54	5.89
10	6.17	5.93	6.70
12	6.76	6.54	7.80
14	7.12	7.00	9.11
16	7.41	7.15	10.25
18	7.56	7.46	11.92
20	7.56	7.45	13.89
22	7.40	7.19	15.80
24	6.52	7.40	16.53
	(9)	(11)	(17)

Euphorbiaceous sp.

Weeks after inoculation	Mean height (cm) Treatment		
	1	2	3
1	1.24	1.08	0.92
2	1.56	1.59	1.47
3	1.93	1.84	1.84
4	2.30	2.26	2.29
5	2.70	2.62	2.72
7	2.94	3.36	3.46
9	3.58	3.86	3.92
11	4.05	4.47	4.28
13	4.30	4.72	4.76
15	4.50	4.86	4.91
17	4.62	4.89	5.41
19	4.70	5.06	6.47
21	4.76	5.07	7.63
23	4.82	5.16	8.29
	(35)	(40)	(39)

All figures: week 1 = 27 Oct. '73; 25 = 15 April '74; 35 = 14 June '74.
Number of plants per treatment at final week are given in brackets.

442   Janos, D. P.

Table 2. Leaf/leaflet length, leaf number, and cotyledon retention for four species of lowland tropical rainforest trees. Treatments as Table 1. Treatment 3 plants are mycorrhizal while plants of the control treatments 1 and 2 are uninfected. Data were taken the indicated number of weeks after inoculation of seedlings. Least significant differences are shown for P = 0.01.

Treatment	Weeks after inoculation	Mean leaf/leaflet length (cm)	LSD	Mean leaf number (leaves)	LSD	Mean Cotyledons retained	LSD
Species:							
Inga cerstediana							
1		6.31		3.63			
2	25	5.93	2.644	3.19	1.599		
3		13.74		8.13			
Sickingia maxonii							
1		8.75		3.17		0.30	
2	36	8.19	2.029	3.23	1.441	0.22	0.4805
3		10.90		5.69		1.14	
Vitex cooperi							
1		1.22		3.00			
2	24	1.23	2.556	4.83	5.132		
3		5.49		11.61			
Euphorbia-ceous sp.							
1		3.69		3.11		1.34	
2	23	3.19	1.003	2.76	1.004	1.47	0.4423
3		7.80		6.43		1.91	

Table 3. Number of surviving *Vitex cooperi*
seedlings. Treatments as in Table 1.

		Weeks after inoculation				
		0	8	16	24	35
	3	24	19	18	17	12
Treatment	2	24	16	16	11	3
	1	24	20	18	9	2

were uniformly attacked and were completely defolia-
ted repeatedly. All plants had some ability to
regenerate new leaves, but as defoliation continued
uninfected plants died.   At the level of herbivory
suffered, significantly more (p = 0.01) mycorrhizal
than uninfected plants survived (Table 3).
    During the course of the experiments the ident-
ity of possible fungal associates was sought because
the chopped cacao root inoculum possibly contained
several endomycorrhizal fungi.   Two Endogonaceous
fungi were discovered fruiting in experimental cans.
Sclerocystis dussii (Pat.) von Höhn formed epigean
fruiting bodies composed of aggregated sporocarps.
Soil wet-sieving (Gerdemann, 1965) revealed numerous
azygospores of an *Acaulospora* sp. in soil samples
from cans of *I. oerstediana*.   Neither fungus was
recovered from cans of mycorrhizal *V. cooperi* or the
Euphorbiaceous species.   It is possible that both of
these fungi are VA mycorrhizal associates causing the
observed growth increases.

DISCUSSION

    Growth increases would be expected in the trop-
ical soils used which are low in available phosphorus
and generally bear plants which are heavily mycorrhizal.   Such plants are likely to be extremely dependent

on VA mycorrhizae for uptake of phosphorus and
perhaps other minerals. Earlier cotyledon abscis-
sion by uninfected plants may have resulted from more
rapid exhaustion of cotyledonary mineral reserves.
The minimal response of *S. maxonii* might have resulted because the endomycorrhizal fungi added in the
cacao root inoculum were less effective with this
species. This seems unlikely in view of the effic-
acy of these fungi with the other three species from
three different families. The large seeds (0.4 gm
dry weight) of *S. maxonii* may simply contain enough
minerals to support continued growth of uninfected
plants for a longer time. The question regarding the
extent to which the observed growth responses to
mycorrhizae affect the fitness of the species involved remains a difficult one, but *V. cooperi,* under the
level of herbivory suffered, could not survive unless
mycorrhizal.

     The restriction of sporulation to inoculated
I. oerstediana has potential implications for heterogeneity in mycorrhiza formation. *I. oerstediana* is
a nodulated legume, and some plants of all treatments
had nodules although only mycorrhizal plants had well
developed nodulation. Thus, improved nitrogen
nutrition of the host is an obvious correlate of
endomycorrhizal fungus sporulation. If we infer
that nitrogen supply can limit fruiting by endomycorrhizal fungi, perhaps an explanation for the paucity
of spores found in wet-forest soils by Mosse and
Bowen (1968) and Redhead (1971) is afforded.

REFERENCES

BAYLIS, G. T. S. (1959). Effect of vesicular-arbus-
    cular mycorrhizas on growth of *Griselinia littoral-
    is* (Cornaceae). *New Phytol.*, <u>58</u>, 274-280.

BAYLIS, G. T. S. (1967). Experiments on the ecologic-
    al significance of phycomycetous mycorrhizas.

New Phytol., <u>66</u>, 231-243.

BOURGEOIS, W. W., COLE, D. W., REIKERK, H., &
GESSEL, S. P. (1972). *Geology and soils of comparative ecosystem study areas in Costa Rica.*   Cont.
no. 11, Tropical Forestry Series, Institute of
Forest Products, Univ. of Washington, Seattle.

GERDEMANN, J. W. (1965). Vesicular-arbuscular mycorrhizae formed on maize and tuliptree by *Endogone
fasciculata.   Mycologia,* <u>57</u>, 562-575.

KHAN, A. G. (1972). The effect of vesicular-arbuscular mycorrhizal associations on the growth of
cereals.   I.   Effects on maize growth.   *New
Phytol.*, <u>71</u>, 613-619.

KLEINSCHMIDT, G. D. & GERDEMANN, J. W. (1972).
Stunting of citrus seedlings in fumigated nursery
soils related to the absence of endomycorrhizae.
Phytopath., <u>62</u>, 1447-1453.

MOSSE, B. & BOWEN, G. D. (1968). The distribution of
Endogone spores in some Australian and New Zealand
soils and in an experimental field soil at
Rothamsted.   *Trans. Br. mycol. Soc.*, <u>51</u>, 485-492.

REDHEAD, J. F. (1968). Mycorrhizal associations in
some Nigerian forest trees.   *Trans. Br. mycol.
Soc.*, <u>51</u>, 377-387.

REDHEAD, J. F. (1971). *Endogone* and endotrophic
mycorrhizae in Nigeria.   *XV IUFRO Congress,*
Section 24.

ROSS, J. P. (1971). Effect of phosphate fertilization
on yield of mycorrhizal and non-mycorrhizal
soybeans.   *Phytopathology,* <u>61</u>, 1400-1403.

ROSS, J. P. & HARPER, J. A. (1970). Effect of
Endogone mycorrhiza on soybean yields.
Phytopathology, <u>60</u>, 1552–1556.

ENDOTROPHIC MYCORRHIZAS IN NIGERIA: SOME ASPECTS OF THE ECOLOGY OF THE ENDOTROPHIC MYCORRHIZAL ASSOCIATION OF KHAYA GRANDIFOLIOLA C.DC.

J. F. REDHEAD

Federal Department of Forestry,
Ibadan, Nigeria

INTRODUCTION

This paper describes a series of experiments designed to test the effects of various environmental factors on the development of vesicular-arbuscular mycorrhiza in order to gain a better understanding of the ecology of the association under Nigerian conditions. It also examines the importance of VA mycorrhiza for the growth of *Khaya grandifoliola,* one of Nigeria's well known mahogany trees and how death of the host affects the number of extra-matrical spores in the soil.

MATERIAL AND METHODS

Plants of *Khaya grandifoliola* C.DC., raised from seed collected from the same parent tree, were potted in soil known to contain *Endogone* spores and mycelium. The incidence of mycorrhizal infection, number of vesicles and number of extra-matrical spores was assessed in plants subjected to different water, light and nutrient regimes.

Water regimes (1)

Eight uniform *Khaya* seedlings were given 250 ml
of water weekly (W-I), twice weekly (W-II), every
second day (W-III), daily (W-IV) or twice daily
(W-V).   Plants subjected to treatment W-V later had
their pots enclosed in polythene bags to ensure
complete water-logging.   After 30 weeks W-I plants
had their water further restricted until wilting
appeared each time.   After 35 weeks the plants were
harvested.

Light intensity (2)

Twenty *Khaya* seedlings were placed under each of
four different light regimes, approximating to those
in an open nursery (full daylight), a shaded nursery
(light and medium shade), and a natural forest environment (heavy shade).   Shading resulted in light
intensities, approximately 15% (light shade), 8%
(medium shade) and 2% (heavy shade) of full daylight,
as measured by a Megatron Light meter.   After 40
weeks, plants were harvested and six from each group
were examined for mycorrhizal infection.

Nutrient supply (3)

Khaya seedlings were grown in a natural soil
enriched with 100 ml of soil from *Nauclea* pot cultures containing many medium-sized yellow vacuolate
spores.   The plants were watered with rain water for
six weeks and then allocated at random in 3 blocks of
a 3^3 factorial layout (81 pots), replicated three
times.
All seedlings received 100 ml of solution daily,
consisting of: water only (level 0); 0.22 g/l N,
0.03 g/l P or 0.16 g/l K once a week (level 1); or
the same nutrient solution daily (level 2).   Nutrients in level 2 were thus sevenfold those in level
1.   Nitrogen consisted of Na NO_3 and NH_4NO_3 given in

the proportion of 4 : 6 molar.   As the soil used
was of low to moderate fertility, it was anticipated
that level 2 might produce nutrient concentrations
detrimental to *Khaya* growth.   Plants were harvested
after 8 months.

The effect of Endogone mycorrhiza on growth (4)

Ten *Khaya* seedlings were subjected to each of
four treatments: F_0 - controls to which no spores
were added; F_1 - with 50 spores surface sterilized
in 0.5% sodium hypochlorite for 10 - 20 secs.   and
rinsed 5 times in sterile distilled water; F_2 - with
50 spores surface sterilised and then individually
squashed with a needle; F_3 - with 50 unsterilised
spores.
Seedlings were pre-inoculated in aluminium
funnels (Gerdemann, 1955) and then transplanted into
an unsterilized nursery soil which contained few
Endogone spores but a high level of VA infection in
the indigenous plants.

The effect of plant death on extra-matrical spores

Newly germinated *Khaya* seedlings grown in an
unsterilized loam in buckets were arranged in thirteen blocks of three on the basis of a sample of the
spore population of each pot, in order to reduce
variability within the blocks.   After a further four
months one plant in each block was cut off below
ground level, one was covered with a black polythene
hood to exclude all light and the third was left as a
control.   All pots were watered to maintain the soil
at a fairly uniform moisture level and spore numbers
counted after 3 months.

Procedure at harvest (1-4)

Roots, stems and leaves were dried at $100^{\circ}C$ and
weighed.   Infection was recorded in five rootlets

selected at random from an evenly spread root sample.
A 5mm segment was cut from each selected root and the
number of cells with and without hyphae were counted
in five microscope fields (x 600) per segment.   The
number of vesicles was counted in all five root
segments.   The soil from each pot was well mixed,
25 ml (ca. 25 g) was wet sieved and the number of
extra-matrical spores above 100 µm was recorded.

RESULTS

Water regimes (1)

All plants appeared healthy and were of similar
height, but after 30 weeks those in treatment W-I had
most leaves, and those in treatment W-V had fewest.

Table 1.   Effect of watering regimes on
growth and mycorrhizal infection of *Khaya
grandifoliola*.

Water regime	Mean dry weight(g)		Mean % infected cells	Mean no. of vesicles in 5 x 5mm root	Mean no. of extra-matrical spores/25 ml soil
	Shoot	Root			
W-I	13	10	71	71	7
W-II	17	14	70	42	16
W-III	35	27	82	67	27
W-IV	31	26	91	17	62
W-V	27	18	84	39	17
L.S.D. 5%	13.0	8.1	10.1	232.6	33.3

Table 1 shows that plant growth responses to water regimes fell into two main groups, W-I to III and W-III to V.   Waterlogging (W-V) particularly reduced root weight.   The percentage infected cells showed a similar response and was least in treatments W-I and W-II, although with 70% infected cells, infection levels were high even under drought conditions.   Presumably water content within the roots remained sufficiently high for continued fungal spread in the root.   Vesicle numbers were very variable (as is shown by the high L.S.D.) and not affected by watering regimes but extra-matrical spores were adversely affected by very dry or waterlogged soil conditions.   Fungal development in the soil may well be more drastically affected by fluctuations in soil moisture than that in the root, where moisture levels probably fluctuate less.

Light intensity (2)

After 40 weeks the plants were all healthy. Shading somewhat reduced plant height (Table 2), less so with heavy shade which may however have caused some etiolation.   Roots of plants in full sunlight, light or medium shade were sturdy and filled available pot space.   Plants in heavy shade had noticeably more slender, less extensive root systems.   In spite of these effects on plant growth differences in light intensity did not affect the incidence of mycorrhizal infection, vesicle or extra-matrical spore production (Table 2).   This is contrary to the findings of Peuss (1958), Khruschcheva (1960) and to results obtained by Hayman (1974) with onions which were less infected at low light intensity.   However, infection decreased most in onions when daylength was also reduced to 6 hr, whereas in the *Khaya* experiments light intensity was reduced, but daylength remained constant.

Redhead, J. F.

Table 2.   Effect of light on growth and mycorrhizal infected cells in *Khaya grandifoliola*.

Treatment	Mean height (cm)	Mean % infected cells	Mean no. of vesicles in 5x5mm root segments	Mean no. of *Endogone* extra-matrical spores/25 ml soil
Full light	20	66	51	50
Light shade	15	65	41	52
Medium shade	14	54	6	22
Heavy shade	17	64	15	11
L.S.D. 5%	3.5	24.0	50.7	118.2

Nutrient Supply (3)

Table 3. The effect of N, P and K at three
levels on growth and mycorrhizal infection
of *Khaya grandifoliola*.

	Nutrient level	N	P	K
Root dry weight	0	+2.36	2.06	1.84
(g)	1	2.13	1.88	2.18
Mean root wt. = 1.99g.	2	1.49	2.04	1.96
Total dry weight	0	++4.42	4.56	++4.02
(g)	1	5.02	4.24	5.06
Mean total wt. = 4.38g.	2	3.70	4.33	4.05
Root/shoot ratio	0	++1.15	0.85	++0.84
Mean root/shoot	1	0.80	0.85	0.82
ratio = 0.89	2	0.71	0.96	0.99
Percentage of infected	0	+85	81	84
cortical cells	1	83	80	84
Mean % = 82	2	77	83	80
Number of vesicles	0	++12.81	3.11	3.15
in 5 x 5 mm root	1	2.65	4.16	5.12
segments. Mean number	2	1.84	4.94	3.88
per plant = 3.97.				
Number of extra-matrical		+		
spores	0	27.3	37.6	37.1
per 25 ml soil	1	25.0	32.1	29.8
Mean No. = 31	2	44.3	25.1	27.4

Differences between levels for this nutrient were
significant

+ $p < 0.05$, ++ $p < 0.01$.

There was a positive N by K interaction on root
and total plant weight (Table 3). Of the three
fertilizers nitrogen clearly had the greatest

effects on dry weight, percentage infected cells
and numbers of vesicles and extra-matrical spores.
Because of the heavy rates of application, nitrogen
at level 2 reduced plant weight (particularly of
roots), percentage infection and vesicle formation
(sevenfold compared to level 0), but there was a
slight increase (just significant) in extra-matrical
spores.   This may be related to death of roots.
K at the middle level increased plant dry weight,
but did not significantly affect fungal development.
P had no significant effect on plant weight or
fungal development.

*The effect of the Endogone mycorrhiza on the growth
of Khaya grandifoliola (4)*

Table 4.   Mean dry weight and percentage
infected cells in *Khaya grandifoliola*
inoculated with Endogone spores.

Treatment	Mean dry weight (g)		Mean % fungal infection
	Shoot	Root	
F_0 control plants	1.7	1.2	26
F_1 sterilised spores	2.4	1.6	42
F_2 sterilised and squashed spores	1.3	0.9	16
F_3 unsterilised spores	3.9	2.9	74
Average L.S.D. 5%	1.6	1.1	27.3

Some uninoculated plants (F_0) and some with squashed spores (F_3) became mycorrhizal (Table 4). As the soil was not sterilised and had supported a strongly mycorrhizal vegetation this was not surprising.   On the other hand, some seedlings inoculated with sterilised spores failed to become infected. It was therefore decided to re-group plants into those strongly mycorrhizal (over 20%) and those lightly infected (less than 20% mycorrhizal roots) (Table 5).

Table 5.   Mean dry weight of strongly mycorrhizal compared with slightly mycorrhizal plants.

Treatment	Total dry weight (g) per plant	
	Heavily Mycorrhizal	Slightly Mycorrhizal
F_0 Control plants	4.65	1.11
F_1 50 sterilised spores	7.11	0.85
F_2 50 sterilised and squashed spores	5.10	0.81
F_3 50 unsterilised spores	6.83	–
Weighted overall mean	6.21	0.91

The coefficient of correlation (+ 0.84) between the percentage incidence of fungal infection and total dry weight of all the plants was highly significant ($P < 0.001$).   While this does not prove that the better plant growth was due to the

mycorrhiza, Table 4 shows that uninoculated plants
(F_0) and those with squashed spores (F_2) only
weighed on average 2.9 and 2.2g respectively, whereas
those with live inoculum (F_1 and F_3) weighed 4.0 and
6.8g.   One may assume that the former group would
contain as many potentially large plants as the
latter and that the better growth of F_1 and F_3 plants
is therefore attributable to their greater degree of
infection.   Also the beneficial results of VA
mycorrhiza on plant growth in some soils are now well
authenticated (Harley, 1968).

*The effect of plant death on the production of
extra-matrical spores (5)*

        As expected the cut plants died quickly.   The
hooded plants gradually lost their leaves over two
months but even after three months the bases of six
plants were still living.
        The initial assessment of spores showed a mean
number of 11.3 per 25 cm^3 of soil, with a coefficient
of variation of 37%.   Mean spore populations at the
end of the experiment are shown in Table 6.

Table 6.   Mean spore populations in pots
containing Khaya plants subject to quick
versus slow death.

Treatment	Mean no. of spores per 25 ml soil
Quick death by cutting	13.9
Slow death by hooding	11.2
Living plants	8.0
L.S.D. 5%	5.53

In this experiment spore production in the soil appeared to increase when the host plant was weakened by growth in darkness, or killed by cutting off its top.   These results conflict with those of Baylis (1969), who found fewer extra-matrical spores after cutting off the plant tops, but it is not clear whether his topped plants died.

DISCUSSION

In several of the experiments (*1, 3* and *4*) mycorrhizal infection of *Khaya grandifoliola* was related to plant size, large healthy plants having most infection.   Seedlings inoculated with *Endogone* spores became strongly mycorrhizal and weighed seven times more than lightly infected ones.   Endotrophic mycorrhiza therefore seem to be important for plants such as *Khaya* and this has implications for forestry practice.   Care should be taken not to destroy mycorrhizal fungi in nurseries by drastic sterilisation with fumigants such as methyl bromide.

Of the environmental factors tested - moisture levels, light and nutrient supply - moisture had the greatest effect on infection, reducing the number of infected cells by 20%, and different light regimes had the least effect. Of the three nutrients N, P and K, nitrogen most strongly affected fungal development in the root, particularly vesicle formation which was reduced sevenfold at the highest nitrogen level. However, even large additions of nitrogen by no means precluded infection.

In general vesicle development appears rather erratic, seemingly related to the physiological condition of individual rootlets rather than to the whole root system.   Many vesicles were found in dead and moribund roots confirming conclusions of Mason (1964) and Mosse and Bowen (1968).

Water regimes, nutrient supply and plant death had quite marked effects on the numbers of extra-

matrical spores.   Amounts of water optimum for plant
growth also resulted in the greatest production of
extra-matrical spores.   On the other hand high
nitrogen levels, injurious particularly to root
growth (Table 3), increased extra-matrical spores by
62% and killing the plant also increased spore numbers by a similar amount.   If extra-matrical spores
form when roots die this might explain why cultivated
and seasonal savannah soils contain more spores than
soils in the moist forest (Mason, 1964; Mosse &
Bowen, 1968).

Numbers of extra-matrical spores were not correlated with root infection in either experiments on
different water regimes or nutrient levels.   This
contrasts with the findings of Hayman (1970) and Daft
and Nicolson (1972).   From their experiments Daft
and Nicolson suggested that spore numbers were an
effective measure of infection.   The results of the
Nigerian experiments suggest that this might not
always be the case if experimental treatments are
severe enough to cause root injury.

REFERENCES

BAYLIS, G. T. S. 1969. Host treatment and spore prod-
    uction by *Endogone*. *N.Z. J. Bot.*, 7, 173-174.

DAFT, M. J. & NICOLSON, T. H. (1972). Effect of
    Endogone mycorrhiza on plant growth.   IV. Quantit-
    ative relationships between the growth of the host
    and the development of the endophyte in tomato and
    maize.   *New Phytol.*, 71, 287-295.

GERDEMANN, J. W. (1955). Relation of a large soil-
    borne spore to phycomycetous mycorrhizal infection.
    Mycologia, 47, 619-632.

HARLEY, J. L. (1968). Fungal symbiosis: Presidential
    address to the British Mycological Society.

Trans. Br. mycol. Soc., <u>51</u>, 1-11。

HAYMAN, D. S. (1970). *Endogone* spore numbers in soil and vesicular-arbuscular mycorrhiza in wheat as influenced by season and soil treatment.   *Trans. Br. mycol. Soc.*, <u>54</u>, 53-63.

HAYMAN, D. S. (1974). Plant growth responses to vesicular-arbuscular mycorrhiza.   VI.   Effect of light and temperature.   *New Phytol.*, <u>73</u>, 71-80.

KHRUSCHCHEVA, E. P. (1960). Conditions favouring formation of maize mycorrhiza.   *Agrobiologiya*, <u>4</u>, 588-593.

MASON, D. T. (1964). A survey of numbers of *Endogone* spores in soil cropped with barley, raspberry and strawberry.   *Hort. Res.*, <u>4</u>, 98-103.

MOSSE, B. & BOWEN, G. D. (1968). The distribution of *Endogone* spores in some Australian and New Zealand soils and in an experimental field at Rothamsted. *Trans. Br. mycol. Soc.*, <u>51</u>, 485-492.

PEUSS, H. (1958). Untersuchungen zur Ökologie und Bedeutung der Tabakmycorrhiza.   *Arch. Mikrobiol.*, <u>29</u>, 112-142.

POTASSIUM UPTAKE BY ENDOTROPHIC MYCORRHIZAS

C. Ll. POWELL*

Botany Department, Otago University, New Zealand

INTRODUCTION

It is well known that mycorrhizas increase phosphorus uptake and plant P concentrations especially in infertile soils (Mosse, 1973). There is conflicting evidence of some other elements having either higher or lower concentration in mycorrhizal compared to non-mycorrhizal plants.

There are two methods for deciding whether or not mycorrhizas specifically increase the uptake of a particular element. One way is to label the soil with a radioactive isotope. Gray and Gerdemann (1973) demonstrated greater uptake of sulphur in mycorrhizal maize and red clover plants by labelling the sand with ^{35}S and finding much greater radioactivity (c.p.m. [35]S/mg root weight) in the mycorrhizal than in the non-mycorrhizal plants. Similarly Jackson et al., (1973) found much greater ^{90}Sr uptake by mycorrhizal soybeans from ^{90}Sr-amended soil. The other way is to grow mycorrhizal and non-mycorrhizal plants in soil deficient only in one element. Using this method Gilmore (1971) demonstrated increased Zn uptake and plant growth by mycorrhizal peach seedlings in Zn-deficient soil well supplied with N,P,K,Ca,Mg,B and Fe EDTA.

---

* Now at Ruakura Agricultural Research Centre, Hamilton, New Zealand.

I have used this second method to measure
growth and potassium uptake by *Griselinia* seedlings
in a sandy loam.   Many workers have reported significantly lower K concentrations in mycorrhizal
than non-mycorrhizal plants (Gerdemann, 1964;
Holevas, 1966; Ross, 1971; Deal *et al*., 1972;
Kleinschmidt & Gerdemann, 1972).   Mosse (1957) and
Baylis (1959) found increased K concentrations in
mycorrhizal plants, and Ross and Harper (1970) found
no significant difference.   All found increased
total K uptake due to the larger size of the mycorrhizal plants.   These experiments did not conclus-
ively indicate enhanced K uptake by mycorrhizas
however, since most of the soils used were very deficient in P, and well supplied with exchangeable K.

Much of the K in soils is present in the cryst-
alline lattice of the primary minerals and secondary
clay minerals.   A small portion is found as exchangeable K, easily brought into solution by exchange with
other cations, and an extremely small portion exists
as soluble K.   Soils contain about 1-2 per cent
total K.   Exchangeable K normally varies from less
than a hundred to several thousand $\mu g \ g^{-1}$ and soluble
K is usually a few $\mu g \ g^{-1}$ of the soil dry weight
(Russell, 1973).   Over a short time, K is supplied
to roots from the soil solution.   Over an intermediate period, exchangeable K will be released into
solution and over a long term, non exchangeable K
may be released from crystal lattices to supply plant
needs.   When roots are taking up K faster than the
mass flow rate can carry it to the root surface, the
supply of exchangeable K depends upon the diffusion
rate through the soil.

The rate of K uptake by the roots is mainly
dependent on the concentration of K in the soil solution and is little affected by the concentrations of
other ions (Russell, 1973).   Plants begin to respond
to K fertilizers when the exchangeable K can no
longer supply the plant's needs, and when the K in
the primary and secondary minerals has a slow rate of

release (Russell, 1973).   I carried out several
experiments using agricultural soils which had shown
a good response to K fertilizer.   Only in one soil
(described below) did I find a response to both K
fertilizer and mycorrhiza.

MATERIALS AND METHODS

The soil used was a yellow brown sandy loam from
under lupins on coastal sand dunes.   It contained
many mycorrhizal lupin rootlets and sporocarps of
Glomus microcarpus.   The pH was 7.9, exchangeable K
measured by rapid extraction by ammonium acetate was
<15 ppm, and $NaHCO_3^-$ extractable P was 9 ppm.   After
steam-sterilizing the soil, the pH was 7.7, and P 12
ppm.
   Griselinia littoralis (Cornaceae) was the test
plant.   It has short thick hairless roots and
depends on mycorrhizas for growth in P-deficient soil
(Baylis, 1959).   Seedlings germinated on damp filter
paper were planted singly in 250 ml pots of sterilized and unsterilized soil.   Filtered washings of
some unsterilized soil were added to all pots.
Plants were given complete Long Ashton nutrient solution*(C), complete solution lacking K (C-K), or
nitrate only (N) at the rate of 8.3 ml per pot, on
three occasions.   The total amount of K applied in
the complete nutrient solution was equivalent to 65
kg K/ha.   Each treatment had eight replicate pots
of sterilized and sixteen of unsterilized soil.
   Plants were given boiled water every two days.
The fully randomized experiment was maintained in a
glasshouse shaded during midsummer, and was dis-
mantled after 140 days growth.   The plants were air-
dried and part of the roots cleared and stained

---

*Hewitt, E. J. (1965) Sand and Water Culture Methods.
Table 30A, p190.

(Phillips & Hayman, 1970).   Root infection was
determined as the number of 1 mm root segments out of
100 found infected.   Roots and shoots from each
treatment were bulked separately and total P and K
analyses carried out.

RESULTS

   All plants in unsterilized soil became mycorrhizal.   No plants in sterilized soil developed any
infections.   With N only added to the soil, non-
mycorrhizal plants were primarily limited by P deficiency, and mycorrhizas increased growth by 300 per
cent, and plant P concentration by 157 per cent
(Fig. 1, Table 1).   Where C - K nutrient was supplied, plants had enough P and were limited only by K
deficiency.

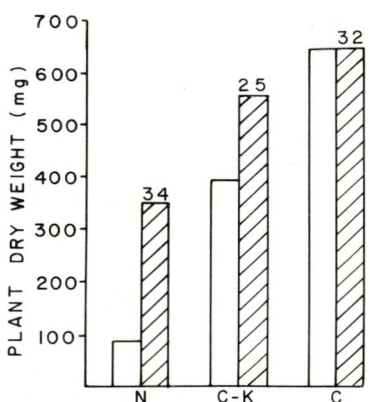

<u>Figure 1.</u>   Effect of mycorrhizas on *Grisel-*
inia growth under three nutrient regimes.
N, nitrate only;   C - K, complete minus potassium; C, complete nutrient.   Cross
hatched columns, unsterilized soil (mycorrhizal plants);   open columns, sterilized
soil (non-mycorrhizal plants).   Infection
percent, as superscript.

Table 1.   P and K composition of *Grisel-*
inia littoralis.

Fert-ilizer app-lied*	Shoot Concentration**				Whole Plant Uptake[‡]			
	P		K		P		K	
	M	NM	M	NM	M	NM	M	NM
N	0.11	0.04	0.57	0.58	0.38	0.03	2.01	0.51
C-K	0.15	0.12	0.89	1.02	0.82	0.45	4.95	4.00
C	0.18	0.12	1.63	1.53	1.13	0.76	10.56	9.91

   * Nutrient abbreviations as in Figure 1.
  ** Percent of shoot dry matter
   ‡ mg per plant

Here, mycorrhizas increased total K uptake by 23 per
cent (Table 1) and plant growth by 42 per cent (Fig.
1).   With K present in the complete solution, there
was a further 65 per cent increase in plant growth
and 149 per cent increase in K uptake, by non-
mycorrhizal plants.   There was no growth response to
mycorrhizal infection.   All growth differences were
significant at $P = 0.01$.

DISCUSSION

In C - K amended soil, mycorrhizal plants had
lower shoot K concentrations than non-mycorrhizal
plants.   This result, also found by most previous
workers, is hard to explain.   It may be that increased K uptake by mycorrhizal plants stimulates an even
greater amount of growth, which in turn dilutes the
overall plant K concentration.   Increased K uptake
by mycorrhizas probably occurs fairly infrequently in
the field, since I had to use a soil of very low

exchangeable K (<15ppm) to demonstrate it at all using single seedlings in pots. It seems probable however, that one plant alone may not use all of the exchangeable K, whereas a whole turf of plants, in the field, will do so and then require additional fertilizer or mycorrhiza for continued growth.

Increased P uptake is without doubt a much more common and important role of mycorrhiza. K uptake is less likely to be enhanced by mycorrhizas, since soluble K is usually maintained at a reasonably high concentration in the soil solution by a diffusion rate 10 to 20 times faster than that of P (rates cited by Newman & Andrews, 1973).

Increased K uptake by mycorrhizas, when it does occur in extremely infertile soils, is probably due to the increased uptake surfaces and more thorough exploration of the soil achieved by mycorrhizal hyphae (Mosse, 1973).

ACKNOWLEDGMENTS

I thank Mr. J. Grigg, Invermay Agricultural Research Centre, Mosgiel, New Zealand for chemical analyses, and Professor G.T.S. Baylis for suggesting the topic and helpful advice.

REFERENCES

BAYLIS, G. T. S. (1959). Effect of vesicular-arbuscular mycorrhizas on growth of *Griselinia littoralis* (Cornaceae). *New Phytol.*, 58, 274-280.

DEAL, D. R., BOOTHROYD, C. W. & MAI, W. F. (1972). Replanting of vineyards and its relationship to vesicular-arbuscular mycorrhiza. *Phytopathology*, 62, 172-175.

GERDEMANN, J. W. (1964). The effect of mycorrhiza on

the growth of maize. *Mycologia,* 61, 342-349.

GILMORE, A. E. (1971). The influence of endotrophic mycorrhizae on the growth of peach seedlings. *J. Am. Soc. hort. Sci.,* 96, 35-38.

GRAY, L. E. & GERDEMANN, J. W. (1973). Uptake of sulphur-35 by vesicular-arbuscular mycorrhizae. *Pl. Soil,* 39, 687-689.

HOLEVAS, C. D. (1966). The effect of a vesicular-arbuscular mycorrhiza on the uptake of soil phosphorus by strawberry (Fragaria sp. var Cambridge favourite). *J. hort. Sci.,* 41, 57-64.

JACKSON, N. E., MILLER, R. H. & FRANKLIN, R. E. (1973) The influence of vesicular-arbuscular mycorrhizae on uptake of ^{90}Sr from soil by soybeans. *Soil Biol. Biochem.,* 5, 205-212.

KLEINSCHMIDT, G. D. & GERDEMANN, J. W. (1972). Stunting of citrus seedlings in fumigated nursery soils related to the absence of endomycorrhizae. *Phytopathology,* 62, 1447-1453.

MOSSE, B. (1957). Growth and chemical composition of mycorrhizal and non-mycorrhizal apples. *Nature,* 179, 922-924.

MOSSE, B. (1973). Advances in the study of vesicular-arbuscular mycorrhiza. *A. Rev. Phytopath.,* 11, 171-196.

NEWMAN, E. I., & ANDREWS, R. E. (1973). Uptake of phosphorus and potassium in relation to root growth and root density. *Pl. Soil,* 38, 49-69.

PHILLIPS, J. M. & HAYMAN, D. S. (1970). Improved procedures for clearing roots and staining parasitic and vesicular-arbuscular mycorrhizal fungi for rapid assessment of infection. *Trans. Br. mycol. Soc.*, <u>55</u>, 158-161.

ROSS, J. P. (1971). Effect of phosphate fertilization on yield of mycorrhizal and non-mycorrhizal soybeans. *Phytopathology*, <u>61</u>, 1400-1403.

ROSS, J. P. & HARPER, J. A. (1970). Effect of *Endogone* mycorrhiza on soybean yields. *Phytopathology*, <u>60</u>, 1552-1556.

RUSSELL, E. W. (1973). *Soil conditions and plant growth*. 10th Ed. Longman, 849 pp.

SPECIFICITY IN VA MYCORRHIZAS

BARBARA MOSSE

Rothamsted Experimental Station, Harpenden, Herts., U.K.

INTRODUCTION

Specificity in host-microorganism relationships covers a range of concepts from host specificity – the restriction of a microorganism to one or several host species – to effectivity – the degree of nutritional or other advantage resulting from the symbiotic association between a particular auto – and heterotroph. In this sense the term effectivity is widely used in studies of symbiotic nitrogen fixation.

Gallaud, in 1905, described two types of vesicular-arbuscular infection differing in the extent and location of arbuscules, but by the mid-1950s it was widely assumed that all these infections were caused by the same or a closely related group of fungi, and that specificity was therefore practically non-existent in VA mycorrhiza. We now know that VA mycorrhiza are caused by a group of phycomycetous fungi sufficiently different taxonomically to make it probable that they belong to different genera. This re-opens the issue of specificity and the questions arise: (1) What determines the distribution of different endophytes? (2) Are some endophytes better adapted to particular conditions? (3) Do endophytes differ in their ability to increase nutrient supply to the host plant, i.e. in effectivity?

REVIEW OF LITERATURE

There is still very little evidence of host specificity and recently VA infections have even been reported in some *Cruciferae* and *Chenopodiaceae* (Ross & Harper, 1973; Williams *et al.*, 1974; Kruckelmann, 1975) increasing the host range to two families previously considered immune to infection. Though complete immunity is rare and mainly confined to plant species that form ectotrophic mycorrhiza, in agricultural situations some plants such as maize (Gerdemann, priv. comm.) and soyabean (Schenck & Kinloch, 1974) appear to be much more heavily infected than others such as potato (Kruckelmann, 1975) or bean (*Phaseolus vulgaris*) (Strzemska, 1975).

There is some evidence that particular endo-phytes are sometimes preferentially associated with particular plant species; Bevege (1971) reported this for *Araucaria* and Porter & Beute (1972) for peanuts and fescue. Fox & Spasoff (1972) found much greater spore replication of *E. gigantea* in assoc-iation with maize, rye and tobacco than with soyabean var. Lee, and Tolle (1958) found surface sterilised mycorrhizal roots of oats and barley to be mutually infective but non-infective with wheat or rye. An endophyte with a fine mycelium *(Rhizophagus tenuis)* may be preferentially associated with high altitude tussock grasses in New Zealand (Crush, 1973) but it can also coexist with the more usual coarse mycelium endophyte in the same root (Ali, 1969; Crush, 1973). Soils, even in adjacent fields, may contain quite different spore populations (Hayman, 1975) but this is not apparently related to host species (Kruckelmann, 1975). On the other hand many inoc-ulation experiments have demonstrated a wide host range of different *Endogone* spore types. For instance yellow vacuolate spores (Mosse & Bowen, 1968) have produced VA infection in twenty plant species belonging to twelve different families (Mosse, 1973).

Evidence is increasing that endophytes can differ in their ability to improve plant growth (Mosse, 1973).   This was first demonstrated in a phosphorus deficient, unsterile field soil in which uninoculated onions became strongly mycorrhizal but nevertheless remained small and phosphorus deficient compared to similar plants inoculated with a non-indigenous endophyte (Mosse, Hayman & Ide, 1969; Mosse and Hayman, 1971).   Benefit from inoculation of an unsterilised soil was also obtained with two citrus species (Kleinschmidt & Gerdemann, 1972). The extent of infection in the non-inoculated seed-lings was not reported, but they grew very much better than those in fumigated soil and presumably this was attributable to infection by the indigenous endophytes.   In other experiments effects of differ-ent endophytes have been compared directly.   These experiments, involving several plant species, indic-ated that: 1. endophytes can differ markedly in the extent to which they improve nutrient uptake (Mosse, 1972a; Gilmore, 1971; Meloh, 1963);   2. the effectivity of a particular endophyte is apparently related to the soil, e.g. the order of effectiveness of seven endophytes differed in two soils, one acid and the other neutral (Mosse, 1972a), and in the same soil before and after liming (Mosse, 1972b); 3. effectivity of a species is not necessarily related to intensity of infection (Mosse, 1972a). 4. contaminating microorganisms associated with different endophytes can also have small but signif-icant effects on growth of the host (Mosse, 1972a). Some of these conclusions have now been confirmed by other work.

INTERACTIONS BETWEEN SOILS AND ENDOPHYTES

Centrosema pubescens inoculated with three different endophytes was grown in two neutral soils with added rock phosphate (Table 1).

Table 1.   Growth and P content of *Centrosema pubescens* inoculated with three different endophytes.   (Mean of 4[+] or 2[++] replicates).

British Honduras soil (pH 7.2)[+]

Endophyte* :	NONE	HON	E₃	YV	*LSD, P=0.05*
Dry weight (g)	0.28	0.29	0.39	0.53	*0.05*
Total P (µg)	265	375	961	1139	*114*
% infection	0	26	97	90	

Ashbridge soil (pH 6.8) + 50% sand[++]

	NONE	HON	E₃	YV	
Dry weight (g)	0.18	0.14	0.22	0.33	*0.11*
Shoot P (µg)	136	130	457	574	*169*
% infection	0	45	98	98	

---

* Endophyte spore types: HON = honey coloured sessile (Mosse & Bowen, 1968); YV = yellow vacuolate (Mosse & Bowen, 1968); E_3 = similar to E_3 (Gilmore, 1968).

In both soils plants infected with the YV endophyte took up most phosphate and grew best, whereas the HON endophyte did not increase growth.   The E_3 spore type in both soils was less beneficial than YV although the percentage infected roots (estimated proportion of the total cortical tissue infected) was equally high for both endophytes.   In another high pH soil (7.5) from Spain, the YV spore type was again much more effective than E_3 (Azcón, Barea & Hayman, priv. comm.).   Lavender plants infected with the YV spore type (60-80% infection) weighed 2 g, those inoculated with E_3 (25-45% infection) weighed 0.44 g and non-infected controls weighed only 0.29 g.   The high effectivity of the YV spore type in neutral or alkaline soils confirms previous experience.

Soil factors other than pH can however also affect the performance of different endophytes. Seedlings of *Paspalum notatum* were grown in three different soils and inoculated with two different endophytes obtained from Brazilian soils (Table 2).

Table 2.   Growth and P content of *Paspalum notatum* grown in three soils inoculated with two different endophytes.   (Mean of 3 replicates).

		Shoot dry wt. (mg)	
	pH	Endophyte A	Endophyte B
Soil 1[o]	5.0	90	82
" 2[o]	4.3	13	76
" 3[oo]	4.0	119	92

[o]   from Brazil

[oo]   from Wareham, England

The consistently poor growth and low infection of plants inoculated with endophyte A in Soil 2 appears to be a specific reaction of this endophyte to the particular soil, rather than a reaction to pH which was even lower in Soil 3.   A similar result obtained by Johnson is quoted by Baylis (1975).   In one New Zealand soil *Coprosma* seedlings inoculated with a *Glomus* sp. weighed ten times more than seedlings inoculated with a *Gigaspora* sp., but in another soil the effectiveness of the two endophytes was reversed and plants inoculated with the *Gigaspora* sp. weighed twice as much.

INDIGENOUS VERSUS INTRODUCED ENDOPHYTES

If introduced endophytes can be more beneficial to plant growth than indigenous ones, this could have important practical applications, especially if it also happened in unsterile soils. This possibility was further examined using a deciduous forest soil (brown earth overlying limestone, pH 5.4; $NaHCO_3$-soluble P, 6 mg/kg soil; $CaCl_2$- soluble P, 0.3 µmoles/l) from Meathop Wood (MW) Lancs. Seedlings of *Viola riviniana* Reichenb., an indigenous species, were transplanted from sand into irradiated (0.8 Mrads)MW soil inoculated with sievings of indigenous fungi (IF) or of E_3, both previously established in pot cultures. Controls were given filtered washings from the IF inoculum. After infection, the seedlings were transplanted into irradiated or unsterile MW soil. Table 3 shows that, although infection levels in E_3 and IF plants in the irradiated soil were similar, E_3 plants weighed three times more and contained more than three times as much P. The non-mycorrhizal control plants (C) weighed least and contained least P, but differences between them and plants inoculated with indigenous endophytes were not statistically significant.

In the unsterile soil control plants (C) became infected from endophytes in the soil. After 10 weeks C plants were very similar to IF inoculated plants in extent of infection, weight and P content, confirming that the IF inoculum was indeed representative of the normal endophyte population in that soil. Both C and IF plants grew very little compared to E_3 inoculated plants. Although E_3 inoculated plants were much larger in the irradiated soil, their relative advantage over IF plants was greater in the unsterile soil.

A similar result was obtained with ash seedlings *(Fraxinus excelsior* L.). As seed was difficult to germinate the plants were raised by rooting cuttings of one-year old seedlings dug up from the

Table 3.   Mean dry weight and P content of
Viola seedlings inoculated with an introduced
and an indigenous endophyte in irradiated and
unsterile Meathop Wood soil.

IRRADIATED SOIL (7 replicates)

Treatments* :	C	IF	E₃	*LSD, P = 0.05*
Dry weight/ plant (mg)	115	157	529	*94*
P content/ plant (µg)	125	317	1021	*214*
Infection	none	medium	medium	

UNSTERILE SOIL (10 replicates)

Dry weight/ plant (mg)	9.4	9.3	59	*14*
P content/ plant (µg)	12	10	95	*9*
% infection	50	50	80	

---

* C = uninoculated;   IF = inoculated with indigenous
  endophytes;   E₃ = inoculated with endophyte
  similar to E₃ (Gilmore, 1968).

===

wood.   This material was rather variable.   Table 4
shows total weight and P content of plants at the
final harvest.   Similar results were obtained at two
intermediate harvests.   The figures include the
original weight and P content of cuttings.   When
these are subtracted, control (C) cuttings in irradiated soil and IF inoculated cuttings in unsterile
soil showed almost no gain in weight or P content.

Table 4.   Mean dry weight and P content of
ash seedlings inoculated with an introduced
and an indigenous endophyte in irradiated
and unsterile Meathop Wood soil.

IRRADIATED SOIL (5 replicates)

Treatments* :	C	IF	E_3	LSD, P = 0.05
Dry weight/ plant (mg)	226	346	857	265
P content/ plant (µg)	135	574	1498	303
% infection	0	58	83	

UNSTERILE SOIL (5 replicates)

Dry weight/ plant (mg)	—	265	454	230
P content/ plant (µg)	—	220	404	235
% infection	—	90	85	

* Treatments C, IF and E_3 as in Table 2.

Although a few new leaves developed older ones
dropped off.   Again the introduced endophyte E_3 was
much more beneficial than the indigenous (IF);   in
the unsterile soil the difference between the two was
significant at the 10% level.   Again E_3 inoculated
plants grew much better in irradiated than in unsterile soil.   While this is not unusual, such
differences are usually smaller and there can also be
marked benefits from inoculation of unsterile soils
(Mosse, 1973).

The establishment of an introduced endophyte in competition with an indigenous one was demonstrated in another experiment with a tropical forage grass Panicum *(Brachiaria* sp.). Plants, taken from their natural habitat, had all their roots and lower sheathing leaves removed and, after thorough washing, were placed in an irradiated soil with and without E_3 inoculum. After ten weeks growth and one inter-mediate harvest, E_3 inoculated plants weighed approx-imately 30% more and contained approximately 25% more P (Table 5), but on examination both sets of plants were found to be strongly mycorrhizal. Infection of the uninoculated plants must have arisen from endo-phyte inoculum lodged in the root base and not removed by washing. Such infections were morpho-logically different and clearly distinguishable from those caused by E_3, which contained many large vesicles.

Table 5. Effects of indigenous (IF) and introduced (E_3) endophytes on growth of *Brachiaria* sp. (4 replicates)

Treatments	IF	IF+E_3	*LSD, P = 0.05*
Dry weight (g/plant)			
1st Harvest	1.11	1.45	*0.48*
2nd Harvest	1.00	2.37	*0.21*
Total P (µg/plant)			
1st Harvest	276	338	*112*
2nd Harvest	347	446	*37*
% infection	100	100	

Morphological evidence of the establishment of an endophyte introduced into an unsterile soil was also obtained in another experiment with *Centrosema*

Table 6. Infection characteristics of *Centrosema* in sterile and unsterile soil with and without an introduced endophyte.

	Plant weight (g)	No. of root segments (out of 20)			
		mycorrhizal	infected throughout cortex	with many vesicles	with much external mycelium
UNSTER. SOIL					
not inocu-* lated	0.85	20	0	0	3
	0.50	20	1	0	0
	0.66	20	0	0	7
inoculated –	0.89	20	5	14	9
E₃	0.91	18	11	18	18
IRRAD. SOIL					
inoculated –	1.89	20	20	17	14
E₃	1.68	20	19	17	18
	1.51	20	16	13	18

* infected from indigenous endophytes

as the host plant. The E_3 endophyte was established
in the seedlings immediately after germination and
they were already mycorrhizal when they were placed
into the unsterile soil. Some were also planted in
irradiated soil. After 10 weeks the plants were
harvested and twenty 1 cm root pieces from each
plant (2 or 3 replicates) were examined microscopically. Virtually all roots of all the plants were
mycorrhizal but infections in non-inoculated plants,
caused by the indigenous endophytes, looked very
different from those caused by E_3 (Table 6).
Characteristics of E_3 infections were more pronounced
in the irradiated soil but still clearly recognisable
in the unsterile soil. The indigenous fungi from
the unsterile soil produced sporadic infections and
the fungus remained confined mainly to the inner
cortex, whereas E_3 spread abundantly throughout the
cortex. The indigenous fungi produced very few
vesicles and this was due more to inherent characteristics of the particular endophyte than to differences in nutritional status of the host, because in
the unsterile soil the E_3 inoculated plants weighed
only a little more than those infected with indigenous endophytes. Very much external mycelium was
attached to E_3 infected roots in irradiated soil,
very little to roots with natural infection, and E_3
inoculated plants in unsterile soil were intermediate.

DISCUSSION

The demonstration that VA endophytes differ in
their ability to improve plant growth raises two
questions:- 1) What causes the differences in effectiveness of different endophytes? 2) Can field inoculation improve yields sufficiently to make it a
practical proposition?
Much evidence suggests that VA endophytes
improve plant P uptake by virtue of the network of
external hyphae that explore the soil beyond the root

hair zone and take up available phosphate that would
not otherwise be accessible to the root.   This
phosphate is translocated within the hyphal network,
reaches the fungus within the root and passes into
the plant either by active transfer out of the arbuscules, by arbuscule breakdown, by leakage, or any
combination of these mechanisms.   Differences in
effectiveness could therefore be related to the
mechanics of the infection, i.e. the extent and
distribution of active mycelium in the soil, the
proportion of external to internal mycelium, the
number of connecting hyphae and the total number of
infected roots.   They could also be due to physiological differences between endophytes in rates of nutrient uptake, translocation and release, in utilisation of plant metabolites, in longevity of soil
hyphae or of arbuscules, or in any particular interaction between endophytes and their soil environment.
We know that certain endophytes do not grow in low pH
soils and that others develop poorly after liming
(Mosse, 1972a & b).   Some endophytes are apparently
specific in their interactions with nematodes
(Schenck *et al.*, 1975), and they are sensitive to
temperature (Schenck & Schroder, 1974; Furlan &
Fortin, 1973).   It is now clear that VA mycorrhiza
can be formed by fungi with very different spores,
fruit bodies and life histories.   According to the
classification of *Endogonaceae* proposed by Gerdemann
and Trappe (1974) four different genera may be
involved.   Such a varied group of fungi are likely
to differ in their optimum growth requirements and in
their reactions to environmental conditions.   Thus,
other specific interactions between VA endophytes and
the environment will be found, which will in turn
affect their symbiotic efficiency.

     The evidence presented in this paper confirms
that non-indigenous endophytes can be introduced into
both irradiated and unsterile soils, that they can
become established in competition with the indigenous
endophytes and can improve plant growth.   This

suggests that the practical possibility of field
inoculation should now be seriously considered.   The
present inability to culture VA endophytes presents
some difficulty, particularly for crops raised from
seed rather than planted out from a nursery, but this
need not prove insurmountable.   Lyophilisation and
seed pelleting are already receiving some attention
(Crush & Pattison, 1975).   Little is known so far
about the persistence of an introduced endophyte in
the soil and its rate of spread.   If it invigorates
its host more than the indigenous endophytes, it
should spread automatically, especially in a pasture
situation.   Such a process of spread has been
observed in citrus nurseries after fumigation, where
islands of green plants gradually enlarged as the
endophytes spread (Kleinschmidt & Gerdemann, 1972).
No one has yet examined to what extent such spread
would occur in an unsterilised soil.   The ability to
recognise some endophytes by the anatomy of the
infected root might be a useful tool for such
studies.   It is improbable that the benefits of
field inoculation will be as great as those of inoculation in irradiated soil in pots where optimum
growth conditions are provided.   But, unless the
introduced endophyte fails to get established or
interacts in some particular way with the existing
soil microflora, some improvement in the plant growth
should result.

REFERENCES

ALI, B. (1969). Occurrence and characteristics of the
    vesicular-arbuscular endophyte of *Nardus stricta.*
    Nova Hedwigia, 17, 409-425.

BAYLIS, G. T. S. (1975). This symposium, p. 373.

BEVEGE, D. I. (1971). Vesicular-arbuscular mycorrhi-
    zas of *Araucaria*: Aspects of their ecology and

physiology and role in nitrogen fixation.   Ph.D.
thesis, University of New England, Armidale, N.S.W.

CRUSH, J. R. (1973). Significance of endomycorrhizas
in tussock grassland in Otago, New Zealand.
N.Z.Jl. Bot., <u>11</u>, 645-660.

CRUSH, J. R. & PATTISON, A. C. (1975). This
symposium, p. 485.

FOX, J. A. & SPASOFF, L. (1972). Host range studies
of *Endogone gigantea*.   *Virginia Journal of
Science*, <u>23</u>, 121, abstr.

FURLAN, V. & FORTIN, J. A. (1973). Formation of endomycorrhizae by *Endogone calospora* on *Allium cepa*
under three temperature regimes.   *Naturaliste
Canadien*, 100, 467-477.

GALLAUD, I. (1905). Etudes sur les mycorrhizes endo-
trophes.   *Revue gén. Bot.*, <u>17</u>, 5-48.

GERDEMANN, J. W. & TRAPPE, J. M. (1974). The *Endogonaceae* in the Pacific Northwest.   *Mycologia Memoir*,
<u>5</u>, 75 pp.

GILMORE, A. E. (1968). Phycomycetous mycorrhizal
organisms collected by open-pot culture methods.
Hilgardia, <u>39</u>, 87-105.

GILMORE, A. E. (1971). The influence of endotrophic
mycorrhizae on the growth of peach seedlings.
J.Am.Soc.Hort.Sci., <u>96</u>, 35-38.

HAYMAN, D. S. (1975). This symposium, p.495.

KLEINSCHMIDT, G. D. & GERDEMANN, J. W. (1972).
Stunting of citrus seedlings in fumigated nursery
soils related to the absence of endomycorrhizae.
Phytopathology, <u>62</u>, 1447-1453.

KRUCKELMANN, H. W. (1975). This symposium, p.511.

MELOH, K. A. (1963). Untersuchungen zur Biologie der endotrophen Mycorrhiza bei *Zea mays* L. und *Avena sativa* L.. *Arch. Mikrobiol.*, 46, 369-381.

MOSSE, B. (1972a). The influence of soil type and *Endogone* strain on the growth of mycorrhizal plants in phosphate deficient soils. *Rev. Ecol. Biol. Sol.*, 9, 529-537.

MOSSE, B. (1972b). Effects of different *Endogone* strains on the growth of *Paspalum notatum. Nature,* London., 239, 221-223.

MOSSE, B. (1973). Advances in the study of vesicular-arbuscular mycorrhiza. *A. Rev. Phytopathol.*, 11, 171-196.

MOSSE, B. & BOWEN, G. D. (1968). A key to the recognition of some *Endogone* spore types. *Trans. Br. mycol. Soc.*, 51, 469-483.

MOSSE, B. & HAYMAN, D. S. (1971). Plant growth responses to vesicular-arbuscular mycorrhiza. II. In unsterilized field soils. *New Phytol.*, 70, 29-34.

MOSSE, B., HAYMAN, D. S. & IDE, G. J. (1969). Growth responses of plants in unsterilized soil to inoculation with vesicular-arbuscular mycorrhiza. *Nature, London.,* 224, 1031-1032.

PORTER, D. M. & BEUTE, M. K. (1972). *Endogone* species in roots of Virginia type peanuts. *Phytopathology,* 62, 783, abstr.

ROSS, J. P. & HARPER, J. A. (1973). Hosts of a vesicular-arbuscular *Endogone* species. *J. Elisha Mitchell Sci. Soc.*, 89, 1-3.

SCHENCK, N. C. & KINLOCH, R. A. (1974). Pathogenic
fungi, parasitic nematodes and endomycorrhizal
fungi, associated with soyabean roots in Florida.
Plant Dis. Reptr., <u>58</u>, 169-173.

SCHENCK, N. C., KINLOCH, R. A. & DICKSON, D. W.
(1975). This symposium, p. 607.

SCHENCK, N. C. & SCHRODER, V. N. (1974). Temperature
response of *Endogone* mycorrhiza on soybean roots.
Mycologia, <u>66</u>, 600-605.

STRZEMSKA, J. (1975). This symposium, p.537.

TOLLE, R. (1958). Untersuchungen über die Pseudomy-
corrhiza von Gramineen.   *Arch. Mikrobiol.*,
<u>30</u>, 285-303.

WILLIAMS, S. E., WOLLUM, A. G., & ALDON, E. F. (1974).
Growth of *Atriplex canescens* (Pursh) Nutt. improved
by formation of vesicular-arbuscular mycorrhizae.
Proc. Soil Sci. Soc. Am., <u>38</u>, 962-965.

PRELIMINARY RESULTS ON THE PRODUCTION OF VESICULAR-ARBUSCULAR MYCORRHIZAL INOCULUM BY FREEZE DRYING

J. R. CRUSH* AND A. C. PATTISON

Rothamsted Experimental Station, Harpenden, Herts., U.K.

INTRODUCTION

Many plant species take up phosphorus more efficiently from phosphate deficient soils when they are mycorrhizal. In sterilised soils, plant growth can be very much increased by inoculation with *Endogone sp.*, and even in unsterilised soils, which usually contain some VA endophytes, the increase can be considerable. This may be because of low inoculum potential in the soil, so that annuals in particular are only lightly infected at the time of maximum vegetative growth and nutrient requirement. Also endophyte strains indigenous in a particular soil may be poor compared with introduced strains (Mosse, 1975). In small field trials inoculation has improved growth of crops such as soyabean (Ross & Harper, 1970), maize, (Khan, 1972), wheat (Khan, 1975), and citrus (Kleinschmidt & Gerdemann, 1972). If such work is to be extended, and particularly if inoculation on a field scale is to become a practical possibility, methods must be found of producing

* Grasslands Division, DSIR, Christchurch, New Zealand.

enough inoculum, of incorporating it with the seed,
or of inoculating during the nursery stage of transplanted crops such as tobacco.

Although *Endogone* still cannot be cultured on
synthetic media, methods of producing mycorrhizal
inoculum have gradually improved; from mixed inocula
of naturally infected soil or roots, to inocula of
known strains of endophytes raised in association
with a host plant grown in a sterilised medium (soil,
sand or agar) inoculated with surface sterilised
spores.   From such cultures one can obtain inoculum
either in the form of infected roots, or of soil
sievings containing extra-matrical spores and mycelium.   Such inocula do not, however, store well.

Preserving fungal cultures by freezing and
desiccation has become standard practice in many
laboratories.   Mazur (1968) reported great variability in survival of strains and species at different
stages of the process.   In general Phycomycetes are
more difficult to lyophilise than septate fungi but,
at the Commonwealth Mycological Institute (Kew,
England) some Mucorales have been successfully lyophilised.   Recently Jackson, Franklin and Miller
(1972) successfully used freeze dried mycorrhizal
soybean roots as an inoculum in pot experiments.
Three week old soybean roots mycorrhizal with *Glomus
mosseae* were frozen at -20°C and dried under vacuum.
The dried roots were ground in a Wiley Mill (20 mesh)
and used as an aqueous suspension.

In this paper we report results of some preliminary trials of seed pelleting with fresh and lyophilised spores and mycorrhizal roots.

MATERIALS AND METHODS

Lyophilisation and Pelleting

Two methods were tested: two-stage (with) and
single-stage (without) preliminary freezing.   In

both methods the material, placed under high vacuum, freezes and ice is sublimed off using P_2O_5 as a drying agent.   In the two-stage process the material is frozen before applying the vacuum.

a)   *Single-stage lyophilisation:* About 0.5g of mycorrhizal roots or sievings were placed in a 0.5 ml glass ampoule (Johnsen and Jörgensen Ltd., London, U.K.) pushed well down to occupy approximately 1cm at the ampoule base and held in place by a loose cotton wool plug.   After constriction the ampoule was evacuated on the manifold of a high vacuum pump (Edwards 2SC30, Edwards High Vacuum Ltd., Crawley, Sussex, U.K.).   Fresh P_2O_5 was placed in the drying chamber and the pump first run with the air ballast system open to speed up removal of water vapour. Within a few minutes the ampoule contents had cooled. Evacuation was continued and the air ballast system closed after 2 hours.   After a further 16 hours under high vacuum (approx. 10 µm Hg) ampoules were sealed with a microburner, removed from the manifold, and stored at -5°C.

b)   *Two-stage lyophilisation:* About 5g of mycorrhizal roots were placed in a McCartney bottle, brought to -30°C in a deep freeze cabinet and quickly transferred to a desiccator at -10°C in a 33% W/W ice-salt mixture.   A little calcium chloride was placed in the base of the desiccator to remove some of the water vapour at this stage.   The vacuum pump was connected to the desiccator and pumping commenced with the air ballast open.   After 5 hours the air ballast was closed and pumping continued under high vacuum for a further 1 hour, during which the specimen temperature rose to ambient (approx 20°C). Air was then slowly re-admitted.   The lyophilised roots were stored in the McCartney bottle at -5°C at atmospheric pressure.

c)   *Pelleting:* Pellets were made by mixing white clover seeds with the inoculum (approximately 5 seeds to 0.5g of sievings or infected roots) in a drop of aqueous high viscosity methyl cellulose and rolling the mixture in pure, finely divided calcium carbonate.   Some inoculum was also pelleted without seed.

Five different strains of endophyte were tested, viz. honey-coloured sessile, yellow-vacuolate, laminate, bulbous-reticulate (Mosse & Bowen, 1968) and strain E_3 (after Gilmore, 1968).

Seedlings of *T. repens* or *T. pratense,* grown in an irradiated field soil (pH 7.0) were used as test plants.

RESULTS

Experiment 1

The infectivity of single-stage lyophilised sievings and of mycorrhizal roots of onion and *Coprosma* was compared with that of fresh material.   The endophyte was E_3.   Seed was either pelleted with the inoculum or sown 2cm above a pellet of inoculum. There were 5 - 10 replicates in each of six treatments, viz. seed pelleted with lyophilised mycorrhizal roots or sievings, with fresh roots or sievings, or unpelleted seed sown above a pellet of fresh sievings or without inoculum.   An inoculum of *Rhizobium trifolii* RCR5* was also given to all plants after emergence. Plants were harvested serially at 4, 6 and 8 weeks when dry weight and extent of infection were recorded.

All plants grown from seed pelleted with fresh or lyophilised sievings, with fresh roots or sown above a pellet of fresh sievings had become infected after 8 weeks, with about 70% of mycorrhizal roots. Only plants from seed pelleted with single-stage lyophilised mycorrhizal roots and uninoculated plants remained non-mycorrhizal.

---

* From the Rhizobium Collection, Soil Microbiology
  Dept., Rothamsted Experimental Station.

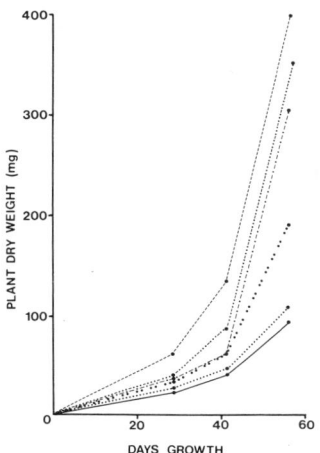

Figure 1. Comparison of dry weight gain of white clover seedlings with fresh & lyophilised spores and roots. ─────── , control (non-mycorrhizal); ············ , seed and single-stage lyophilised mycorrhizas; ── ── ── , seed and single-stage lyophilised spores; ─·─·─·─·─ , seed and fresh roots; ── ·· ── ·· ── ·· ── , seed and fresh spores; ─────── , fresh spores below seed.

In the soil used, growth of *T. repens* is stimulated by VA mycorrhizas (Crush, 1974). Figure 1 shows that pellets of fresh sievings placed 2cm below the seed had the best and most rapid effects on seedling growth. Growth improvements appeared more slowly after pelleting with fresh sievings and were slowest after pelleting with lyophilised sievings. Uninoculated plants grew least and were not significantly different from those grown from seed pelleted with lyophilised roots. Pelleting therefore was not injurious to the inoculum but mycorrhizal roots were non-infective after single-stage lyophilisation.

Experiment 2

Roots of six-week-old soyabeans infected with

one of five different endophytes were lyophilised by
the single or two-stage process.   They were stored
for 3 days at $-5^{O}C$ and then used as inoculum either
in the form of root segments, or as a ground powder
placed 2cm below young seedlings of *T. repens*.
There were six replicates of each treatment and
infection was recorded after eight weeks.

All roots lost their infectivity after single-
stage lyophilisation and those infected with honey-
coloured, laminate or bulbous reticulate endophytes
were also non-infective after two-stage lyophilisation.   However, 1cm root segments infected with
the yellow vacuolate endophyte infected all six test
plants, producing 60-80% infection.   Root segments
infected with E_3 also retained some infectivity and,
after eight weeks produced up to 75% infection in
T. repens.   All infectivity was lost if the lyophilised roots were ground or stored for 9 weeks.   In
another experiment a little infectivity was retained
in E_3 infected clover roots lyophilised by the two-
stage process and stored for 9 weeks.

Experiment 3

Roots of eight-week-old soya beans, heavily
infected (55-95%) with one of five endophytes, were
lyophilised by the two-stage process and an inoculum
of the intact lyophilised roots, stored for 3 days
at $-5^{O}C$, was placed below young seedlings of *T. pratense*.   There were six replicates and plants were
harvested after eleven weeks.

Table 1 shows that infectivity was quite well
preserved after two-stage lyophilisation with 1-40%
mycorrhizal infection in inoculated plants.   As
before, roots infected with the yellow vacuolate
endophyte survived lyophilisation well, and those
infected with E_3 retained some infectivity.   The
honey-coloured endophyte also survived well, contrasting with loss of infectivity in experiment 2.
No infectivity was retained in roots infected with

Table 1.   Infectivity of two-stage lyophilised soya bean roots on *Trifolium pratense*
eleven weeks after inoculation.   (Six
replicates).

Endophyte strain	No. of infected plants	% Infect-ion	No. of plants with at least 5% infection
Honey-coloured, sessile	5	1 - 10	5
Yellow-vacuolate	6	1 - 30	5
E_3	5	1 - 40	3
Laminate	6	2	0
Bulbous reticulate	0	0	0
Control (non-inoculated)	1	2	0

the bulbous reticulate endophyte;   as fresh inocula
of this endophyte are not always successful this lack
of infectivity may not be entirely attributable to
lyophilisation.   One of the uninoculated control
plants in this experiment also became infected (2%).
Plants with only slight infection are therefore suspect and are omitted from the last column in Table 1.

DISCUSSION

These experiments confirm that mycorrhizal inoculum either in the form of sievings or infected roots
can survive lyophilisation.   Sievings survived both
single- and two-stage lyophilisation but roots only
survived the latter.   The better survival of sievings containing spores is not surprising because
resting spores are thick-walled and naturally adapted
for survival.   They contain much lipid and are
probably less hydrated than mycorrhizal roots.   It
seems that the early stages of lyophilisation,

i.e. the initial freezing may be critical for survival and might repay more detailed study.

Vesicles in the root may perform similar storage functions as the extra-matrical spores and resemble them in their contents and structure.   It is therefore possible that roots with many vesicles might be better adapted to lyophilisation.   Such roots can be produced by appropriate feeding of the source plants. Contrary to the findings of Jackson, Franklin and Miller (1972) grinding the lyophilised roots destroyed their infectivity, possibly by damaging the vesicles.

Seed pelleting showed some degree of success although sowing seed above pelleted inoculum was preferable.   This is not surprising because emerging roots would very quickly grow past the pellet of inoculum around the seed coat.   Possibly larger pellets with inoculum and seed dispersed in an inert medium like powdered lignite (such as are being developed for *Rhizobium* inoculum) might offer advantages.   Alternative methods of establishing infection in the field would be direct inoculation of rows, or pre-inoculation of plants that are raised in nurseries or small containers.

Clearly much remains to be found out about lyophilisation of mycorrhizal inocula but, such inocula have potential advantages over fresh material in longevity, ease of storage and handling.

ACKNOWLEDGMENTS

Thanks are due to Mr. K. Richards for technical help.

REFERENCES

CRUSH, J. R. (1974). Plant growth responses to vesicular-arbuscular mycorrhiza VII.   Growth and nodulation of some herbage legumes.   *New Phytol.*, 73, 743-749.

GILMORE, A. E. (1968). Phycomycetous mycorrhizal organisms collected by open-pot culture methods. *Hilgardia,* 39, 87-105.

JACKSON, N. E., FRANKLIN, R. E. & MILLER, R. H. (1972). Effects of vesicular-arbuscular mycorrhizae on growth and phosphorus content of three agronomic crops. *Proc. Soil. Sci. Soc. Am.,* 36, 64-67.

KHAN, A. G. (1972). The effect of vesicular-arbuscular mycorrhizal associations on growth of cereals 1. Effects on maize growth. *New Phytol.,* 71, 613-619.

KHAN, A. G. (1975). The effect of vesicular-arbuscular mycorrhizal associations on growth of cereals II.   Effects on wheat growth *J. appl. Biol.* (in press).

KLEINSCHMIDT, G. D. & GERDEMANN, J. W. (1972). Stunting of citrus seedlings in fumigated nursery soils related to the absence of endomycorrhizae. *Phytopathology,* 62, 1447-1453.

MAZUR, P. (1968). Survival of fungi after freezing and desiccation.   In *The Fungi, An Advanced Treatise,* pp 325-394.   Ainsworth, G. C. & Sussmann, A. S. eds., Academic Press, London.

MOSSE, B.   This symposium. p.469.

MOSSE, B. & BOWEN, G. D. (1968). A key to the recognition of some *Endogone* spore types.   *Trans. Brit. mycol. Soc.,* 51, 469-483.

ROSS, J. P. & HARPER, J. A. (1970). Effect of *Endogone* mycorrhiza on soybean yields. *Phytopathology,* 60, 1552-1556.

THE OCCURRENCE OF MYCORRHIZA IN CROPS AS AFFECTED BY SOIL FERTILITY

D. S. HAYMAN

Rothamsted Experimental Station, Harpenden, Herts., U.K.

INTRODUCTION

There are many reports on the occurrence of vesicular-arbuscular (VA) mycorrhiza in diverse crops throughout the world.   However, quantitative studies on the effects of different fertilizer treatments on mycorrhizal propagules and infection in the field are few.   Fertilizer effects have been studied more in the glasshouse, but conditions in pots, where large plant growth responses to mycorrhiza have been reported, are rather different from conditions in the field.   For example, the volume of soil accessible to roots is smaller in pots and the test soils have usually been sterilized.   Field observations are therefore important for a realistic assessment of the significance of mycorrhiza in agriculture.

This paper summarizes quantitative studies at Rothamsted on the effects of fertilizers on the occurrence of VA mycorrhiza and *Endogone* spores under agricultural conditions.   Quantitative field observations from elsewhere and results from glasshouse and laboratory experiments are discussed, and the type of information that should be obtained in future field studies is considered.

FIELD OBSERVATIONS AT ROTHAMSTED

Spores and roots were recovered by wet-sieving and decanting from soil samples collected from fertilizer experiments on the silty clay loam at the Rothamsted farm (Hayman, 1970; Hayman *et al.,* 1975) and the sandy loam at Woburn. Spore numbers and root infections were assessed as described elsewhere (Hayman, 1970). Spores were identified according to Mosse and Bowen (1968a).

General fertilizers

Table 1 gives the numbers of *Endogone* spores and the amount of VA root infection found in September in samples from plots which had grown wheat and received the same fertilizer treatments annually since 1843 (see Rep. Rothamsted exp. Stn. for 1968, Part 2). Two of the most striking results here are the much larger mycorrhizal population in soil never given fertilizer and the presence of several *Endogone* spore types in plots having grown the same crop for so long. The population of white reticulate spores was the most affected by the addition of fertilizer.

Nitrogen fertilizers

Wheat plots with or without nitrogen fertilizer and formalin treatment were sampled monthly. They showed a large negative effect of nitrogen fertilizer (188 kg N/ha as 'Nitro-Chalk') on the mycorrhizal population. Table 2 gives the figures for September. Plots not given N contained 2 to 7 times more *Endogone* spores and 2 to 4 times more VA infection, depending on their formalin treatment, than plots given N. All spores were of the laminate type although this field grew the same crop and was close to Broadbalk which contained at least four spore types.

Since the negative effect of N fertilizer on mycorrhiza in field-grown wheat was so pronounced in

Table 1. Mycorrhiza in Broadbalk wheat plots given different fertilizer treatments.

Fertilizer treatment	Amount of VA infection[+]		Numbers of *Endogone* spores per 100 g soil[+]				
	% root segments with any VA infection	% length of VA-infected roots	White Reticulate	Bulbous Reticulate	Laminate	Yellow Vacuolate	Total*
No fertilizer	61	36	72	66	30	28	378
N	33	13	16	60	18	20	260
NP	17	5	24	70	18	24	238
NPK Na Mg	22	10	14	36	32	14	238

* including *Endogone* spores not clearly assignable to type

+ Means of 3 bulked samples.

Table 2.   Mycorrhiza in Little Knott wheat plots with and without nitrogen fertilizer.

Treatment		Amount of VA infection[*]		Numbers of laminate Endogone spores per 100 g soil[+]
		% root segments with any VA infection	% length of VA-infected roots	
No formalin in previous year	No N	57	33	180
	+ N	31	12	52
Formalin in previous year only	No N	39	17	100
	+ N	15	4	14
Formalin in previous and other years	No N	45	22	154
	+ N	33	11	78

[*] In the complete date for Little Knott (Hayman, 1970, text-figures 2 and 3), the histogram captions for % roots with any infection and % root length infected were reversed after the return of the galley proofs.

[+] Means of 3 bulked samples.

the heavy soil at Rothamsted, plots of wheat given different amounts of N fertilizer in the light soil at Woburn were examined for comparison. Samples were collected in October from Stackyard field which is on a 5 year rotation with wheat following potatoes following an arable or pasture legume. Table 3 shows that N fertilizer (63 and 188 kg N/ha as 'Nitro-Chalk') decreased spore numbers even more than at Rothamsted (Table 2) and this effect was consistent throughout plots given different subsidiary treatments. Mainly spores of the reticulate type, both white and bulbous, were found in these plots, plus a few laminate spores. The preponderance of white reticulate spores in this field may partly account for the greater susceptibility of the *Endogone* spore population to additions of N fertilizer (cf. Table 1).

Table 3. Population of *Endogone* spores in Woburn Stackyard wheat plots given different amounts of nitrogen fertilizer.

Treatment (kg N/ha as 'Nitro-Chalk')	Numbers of *Endogone* spores per 100 g soil[+]		
	White Reticulate	Bulbous Reticulate	Total *Endogone* spores
0	421	40	595
63	84	16	183
188	43	10	77

[+] Means of 12 bulked samples.

The effects of N fertilizer on mycorrhiza at
Woburn were also examined in a crop which produces
its own N, viz. field beans (*Vicia faba*).   Samples
collected in October from plots in Butt Furlong field
yielded spores of the laminate type only and numbers
ranged from an average of 136/100 g soil in the no N
plots to an average of 147/100 g soil in plots given
126 kg N/ha as 'Nitro-Chalk' and 203/100 g soil in
plots given 252 kg N/ha as 'Nitro-Chalk'.   Spore
counts were very variable and these figures are not
statistically different, but it seemed that N fertilizer had no clearcut effect on spore numbers associated with this crop.

Phosphate fertilizers

Figures in Table 1 suggested a slightly negative
effect of phosphate fertilizer.   To expand this
observation Great Field IV at Rothamsted, which
embraces a wide range of phosphate treatments and
grows potatoes, barley and swedes on rotation, was
examined for mycorrhiza.   Samples were collected in
September in two successive years.   Table 4 shows
that the effects of P fertilizer were not as marked
as the effects of N fertilizer (compare Tables 2 and
3).   Two trends were apparent however, namely that
most VA infection was present in plots given least
phosphate (no phosphate and rock phosphate) and most
spores occurred in plots given intermediate amounts
of P (a total of 330 kg P/ha applied as superphosphate over 12 years).   Swede roots had no VA
infection.   The fact that spore numbers did not fall
drastically in the swede plots suggests that the
spores remained viable for at least a year because
the weeds in these plots were mostly non-mycorrhizal
species.

Table 4. Mycorrhiza in Great Field IV plots given different phosphate treatments.

Total kg P/ha superphosphate over 12 years	ppm NaHCO$_3$⁻ soluble P	Crop	Amount of VA mycorrhiza (% root length infected)[+]	Numbers of laminate *Endogone* spores per 100 g soil[+]
0	8	Potatoes	20	100
	8	Barley	21	148
	8	Swedes	0	109
165 (Rock P, once)	12	Potatoes	25	105
	13	Barley	8	159
	10	Swedes	0	134
330 (2 rotavations)	13	Potatoes	17	173
	15	Barley	9	251
	14	Swedes	0	123
330 (triennial)	24	Potatoes	3	148
	29	Barley	5	231
	24	Swedes	0	112
330 (annual)	23	Potatoes	8	113
	24	Barley	5	250
	21	Swedes	0	131
660 (annual)	39	Potatoes	3	111
	34	Barley	2	176
	30	Swedes	0	132

+ Means of 3 bulked samples

FIELD OBSERVATIONS ELSEWHERE

A detailed account of the effects of different
fertilizer and manurial treatments in some German
soils is given by Kruckelmann (1973).   In a loamy
sand he found fewest spores in plots given no fertilizer and most in plots given NPK plus farmyard
manure (FYM).   This contrasts with results at
Rothamsted and the widely held view that VA mycorrhiza tends to be most abundant in soils with a low or
unbalanced nutrient content.   He also found more
Endogone spores with increased applications of N,
although different amounts of P or FYM applied to
certain soils did not significantly affect spore
numbers.   On the other hand Khan (1972) and
Krushcheva (1960) reported decreased mycorrhizal
development in the field when phosphate fertilizer
was added and Porter and Beute (1972) found that
peanuts grown in soils containing little N had more
Endogone spores than in soils containing much N.
In field plots of hoop pine *(Araucaria cunninghamii)*
given different proportions of N and P, Bevege (1971)
found a trend towards increased VA mycorrhizal
infection with increasing application of N (448 to
4032 kg N/ha as urea), especially at intermediate
levels of P (37 and 74 kg P/ha as super-phosphate),
but with 148 kg P/ha infection decreased with
increasing N.   There were most *Endogone* spores at
low and high levels of N, and fewer spores with no
P than with various amounts of P.   Clearly results
may be very clearcut at individual sites but they
vary in different soils and this makes generalizations difficult.   The initial fertility of a soil
before the addition of fertilizers must be a major
factor determining the mycorrhizal response.

A further complication is that certain crop
plants are heavily mycorrhizal even in very fertile
soils.   This is so in the American Mid-West
(Gerdemann, 1970) and in southern Spain.   Observations at several sites in southern Spain (Hayman,

Barea and Azcón, 1975) showed no correlation between the amount of mycorrhiza at a particular site and the fertility level of the soil there, and mycorrhiza were often abundant in both poor and rich soils. This indicates that low soil fertility is not always a prerequisite for extensive mycorrhizal development. Another reason for apparently conflicting results is that sites may differ in initial inoculum density. Other variables are differences in soil structure, moisture levels, pH, seasonal effects, crop differences in susceptibility to mycorrhizal infection, and the variability even within a small site.

Organic manures often, but not always, enhance mycorrhizal development compared to mineral fertilizers or no fertilizer (see Johnston, 1949; Reed and Frémont, 1935), and may favour particular spore types, e.g. yellow vacuolate spores (Mosse and Bowen, 1968b). In citrus roots FYM favoured arbuscular development, compared to no fertilizer or mineral fertilizer treatments where infection was less and consisted mainly of intercellular hyphae lacking arbuscules (Reed and Frémont, 1935). In alpine plants, Peyronel (1940) found that humus mitigated the negative effect of low light intensity on VA infection.

GLASSHOUSE AND LABORATORY EXPERIMENTS

Although no substitute for field observations, pot experiments give more controlled conditions and permit clearer interpretations of data. Results have generally shown that VA infection in pot-grown plants is usually decreased by the addition of nutrient solution and especially by N and P. Table 5 gives some examples.

In her detailed examination of phosphate effects in soils Mosse (1973) concluded that the decline of infection with increasing addition of P to onions grown in pots was probably connected with the high

Table 5. Mycorrhiza in plants given different nutrients in pots.

Plant	Length of expt. (weeks)	Treatment	% VA infection	Reference
Coprosma robusta	15	Forest soil + no fertilizer	70	Baylis (1967)
		Forest soil + P (as KH_2PO_4)	36	
		Forest soil + Complete nutrients	5	
Maize	10	Sand + complete nutrients (P as Ca_3PO_4)	64	Daft and Nicolson (1969)
		Sand + complete nutrients (P as KH_2PO_4)	27	
Tomato	7	Sand + no fertilizer	33	Koch (1961)
		Sand + complete nutrients	2	
Onion	12	Wheatfield soil + no fertilizer	65	Mosse (1973)
		Wheatfield soil + Ca $(H_2PO_4)2$ 23 mg P/Kg soil	40	
		Wheatfield soil + Ca $(H_2PO_4)2$ 115 mg P/Kg soil	10	

phosphate concentrations attained in the host tissues. The balance between N and P was important (cf. Bevege, 1971) because adding extra N to the soil slowed down the decline of infection, probably because it encouraged growth and hence decreased the concentration of P in the plant.   Experiments in test-tubes further indicate the importance of the balance between N and P. Mosse and Phillips (1971) showed that clover seedlings in a medium containing much P only became infected when N was lacking, whereas with much less P infection occurred readily in the presence of adequate N.   These results point to the desirability of obtaining more chemical measurements in the field to improve the value of observations on mycorrhizal populations.

CONCLUSIONS

A valuable feature of quantitative field observations is that they point to the possibility of manipulating VA mycorrhiza under field conditions. The findings that, in mixed field populations of *Endogone*, white reticulate spores were selectively decreased by N and yellow vacuolate spores selectively increased by FYM make it feasible to consider soil amendments that will encourage the proliferation of selected spore types.   This is important because specificity between host plant, *Endogone* spore type and soil affects the degree of stimulation of plant growth by mycorrhiza (see Mosse, 1972). Also organic matter seems to favour arbuscular development, probably by directly affecting the host-fungus interaction rather than by selectively stimulating an "arbuscular strain", and total arbuscular infection may sometimes be more important than total infection *per se* (see Hayman, 1974). The abundance of mycorrhiza in some quite fertile soils further indicates the feasibility of maintaining an appreciable mycorrhizal population in

cultivated soils.   The most efficient strains of
Endogone (see Mosse, 1972) should be used for this
purpose.

Most field studies have indicated an effect of
fertilizers on the mycorrhizal population.   Although
many locations (and pot experiments) show this effect
to be a negative one, some locations indicate a
positive effect.   Consequently, generalizations are
difficult.   There are many variables in the field;
the original fertility of the soil before fertilizers
are added and the nutrient content of the host plant
are probably major factors to consider.   Therefore,
to understand the field situation more fully and to
facilitate comparisons between different locations,
it is suggested that samples be mostly collected in
late summer when the mycorrhizal population is highest, and that additional data be obtained, including
N and P contents of the plant tissues and N, P,
organic matter contents and pH of the soil, measured
according to universally accepted techniques.

REFERENCES

BAYLIS, G. T. S. (1967). Experiments on the ecological significance of phycomycetous mycorrhizas.   *New
Phytol.*, 66, 231-243.

BEVEGE, D. I. (1971). Vesicular-arbuscular mycorrhizas of Araucaria: aspects of their ecology and
physiology and role in nitrogen fixation.   Thesis,
University of New England, Armidale, N.S.W.,
Australia.

DAFT, M. J., & NICHOLSON, T. H. (1969). Effect of
Endogone mycorrhiza on plant growth.   II. Influence
of soluble phosphate on endophyte and host in
maize.   *New Phytol.*, 68, 945-952.

GERDEMANN, J. W. (1970).   The significance of

vesicular-arbuscular mycorrhizae in plant nutrition.
In: *Root Diseases and Soil-borne Plant Pathogens*.
Eds. Toussoun, T. A., Bega, R. V., & Nelson, P. E.
University of California Press, Berkeley.

HAYMAN, D. S. (1970). *Endogone* spore numbers in soil
and vesicular-arbuscular mycorrhiza in wheat as
influenced by season and soil treatment.   *Trans.
Br. mycol. Soc.*, 54, 53-63.

HAYMAN, D. S. (1974). Plant growth responses to
vesicular-arbuscular mycorrhiza. VI.   Effect of
light and temperature.   *New Phytol.*, 73, 71-80.

HAYMAN, D. S., BAREA, J. M., & AZCÓN, R. (1975).
Vesicular-arbuscular mycorrhiza in southern   Spain:
its occurrence in crops growing in soils of
different fertility.   (In press).

HAYMAN, D. S., JOHNSON, A. M., & RUDDLESDIN, I. (1975).
The influence of phosphate and crop species on
Endogone spores and vesicular-arbuscular mycorrhiza
under field conditions.   (In press).

JOHNSTON, A. (1949). Vesicular-arbuscular mycorrhiza
in Sea Island cotton and other tropical plants.
Trop. Agric., Trin., 26, 118-121.

KHAN, A. G. (1972). The effect of vesicular-arbuscular
mycorrhizal associations on growth of cereals. I.
Effects on maize growth.   *New Phytol.*, 71, 613-619.

KHRUSHCHEVA, E. R. (1960). Conditions favourable for
the formation of mycorrhiza of maize.   (In Russian).
Agrobiologiya, 4, 588-593.

KOCH, H. (1961). Untersuchungen über die Mykorrhiza
der Kulturpflanzen unter besonderer Berücksichtigung
von *Althaea officinalis* L., *Atropa belladonna* L.,
Helianthus annuus L. und *Solanum lycopersicum* L.

508 Hayman, D. S.

Gartenbauwissenschaft, 26, 5-32.

KRUCKELMANN, H. W. (1973). Die vesikulär-arbuskuläre
Mykorrhiza und ihre Beeinflussung in landwirt-
schaftlichen Kulturen. Thesis, Universität
Braunschweig, Germany.

MOSSE, B. (1972). The influence of soil type and
Endogone strain on the growth of mycorrhizal plants
in phosphate-deficient soils. *Rev. Ecol. Biol.
Sol,* 9, 529-537.

MOSSE, B. (1973). Plant growth responses to vesicular-arbuscular mycorrhiza. IV. In soil given
additional phosphate. *New Phytol.,* 72, 127-136.

MOSSE, B., & BOWEN, G. D. (1968a). A key to the
recognition of some *Endogone* spore types. *Trans.
Br. mycol. Soc.,* 51, 469-483.

MOSSE, B., & BOWEN, G. D. (1968b). The distribution
of *Endogone* spores in some Australian and New
Zealand soils, and in an experimental field soil at
Rothamsted. *Trans. Br. mycol. Soc.,* 51, 485-492.

MOSSE, B., & PHILLIPS, J. M. (1971). The influence of
phosphate and other nutrients on the development of
vesicular-arbuscular mycorrhiza in culture. *J.
gen. Microbiol.,* 69, 157-166.

PEYRONEL, B. (1940). Prime osservazioni sui rapporti
tra luce e simbiosi micorrizica. Estratto dell'
Annuario N.4 del Laboratorio della Chanousia,
Giardino Botanico Alpino dell'Ordine Mauriziano al
Piccolo San Bernardo, 4, 1-19.

PORTER, D. M., & BEUTE, M. K. (1972). *Endogone*
species in roots of Virginia type peanuts.
Phytopathology, 62, 783. (Abstr.).

REED, H., & FREMONT, T. (1935). Factors that influence
the formation and development of mycorrhizal associations in citrus roots.   *Phytopathology*, 25,
645-647.

EFFECTS OF FERTILIZERS, SOILS, SOIL TILLAGE, AND PLANT SPECIES ON THE FREQUENCY OF ENDOGONE CHLAMYDOSPORES AND MYCORRHIZAL INFECTION IN ARABLE SOILS

H. W. KRUCKELMANN

Forschungsanstalt für Landwirtschaft, Braunschweig, W. Germany

INTRODUCTION

Improved growth of crop plants in the presence of vesicular-arbuscular mycorrhiza has been demonstrated many times (Mosse, 1973). Therefore it should be of interest to get information on the factors controlling the mycorrhiza in arable soils.

This paper reports on effects of mineral and organic fertilizers and of different soils and soil tillage, on the frequency of *Endogone* chlamydospores. Whether crops and weeds are infected to different extents, whether they can influence the frequency of *Endogone* chlamydospores in arable soils and whether certain endophytes are preferentially associated with particular host species was also examined.

SOILS AND METHODS

The soil samples for the determination of spore numbers were collected on several experimental fields in the Forschungsanstalt für Landwirtschaft (FAL) in Braunschweig, FRG, and on Broadbalk in Rothamsted Experimental Station, Harpenden, Hertfordshire,

England.   Data on the different soils are listed in
Ahrens (1970), Avery and Bullock (1968), Johnston
(1968), Sauerlandt and Tietjen (1970).

The effects of crop rotations on the frequency
of *Endogone* chlamydospores was studied in the field
and was compared with the extent of mycorrhizal infection in different crop and weed species grown in
pots.

In the crop rotation studies three soil samples,
each consisting of twenty randomly selected subsamples, were examined from each field.   The
samples were collected after harvest in 1970 and
1972 and were taken from the top 7 cm of soil.
Endogone spores were extracted from the samples by a
wet sieving and decanting method (Gerdemann and
Nicolson, 1963; Kruckelmann, 1973).   They were
recorded per 50 g dry soil as determined by the
"Ultramat" soil moisture apparatus.   Spores were
examined microscopically and determined according to
the nomenclature of Mosse and Bowen (1968).

A comparison of infection in various weed and
crop species was made in plants grown in 8 cm pots.
The rooting medium consisted of 13 parts sand: 13
parts inorganic sediment from wet sieving, inoculated
with 1 part (by volume) of the 125-250 µm organic
fraction also obtained by wet sieving from a field
soil.   After two months plants were harvested.
Twenty to fifty, 1 cm root segments from the middle
part of each root system were cleared overnight in
10% KOH, rinsed in water, stained for some days in
0.1% trypan blue in lactophenol and contrasted in
lactic acid for some days.   There were three replicates for each plant species.

Differences in spore numbers and infection
ratings were analysed by the t-test.

FACTORS INFLUENCING <u>ENDOGONE</u> CHLAMYDOSPORES

Fertilizers

Application of different amounts of phosphate
(Thomasslag) from 0 to 220 kg P/ha for 7 consecutive
years from 1960 onwards (124 kg K and 60 kg N/ha were
also given annually), did not significantly influence
the frequency of *Endogone* chlamydospores in the
experimental field eleven years later, although soluble phosphate and yields of rye still showed significant after-effects of the fertilizer treatments
(Table 1).   There was a trend towards fewer spores
in soil given 110 and 220 kg P/ha.   The relatively
low number of spores in this field was typical for
Secale cereale following *Solanum tuberosum* (Kruckelmann, 1973).

Table 1.   Frequency of *Endogone* chlamydospores in soils with increasing phosphate
content.

Fertilization kg P/ha/year 1960-67	In the year of investigation (1971)		
	mg lactate soluble P/ 100 g soil	Yield of rye t/ha	Chlamydospores/ 50 g soil
0	1.8	3.3	28.7 ± 8.1
55	2.1	3.4	30.0 ± 14.9
110	4.3	4.4	13.3 ± 1.2
220	6.8	5.8	21.3 ± 3.2

Calcium ammonium nitrate (from 0 to 200 kg N/ha)
applied annually for 11 years, together with 124 kg K
and 21 kg P/ha, did not influence spore numbers in an
experimental field cropped with *Beta vulgaris* after
Avena sativa.   Biennial applications of about 25.5 t

farmyard manure/ha (equivalent to 2.0 t organic material) for 18 years, together with an annual application of 100 kg K and 60 kg N/ha also had no effect
on spore numbers in an experimental field cropped
with *Avena sativa* following *Secale cereale*.

Fig. 1 shows a comparison of the effects of
mineral and organic fertilizer and no fertilizer on
the frequency of *Endogone* chlamydospores in two different soils.   The amounts of fertilizer/year/ha were
equivalent to 96 kg N, 33 kg P, 90 kg K, 16 kg Na,
and 31 kg Mg in Broadbalk.   Farmyard manure (FYM)
was calculated to be equivalent to 224 kg N.   Plots
of FAL were fertilized/year/ha with 40 kg N, 49 kg P,
104 kg K, compost of farmyard manure and municipal
refuse respectively, equivalent to 60 t organic
matter.

Spore numbers were greatly influenced not only
by the different fertilizers but also by the different soils.   In Rothamsted soil (silty clay loam)
most spores were found in the unmanured plot whereas
at FAL (loamy sand) fewest spores occurred in the
unmanured plot.   In the sandy soil all fertilizer
applications increased spore numbers.   The response
was more pronounced with added compost than with
mineral fertilizers (NPK).   In the silty clay loam
the number of spores was decreased by all fertilizers,
and the negative effect was more pronounced with mineral fertilizers (NPKMgNa) and with nitrogen + FYM.
Farmyard manure alone caused a smaller decrease in
the spore population.   Except for the compost from
municipal refuse all differences between spore freq-
uencies in different plots are statistically significant at $P<0.001$.

There was no apparent relationship between total
carbon, potassium and phosphate content and the
frequency of spores, but spore numbers appear to be
related to pH.

Soils

Four soils brought to FAL in 1952 from different

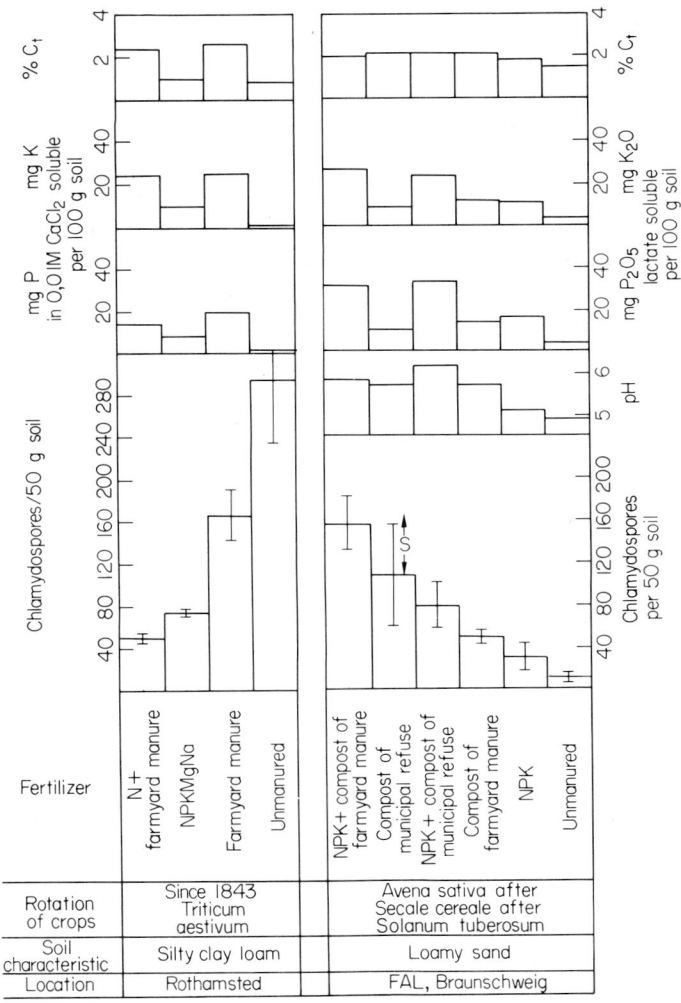

Figure 1. Frequency of *Endogone* chlamydospores
as influenced by different fertilizers in two
soils.

Kruckelmann, H. W.

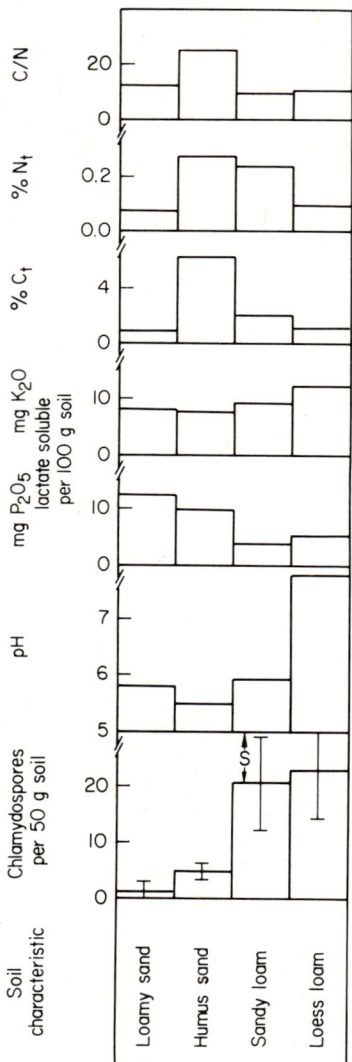

<u>Figure 2</u>. Frequency of *Endogone* chlamy-dospores as influenced by different soils in the cropping sequence of *Solanum tuberosum* after *Avena sativa*.

locations were investigated for their influence on
the frequency of spores (Fig. 2). The loam soils
contained significantly more spores (P<0.001) than
the sands. Similar differences were recorded
between the loess loam of Karlshof in Stuttgart-
Hohenheim and the loamy sand in FAL (Kruckelmann,
1973). The silty clay loam at Broadbalk (Rothamsted)
also contained more spores than the sandy FAL soil
(Fig. 1).

Again there was no apparent relationship between
the frequency of spores and the contents of potassium,
carbon and nitrogen in the soils but spores increased
with higher pH values of the soils and decreased with
increasing phosphate contents. These results are in
agreement with findings of Mosse (1966, 1972),
Strzemska (1949) and Daft and Nicolson (1972).

Soil Tillage

Effects of soil tillage on mycorrhizal development have not been considered before. Spore counts
were made in a field that had been differentially
tilled for the last 10 seasons after harvest (Fig. 3).
Chopped straw was applied before tillage in half the
field each year.

Compared with no-tillage, shallow ploughing
tended to increase spore frequency, while tilling by
rotary hoe tended to decrease it. Chopped straw
increased the differences between the treatments
(P<0.001) and greatly increased spore numbers in the
shallow ploughed soil.

*Effect of plant species on frequency and type of
Endogone spores*

Rotation of crops had no effect on the frequency
of spore types in the soil (Table 2). The yellow
vacuolate (YV) spores predominated over all others.
In the loamy sand most white reticulate (WR) and
white crenulate (WC) spores were found in two fields

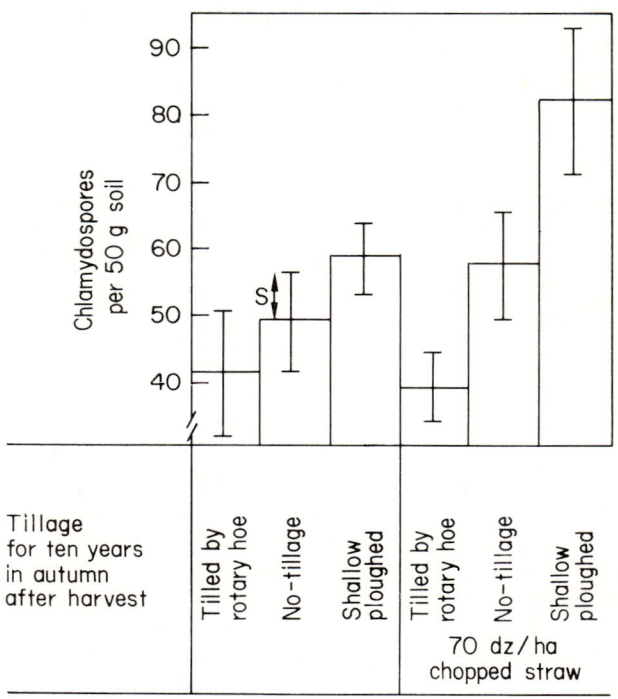

Figure 3.   Frequency of *Endogone* chlamydospores as influenced by tillage and chopped straw in the cropping sequence of *Beta vulgaris* after *Avena sativa*.

with *Beta vulgaris,* but all attempts to infect *Beta vulgaris* with these spores in pot cultures were unsuccessful (Kruckelmann, 1973) and only two percent WR and WC spores occurred after 16 years monoculture with *Beta vulgaris* in a loess soil (Table 2).   In

Table 2. Frequency of different *Endogone* spore types in fields with different crop rotations

Soil	Crop rotations Crop/pre-crop/pre-pre-crop	pH	P	K	Mg	Total no.of spores	YV	RL	WR WC	HS	BV BR
	Avena/Secale/Solanum*	5.0	3	15	2	1745	18	2	18	56	6
	Avena/Secale/Solanum	5.5	8	12	4	1302	90	1	3	3	3
	Avena/Beta/Secale	6.0	11	12	2	2160	90	3	3	2	2
	Solanum/Beta/Zea					283	86	6	7	1	1
Loamy sand	Solanum/Solanum/Triticum	6.0	14	12	2	279	62	6	20	6	6
	Triticum/Avena/Beta	6.2	16	13	2	216	77	3	16	1	3
	Secale/Solanum/Avena	5.7	2	10	3	515	57	1	31	3	8
	Beta/Avena/Beta	6.2	14	13	2	324	50	4	40	3	3
	Beta/Avena/Solanum	6.3	3	15	2	2083	37	6	45	7	5
Loess	)	(7.7	2	10	4	139	84	2	3	9	2
Sandy loam	) Solanum/Avena/Solanum	(5.9	1	7	4	125	77	4	8	8	3
Humic sand	)	(5.5	4	7	4	29+					
Loamy sand	)	(5.8	5	7	3	9+					
	16 yrs Solanum tuberosum	6.4	6	13		11			0		
	16 yrs Zea mays	6.6	5	15		108			0		
Loess	16 yrs Beta vulgaris	6.4	4	10		39			2		
	16 yrs Avena sativa	6.5	4	13		314			6		
	16 yrs Triticum aestivum	6.5	4	17		392			34		
	16 yrs Secale cereale	6.5	4	14		141			70		

* Full names in lower section

Spore types: YV = yellow vacuolate, RL = red-brown laminate, WR = white reticulate, WC = white crenulate,
HS = honey-coloured sessile, BV = bulbous vacuolate, BR = bulbous reticulate

+ too few spores identified to make percentages useful.

the same soil 34 and 70% WC/WR spores were found
after 16 years monoculture with *Triticum aestivum* and
Secale cereale respectively.

Generally, fewer than 10% honey-coloured sessile
(HS) spores were found, but in one field there were
56% (Table 2).   This was not apparently connected
with the crop rotation because in a neighbouring
field with the same crop rotation only 3% of HS
spores occurred;   the soil with 56% HS spores also
had the lowest pH (5.0).   These observations agree
with those of Mosse (1972).   The red-brown laminate
(RL), bulbous vacuolate (BV) and bulbous reticulate
(BR) spores were rare in all plots.

The frequency of *Endogone* spores was clearly
influenced by monocropping for 16 years (Table 2,
Fig. 4).   Most differences are significant at the

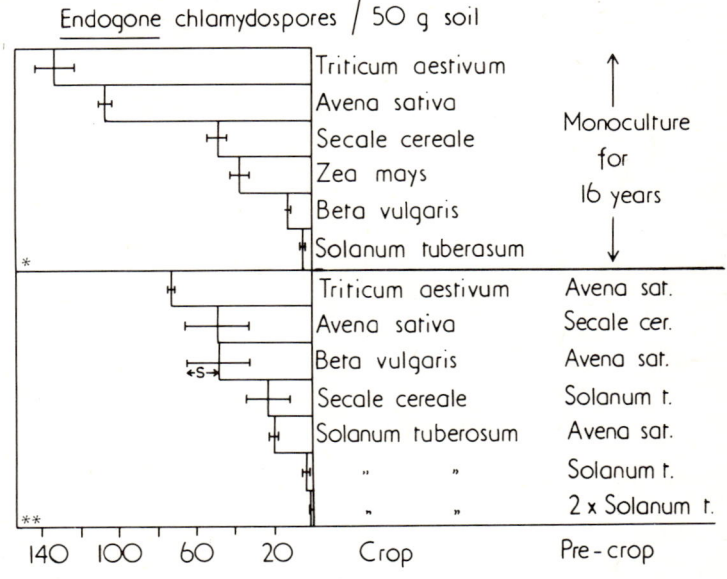

* Loess, at Karlshof, Stuttgart.   ** Loamy sand, at FAL, Braunschweig.

Figure 4.   Frequency of *Endogone* chlamydospores as influenced by crop rotations.

0.1% level.

The effect of plant species on spore numbers was also investigated in fields with different crop rotations (Fig. 4).   Relatively many spores were found in fields of *Solanum tuberosum* and *Beta vulgaris* when *Avena sativa* was the previous crop.

Altogether these results indicate some differential effects of host species on *Endogone* spore numbers.

MYCORRHIZAL INFECTION IN INOCULATED CROP AND WEED SPECIES

Different plant species became infected to different extents after inoculation (Figs. 5 and 6).

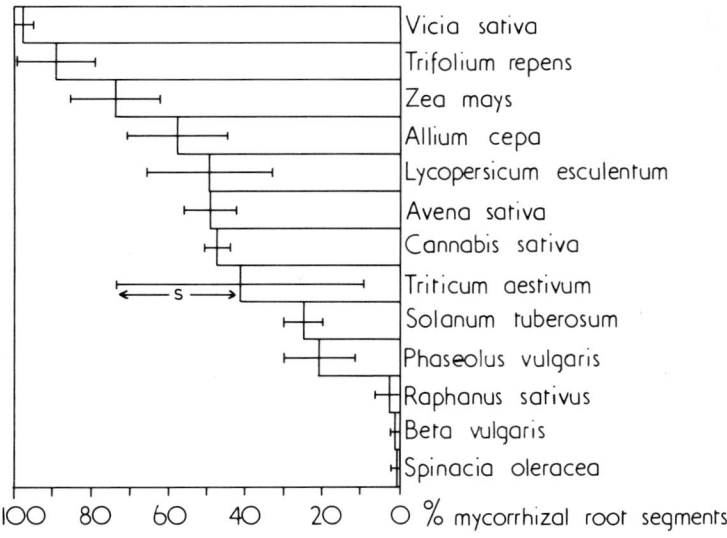

Figure 5.   Mycorrhizal infection in seedlings of different crop plants grown for two months in a sand:soil mixture inoculated with sievings from a field soil.   (Mean of three replicates).

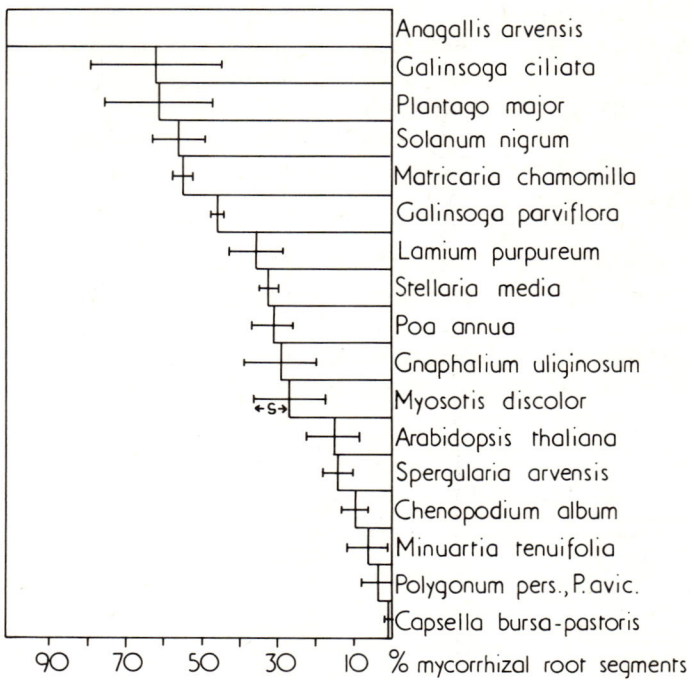

Figure 6.   Mycorrhizal infection in weed
seedlings grown for two months in a sand:
soil mixture inoculated with sievings from
a field soil.   (Mean of three replicates).

Most of the differences are signficant at the 0.1%
level.   The extent of infection in the different
crop plants (*Beta, Solanum, Avena* and *Zea*) does not
follow the pattern of spore production in the field
after 16 years monoculture with these crops (Fig. 4).
     Weed species such as *Stellaria media, Spergul-
aria arvensia, Minuartia tenuifolia,* (Caryophyllac-
eae), *Arabidopsis thaliana* (Cruciferae), *Chenopodium
album* (Chenopodiaceae), *Polygonum persica* and *Polygonum aviculare* (Polygonaceae) were clearly mycorrhi-

zal, although, with the possible exception of *Stell-aria,* infection was slight.   It seems that there is some susceptibility to mycorrhizal infection in some plant families previously reported to contain few or no mycorrhizal species (Gerdemann, 1968).

CONCLUSIONS

These results show that *Endogone* chlamydospores were more frequent in loamy soils than in sandy ones and that spore numbers can be influenced by agricult-ural practices such as manuring and soil tillage. In a silty clay loam most spores occurred in the unfertilized plots and numbers were reduced by long term fertilizer treatments but in a sandy soil unfertilized plots contained least spores and numbers increased progressively in plots manured for 50 years with NPK, compost and NPK + compost.   Results of fertilizer applications are not always predictable but spore numbers appeared to be more closely related to pH than to organic matter, potassium or phosphate content of the soil.   Where spores were few, as with *Beta vulgaris,* and in the rye experiment (Table 2), no significant effects of fertilizer treatments could be demonstrated possibly because of the great varia-bility of spore counts in soil.   The effects of soil tillage on spore numbers are of some interest in relation to the presently increasing practice of direct drilling.
An effect of crops on the frequency of partic-ular chlamydospore types of the obligate biotrophic *Endogone* spp. in arable soils could not be proved. The total population of *Endogone* chlamydospores was however greatly influenced by different crops.   In pot experiments mycorrhizal infection ranged from 1% to 100% in 14 crop and 17 weed species some of which belonged to families previously reported to show little or no mycorrhizal infection.

524

Kruckelmann, H. W.

ACKNOWLEDGMENT

I express my gratitude to Dr. B. Mosse for her
inspiring help and for help received from friends in
the Forschungsanstalt für Landwirtschaft, Braunschweig, in bringing this publication to completion.

REFERENCES

AHRENS, E. (1970). Untersuchungen an Böden, die in
klimatisch unterschiedliche Standorte verlagert
wurden, mit besonderer Berücksichtigung ihrer
mikrobiellen Eigenschaften (Klimaversuch). Habil.,
Universität Gießen.

AVERY, B. W. & BULLOCK, P. (1968). Morphology and
classification of Broadbalk soils. *Rep. Rothamsted exp. Stn.*(1968), pt. 2, p. 63-81.

DAFT, M. J. & NICOLSON, T. H. (1972). Effect of
Endogone on plant growth. IV. Quantitative
relationship between the growth of the host and
the development of the endophyte in tomato and
maize. *New Phytol.*, 71, 287-295.

GERDEMANN, J. W. (1968). Vesicular-arbuscular mycorrhiza and plant growth. *A. Rev. Phytopath.*,
6, 397-418.

GERDEMANN, J. W. & NICOLSON, T. H. (1963). Spores of
mycorrhizal *Endogone* species extracted from soil
by wet sieving and decanting. *Trans. Br. mycol.
Soc.*, 46, 235-244.

JOHNSTON, A. E. (1968). Plant nutrients in Broadbalk
soils. *Rep. Rothamsted exp. Stn.*, (1968), pt. 2,
p. 12-25.

KRUCKELMANN, H. W. (1973). Die vesikulär-arbuskuläre
Mykorrhiza und ihre Beeinflussung in landwirtschaft-
lichen Kulturen. Dissertation, Universität Braunschweig.

MOSSE, B. (1966). Effect of host nutrient status on
mycorrhizal infection. *Rep. Rothamsted exp. Stn.*,
(1966), p. 76.

MOSSE, B. (1972). Effect of different *Endogone*
strains on the growth of *Paspalum notatum*. *Nature,
London,* 239, 221-223.

MOSSE, B. (1973). Advances in the study of vesicular-
arbuscular mycorrhiza. *A. Rev. Phytopath.*, 11,
171-196.

MOSSE, B. & BOWEN, G. D. (1968). A key to the
recognition of some *Endogone* spore types. *Trans.
Br. mycol. Soc.*, 51, 469-483.

MYCORRHIZA IN FARM CROPS GROWN IN MONOCULTURE

J. STRZEMSKA

Laboratory of Soil Microbiology,
Academy of Agriculture, Olsztyn, Poland

INTRODUCTION

This work has been done in the framework of the program of the Polish Academy of Sciences "Ecological Effects of Intensified Land Cultivation", and of the program of the Institute of Soil Science and Plant Cultivation in Pulawy "Growing of Cereals in Monoculture".

In modern agriculture there are tendencies towards simplification of crop rotations and towards monoculture. Current experiments are testing how long, after applying different agrotechnical methods, particular plant species may be grown in monoculture without destroying the biological equilibrium. One factor in such ecosystems is the symbiosis of cultivated plants with soil fungi in the form of mycorrhiza. This paper reports results of a study of mycorrhizal infection in crops grown in monoculture in three experiments.

MATERIAL AND METHODS

Experiment 1. Wheat, barley and horse bean were grown for two years in a loamy sand (pH in KCl = 6.5) with the following fertilizer treatments in kg/ha;

wheat NPK = 70/31/83; barley NPK = 50/31/33; horse
bean NPK = 20/35/116.*

Experiment 2.   In this experiment rye, wheat,
barley, oats and horse bean were grown for five years
in a sol lessivé derived from clay (pH in KCl =
5.5 - 6.0) with fertilizer applied at two levels.
For the first four crops NPK was applied at 60/26/50
and 120/52/100 kg/ha.   For horse beans it was
applied at 20/35/83 and 40/70/166 kg/ha.   In 1971
all plots also received 3 t/ha of dolomite.**

Experiment 3.   Wheat and rye, in the 12th and 13th
year of monoculture respectively, were grown with the
following six fertilizer treatments: A - nil;
B - farm yard manure (FYM) at 15 t/ha; C - FYM at
30 t/ha; D - commercial fertilizer (NPK) at
35/26/66 kg/ha; E - NPK at 70/52/133 kg/ha; and
F - FYM at 15 t/ha + NPK at 35/18/50 kg/ha.   The rye
was grown in a sol lessivé derived from clay
(pH in KCl = 6.0), the wheat in a black earth derived
from clay (pH in KCl = 6.8).***

Assessment of Mycorrhizal infection

Roots were sampled at the time of crop ripening
and were fixed in Navashin's solution.   100 roots
from each treatment were sectioned transversely and
stained in Thionine + Orange G.   The number of roots

*   The experiment was conducted at the Experimental
      Station at Zelislawki, nr. Gdansk, and I am
      grateful to Professor Dr. S. Nawrocki for the
      material.
**   The experiment was conducted at the Experimental
      Farm at Balcyny, nr. Olsztyn, and I am grateful
      to Professor Dr. W. Niewiadomski for the material.
***   The experiment was conducted at the Experimental
      Station Chylice, nr. Warsaw, and I am grateful
      to Dr. A. Gawronska for the material.

with mycorrhiza was counted and the intensity of
infection in each infected root assessed in six
categories (Fig. 1).   For the purposes of this paper,
each category of infection was given a corresponding
numerical rating (1-6) and multiplied by the number
of roots in that category, giving a maximum rating of

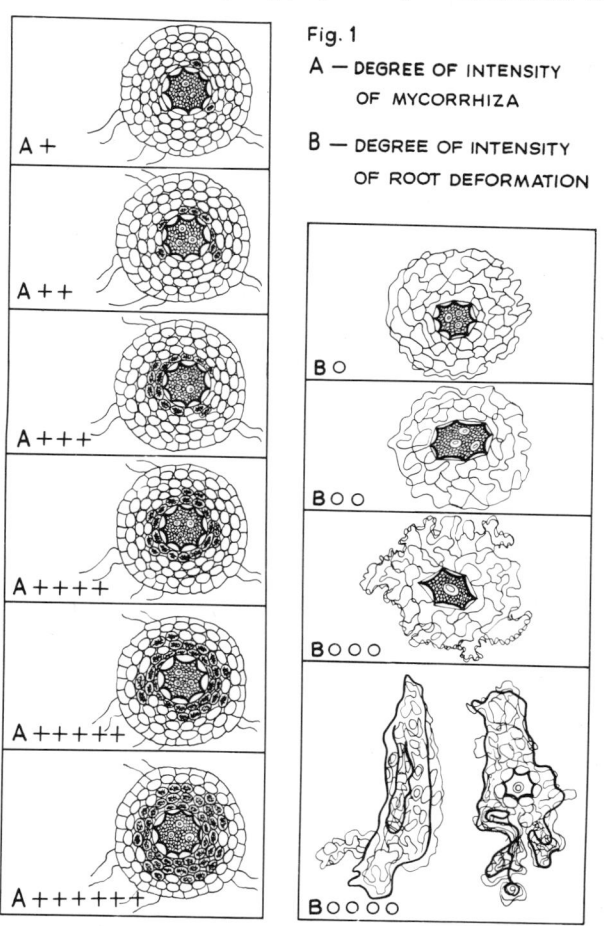

Fig. 1

A — DEGREE OF INTENSITY
OF MYCORRHIZA

B — DEGREE OF INTENSITY
OF ROOT DEFORMATION

Figure 1.   Explanation of the system of
rating mycorrhizal infection intensity and
root deformation.

600 (all 100 roots in category 6) and a minimum of
nil (no infected roots). The degree of deformation
in the uninfected roots was also recorded on a similar category basis.

RESULTS

Experiment 1.   (Two years of monoculture)

All three crop plants (wheat, barley and horse
beans) formed endotrophic mycorrhiza. There was no
marked change in infection rating of barley and
horse bean in the second year of monoculture and that
of wheat was slightly greater. Roots without
mycorrhiza were generally deformed. Yields of wheat
and horse bean were slightly less in the second year,
that of barley was 30% less.

Table 1.   Infection rating and yield in the
first and second years of monoculture.

| Crop | Yield, t/ha | | Infection rating* | |
	Year: 1	2	1	2
Wheat	3.4	3.2	263	302
Barley	5.0	3.4	198	205
Horse bean	2.4	2.1	279	254

* Determined as described in Materials & Methods.

Experiment 2.   (Five years of monoculture)

Table 2.   Infection rating and yield in the
first and fifth years of monoculture with
two levels of fertilizer.

Crop	N/P/K kg/ha	Yield t/ha		Infection rating*	
		Year: 1	5	1	5
Rye	60/26/50	2.0	3.9	173	78
Wheat	" " "	2.3	2.0	175	269
Barley	" " "	3.6	3.5	248	222
Oats	" " "	3.4	3.1	25	176
Bean	20/35/83	3.9	2.3	194	196
Rye	120/52/100	2.0	3.5	46	11
Wheat	" " "	2.2	2.0	115	103
Barley	" " "	3.3	3.2	220	66
Oats	" " "	2.9	2.7	48	54
Bean	40/70/166	3.3	2.2	116	205

* Determined as described in Materials & Methods.

    Table 2 shows the results of this experiment.
Yields of all five crops were slightly reduced by the
higher rate of fertilizer.   Wheat, barley and oat
yields were also slightly reduced by five years of
monoculture.   However, in the 5th year winter - and
in the 1st - spring rye and wheat were used which
produce slightly lower yields.   Rye yields increased
and bean yields decreased in the fifth year at both
fertilizer levels.   In rye, the increased yield was
associated with less mycorrhizal infection and in
bean the decreased yield was associated with more
infection at the high but not the low fertilizer
level.   In wheat infection was less with high than
with low fertilizer, and with low fertilizer was 50%
greater in the 5th than in the 1st year.   A prop-

ortionally even greater (sevenfold) build-up of
infection occurred in oats but again only at the
lower fertilizer level.   Barley had the highest
infection rating of all five crops at the low fertilizer level but at the higher level infection was
73% less in the 5th year of monoculture.   Except
for beans infection ratings were less in the fifth
year at high fertilizer levels than at low.
Clearly mycorrhizal responses to monoculture depend
on crops and fertilizer levels.

In general few mycorrhizal roots were deformed,
but in horse beans after five years monoculture with
high fertilizer additions many were.

Experiment 3

After 12 and 13 years monoculture respectively
both wheat and rye yielded less in unfertilised
plots (A) and those with low farm yard manure additions (B) than in plots with other fertilizer treatments (high FYM (C);   low and high NPK (D + E);   and
FYM + NPK (F)).   In rye but not wheat mycorrhizal
infection was reduced by these four treatments
(C, D, E & F) compared to treatments A and B.

As in Experiment 2 therefore, mycorrhizal
response to monoculture and fertilizer treatment
differed with the crop.   The wheat was grown in a
black earth and the rye in a sol lessivé and this
may also have affected the mycorrhizal responses.

Non-mycorrhizal roots of wheat and rye were
almost always deformed.

DISCUSSION

Although there have been several *ad hoc* studies
of mycorrhizal infection in agricultural crops this
is the first study of the long-term effects of monoculture and fertilizer treatments.   After five years
monoculture (Experiment 2) there were marked reduct-

Table 3.   Infection rating and yield after twelve and thirteen
years of monoculture under different fertilizer treatments.

	A NIL	B FYM(low)	C FYM(high)	D NPK(low)	E NPK(high)	F FYM + NPK
Rye (13 years)						
Yield t/ha	2.5	3.6	4.6	4.4	4.3	4.5
Infection rating*	272	214	85	85	90	206
Wheat (12 years)						
Yield t/ha	0.5	0.6	0.9	0.8	0.8	1.0
Infection rating	248	244	233	295	207	206

ions in mycorrhizal infection of rye, wheat, barley
and oats with high levels of fertilizer. These
fertilizer additions, however, seemed to be supra-
optimal and did not increase crop yield. A similar
reduction in mycorrhizal infection, from rating 295
to 207, occurred in wheat in Experiment 3 with high
compared to low NPK additions, but again the higher
NPK dosage did not increase yield. Generally the
fertilizer treatments in Experiment 3 increased
yield and in wheat did not decrease infection. In
rye, on the other hand, all fertilizer additions,
with the exception of low FYM (B), markedly decreased
infection after 13 years of monoculture. In general
mycorrhizal infection was reduced by high but not by
low fertilizer additions. Whether such reduction in
infection occurred before or after fertilizer levels
became supra-optimal for yield seemed to depend on
the crop.

Mycorrhizal infection in beans (Experiment 2)
was not reduced by high fertilizer levels, in fact it
appeared to increase. This supports Hayman's (1975)
results which showed a marked reduction in mycorrhizal infection after heavy nitrogen applications in
wheat, but not in beans. In Experiment 2, bean
yields were less after 5 years monoculture irrespect-
ive of fertilizer treatment. Such a reduction after
several years monoculture is not unusual in beans and
may be due to a disease build-up.

It is generally agreed that mycorrhiza usually
die out in rich garden soils (Gerdemann, 1968). In
the experiments described here levels of fertilizer
that might normally be given to agricultural crops
were not necessarily deleterious to mycorrhiza formation. There was no clear relationship between crop
yield and mycorrhizal infection, and many non-
mycorrhizal roots appeared to be deformed.

REFERENCES

GERDEMANN, J. W. (1968). Vesicular-arbuscular
  mycorrhiza and plant growth.   A. *Rev. Phytopathol.*
  397–418.

HAYMAN, D. S. (1975). This symposium, p. 495.

STRZEMSKA, J. (1975). This symposium, p. 537.

OCCURRENCE AND INTENSITY OF MYCORRHIZA AND DEFORMATION OF ROOTS WITHOUT MYCORRHIZA IN CULTIVATED PLANTS

J. STRZEMSKA

Laboratory of Soil Microbiology,
Academy of Agriculture, Olsztyn, Poland

INTRODUCTION

Much more attention has been paid to ectotrophic than to endotrophic mycorrhiza and comparatively few studies of the latter have been concerned with cultivated plants in field ecosystems. In recent years interest in mycorrhiza of cultivated plants has grown, and attention has been paid to the role the fungi play in increasing the supply of nutrients, particularly phosphorous, to crops. To estimate the importance of symbiotic fungi in the life of cultivated plants one must have some quantitative data on the occurrence of mycorrhiza, which at present is scarce.

Since 1948 the author has begun regular studies on the occurrence and intensity of mycorrhizal infection in cultivated plants which are reported in this paper.

MATERIALS AND METHODS

Root samples of 68 plant species belonging to 13 families have been collected, fixed in Navashin's fixative and stored. Sections were stained in

Thionine and Orange G, and the occurrence and intensity of mycorrhizal infection were recorded as described (Strzemska, 1975).

The first family studied was the Gramineae.
Rye, wheat, barley and oat roots were collected in
1948 at different growth stages and from different
soil types with pH values between 4.0 and 8.0.   A
total of 1500 roots have been examined.

Based on the experience gained with cereals,
other plant species were taken from only one soil
type per species, in three separate years, at a particular vegetative period viz. at or after flowering.
In the Papilionaceae, 20 species have been examined
and 6 more are still outstanding.   Root samples were
collected in 1957, 1958 and 1960 at flowering.   75
samples of each species were examined for each year
and infection assessed as described (Strzemska, 1975)
with a maximum rating of 450 (6 x 75) for each year.

RESULTS AND DISCUSSION

Mycorrhiza in cereals

Table 1 shows the mean occurrence and intensity
of infection in barley, oats, wheat and rye.

Table 1.   Occurrence and intensity of
mycorrhiza in cereals.

Occurrence % infected roots		Intensity mean of categories 1-6	
Barley	81.7	Wheat	2.9
Oats	78.3	Oats	2.6
Wheat	76.2	Barley	2.5
Rye	41.0	Rye	2.4

Mycorrhiza were common in cereals and increased with
plant age.   Soil type and pH had no determinable
effects on infection.   Non-mycorrhizal roots were
generally deformed; this also was unrelated to soil
type and pH.

Mycorrhiza in Papilionaceae

Results summarised in Table 2 show that, although
there were marked annual fluctuations in infection, no
year was consistently low or high for all species but
there were consistent differences in infection ratings
of different species.

The species formed a continuous series from
strongly mycotrophic ones like common seradella, red
clover, common vetch, horse bean and field pea to
weakly mycotrophic ones like common bean, black
medic, fenugreek, greater birds foot trefoil and
melilot.   In common bean and fenugreek no mycorrhizal roots were found in 1958.

In most species non-mycorrhizal roots tended to
be deformed, but in some with little mycorrhizal
infection viz. common bean, fenugreek and broad bean
25-30% of non-mycorrhizal roots were normal and in
white clover 60% were normal.   In Swedish clover,
black medic and common vetch 25-30% of mycorrhizal
roots also showed deformation.

These results show that within a single family
(Papilionaceae) individual species can differ markedly in the extent of mycorrhizal infection.   Some
further references to this work are attached.

Table 2. Plant species, soil and infection rating for 20 species in the *Papilionaceae*.

Plant sp.	Soil	pH in KCl	Infection rating*			
			1957	1958	1960	Total
Ornithopus sativus L. Common serradella	Grey, light loamy sand	4.5 – 5.5	166	226	233	625
Trifolium pratense L. Red clover	Black earth silt deposit	about 5.5	170	135	255	560
Vicia sativa L. Common vetch	Brown, loamy sand	about 6.5	158	129	205	492
Trifolium hybridum L. Swedish clover	Brown silt deposit	about 5.9	136	122	187	445
Vicia faba L.var. minor Horse bean	Grey, light loamy sand	about 5.0	97	185	234	516
Lotus corniculatus L. Birdsfoot trefoil	Brown soil derived from sand	5.0 – 6.0	163	243	81	487
Pisum sativum L. Pea	Brown loamy sand	about 6.5	277	155	85	517
Pisum arvense L. Field pea	Brown, loamy sand	about 6.5	145	177	199	521
Vicia villosa Roth. Hairy vetch	Brown, loamy sand	about 6.5	148	156	155	459

* Maximum rating/year = 450   No. roots examined/year/sp. = 75

Table 2. (Contd.)

Plant sp.	Soil	pH in KCl	Infection rating*			
			1957	1958	1960	Total
Glycine hispida (Moench) Maxim. Soy bean	Grey, light loamy sand	4.5 - 5.5	13	202	212	427
Lens esculenta Mnch. Lentil	Grey, light loamy sand	4.5 - 5.5	117	132	147	396
Anthyllis vulneraria L. Kidney vetch	Brown, light loamy soil	about 6.5	163	62	75	300
Vicia faba L.var.major Broad bean	Black earth - silty, light loam	about 6.5	124	88	115	327
Onobrychis sativa Lam. Sainfoin	Grey, light loamy sand	4.5 - 5.5	65	52	157	274
Phaseolus vulgaris L. Common bean	Black earth derived from sand	about 6.5	16	0	96	112
Medicago lupulina L. Black medic	Brown soil derived from sand	about 6.5	84	48	3	135
Trifolium repens L. White clover	Grey, loamy sand	about 5.7	46	42	89	177

Strzemska, J.

Table 2. (Contd.)

Plant sp.	Soil	pH in KCl	Infection rating*			
			1957	1958	1960	Total
Trigonella foenum-graecum L. Fenugreek	Grey, light loamy sand	4.5 – 5.5	139	0	23	162
Lotus uliginosus Schk. Greater birdsfoot trefoil	Black earth derived from sand	5.0 – 7.0	78	26	48	152
Melilotus albus Desr. Melilot	Brown soil derived from sand	about 6.5	57	11	7	75

REFERENCES

STRZEMSKA, J. Zagadnienie mykorizy u zbóz (The mycorrhiza of corn plants).
- (1949). Cz.I.Owies (Part I. Oats). *Annales UMCS*, 4, 359 - 372.
- (1952). Cz.II.Zyto (Part II. Rye). *Acta microbiol. pol.*, 1, 24 - 35.
- (1953). Cz.III. Pszenica (Part III. Wheat). *Acta microb. Pol.*, 2, 297 - 306.
- (1955). Cz.IV. Jeczmień (Part IV. Barley). *Acta microbiol. pol.*, 4, 183 - 189.

STRZEMSKA, J. (1955). Wyniki badań nad mykoriza u zbóz (Investigations on the mycorrhiza in corn plants). *Acta microbiol. pol.*, 4, 191 - 204.

STRZEMSKA J. Mycorrhiza of cultivated plants of the Papilionaceae Family.
- (1969). Part I. Broad Bean (Vicia faba ssp. major), Horse Bean (Vicia faba ssp. minor). *Polish J. Soil Sci.*, 2, 43 - 50.
- (1969). Part II. Red Clover (Trifolium pratense), Swedish Clover (Trifolium hybridum) and White Clover (Trifolium repens). *Polish J.Soil Sci.*, 2, 137 - 143.
- (1970). Part III. Pea (Pisum sativum) and Field Pea (Pisum arvense). *Polish J.Soil Sci.*, 3, 25 - 29.
- (1973). Part IV. Fenugreek (Trigonella foenum graecum) and Soy-Bean (Glycine hispida). *Polish J. Soil Sci.*, 6, 57 - 62.

MYCORRHIZAS IN PENNINE GRASSLAND

G. P. SPARLING AND P. B. TINKER
Department of Plant Sciences,
University of Leeds, U.K.

INTRODUCTION

Despite increasing awareness of the widespread occurrence of endotrophic mycorrhizal associations in natural (or semi-natural) ecosystems, most work, (see Mosse, 1973a) has concentrated on agricultural crops. However, Ali (1969), Crush (1973a, b), Mejestrik (1972) and Nicolson (1958, 1959, 1960) attempted to measure the extent and influence of mycorrhizal infection in grassland.   This paper presents preliminary findings on the distribution and level of mycorrhizal infection in three grassland areas in the Yorkshire Pennines.

The main agricultural use of the Pennines is for permanent pastures and rough grazings.   The climate is windy, humid and cloudy with a low yearly average temperature of 7° (Manley, 1955).   Rainfall is 1 - 2 m per annum, with evapo-transpiration of only 500 mm (Bullock, 1964).   The soils of the area are consequently highly leached, predominantly acid and nutrient deficient (Bullock, 1964).   The low Olsen phosphate levels (<5 µg/g) of the soils suggested that a mycorrhizal response could be expected.

PRELIMINARY SURVEY

Fifteen sites in the Grassington area (N.G.R.

SE06 0064) were sampled in autumn 1972. Despite site differences of vegetation, soil type, soil pH and altitude, all plants sampled showed moderate to heavy mycorrhizal infection. Cleared and stained roots (Phillips and Hayman, 1970) had endotrophic mycorrhizal infections along 50 - 80% of the total root length.

Three sites were chosen for more detailed work. These were characteristic of their type and representative of larger Pennine areas, (pers. comm. Soil Survey of England and Wales). The three sites chosen belonged to the Malham, (Bullock 1964), Roddlesworth and Brickfield Series (Crompton 1966) and were respectively an acid brown earth, a peaty gley and a surface water gley soil over sandstone.

METHODS

Roots were sampled from the field by taking soil cores with a Dutch auger or cheesecore auger. Roots were washed out from the cores, cleared and stained to assess mycorrhizal infection. Spores were collected by the wet sieving technique (Gerdemann & Nicolson 1963). Soil pH was measured in a 2 : 1 distilled water : soil suspension. Available phosphate was determined by the Olsen method (Olsen *et al.*, 1954). This sampling method did not allow the infection in individual plants to be assessed, but in any event it proved impossible to extract individual root systems, with their heavily infected fine lateral roots, (Ali, 1969) from turf.

Within site variation was high (Table 1) and to reduce standard error in later samples thirty cores were bulked and subsamples taken. Collections were made from the field every three months.

Infection was assessed by a simple modification to the point intersection method, (Newman, 1966). A randomly selected, cleared and stained root sample was scored as mycorrhizal or non-mycorrhizal at each

Table 1. Root fresh weight, percent mycorrhizal root length and soil phosphate at
different depths.   Brickfield series site[+].
(Number of cores = 8).

Depth (cm)	Root fresh weight (g/l)	Mycorrhizal Root (%)	Olsen phosphate (µg P/g)
5	91.1 ± 13.06*	29.9 ± 4.48	15.5 ± 1.71
10	24.8 ± 5.23	41.9 ± 4.45	4.5 ± 0.70
25	7.0 ± 1.74	42.7 ± 2.20	3.4 ± 0.43

* Standard error
+ This site received a basic slag phosphate
  dressing 6 months previously.

intersection.   The ration of mycorrhizal to non-
mycorrhizal intersections gave the proportion of infected root, and the fraction of root lacking cortex
was also calculated in a similar way.   The normal
procedure for the method gave the total root length.
The method proved very suitable for the fine, much
branched roots of grasses.   The effect of fertiliser
on the infection level was checked by top-dressing
non-replicated three by three metre square plots with
various fertiliser treatments (Table 2) in May every
year.

Agrostis tenuis, Anthoxanthum odoratum, Festuca
rubra, Cynosurus cristatus and Nardus stricta were
grown in γ-irradiated (0.8 M rad.) soils from the
three sites.   Plants were grown in a glasshouse with
additional lighting and heating during the winter
months and were either mycorrhizal, non-mycorrhizal
or non-mycorrhizal receiving phosphate fertiliser.
Sievings from field soil were used as inoculum in the
mycorrhizal treatments.   Phosphate (80 mg P/kg soil)
was supplied as potassium dihydrogen phosphate solution.   All plants originally received a soil leachate

Table 2. Effect of fertiliser dressings on soil pH, available phosphate, mycorrhizal infection and yield.

Malham Series (Brown earth) samples from top 15 cms.

Treatment (One year's application)	Soil pH	Olsen P µg/g	% Infection	% Yield Increase
No additions (control)	5.05	0.2	61.3	–
Calcium Carbonate 5.02 tonnes/ha	5.40	0.2	58.6	14.3
Calcium Carbonate 5.02 tonnes/ha	6.34	0.2	56.8	25.3
Nitrogen (Nitram) 125 kg/ha	5.43	0.2	59.8	27.6
Potassium (Muriate) 125 kg/ha	5.01	0.2	60.4	16.1
Phosphate (Basic Slag) 250 kg P/ha	5.32	0.5	44.1	1.0
Phosphate (Triple Super) 250 kg P/ha	5.20	0.5	42.2	7.4
NPK (Combination of Nitrogen, Super-phosphate & Potassium treatments) 125 : 250 : 125 kg/ha	5.25	3.0	46.3	77.6

and were watered to saturation with 1×10^{-3} M potassium nitrate solution every week. The plants were harvested by clipping the foliage down to pot level.

RESULTS

Root distribution and level infection in the field

Results are presented for the Brickfield series

(Table 1) and in more detail (Fig. 1) for the Malham
series soils.   Similar patterns were obtained on all

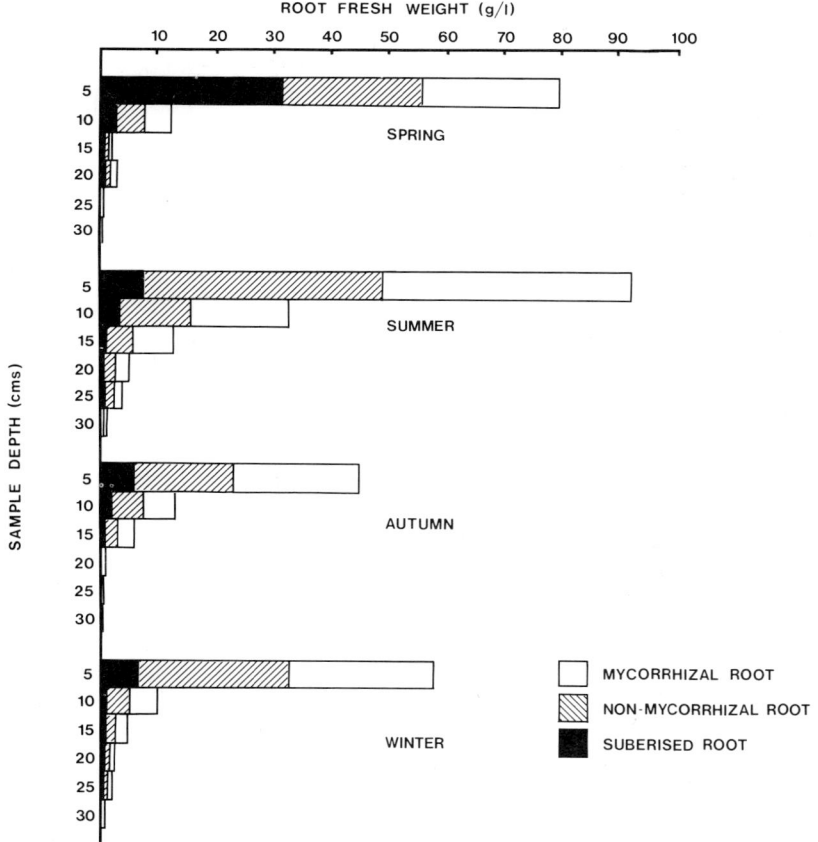

Figure 1.   Seasonal fluctuations in mycorrhizal root.

three sites.   Root weight declined sharply with
depth, so that over 95% of the total root weight was
in the top 15 cm of soil.   The root weight was highest in the summer.   The percentage infection varied

little between the sampling times, so the increased
root growth was matched by a corresponding increase
in infection.   The percentage infection varied
little above 25 cm, but declined slightly below this.
(Table 3).

Table 3.   Variation in the percentage
length of infected root with depth and
season.

(Malham Series (Brown earth) - suberised root omitted)

Depth (cms)	Spring	Summer	Autumn	Winter	Average
5	51.3	51.0	55.8	48.6	51.7
10	50.2	58.1	52.7	58.4	54.8
15	47.2	50.8	65.9	60.5	56.1
20	62.9	51.4	51.4	54.8	55.1
25	53.6	44.6	42.9	55.7	49.2
30	30.5	40.7	33.3	45.7	37.5

*Effect of fertiliser applications on growth and level
of infection in the field*

Only dressings incorporating phosphorus reduced
infection, (Table 2).   Nitrogen had no effect on the
infection level.   Phosphorus alone gave only a small
yield increase whereas nitrogen had a great effect on
dry matter production, but the use of single plots
did not allow significance testing.   The lime and
potassium applications increased yield but had no
effect on infection.   Yields with NPK were greater
than those of the nutrients added separately but the
decrease in infection was the same as that of the
phosphate alone.

The fungal endophytes

Both arbuscular and vesicular-arbuscular mycorrhizal types were present as well as non-mycorrhizal endophytes (Nicolson 1959). The mycorrhizal endophytes were confined to the inner cortical cells and did not invade the stele or meristems. Extra-matrical mycelium was common.

Fine endophytes with hyphae 1 - 3 μm diameter were present, as well as the more usual "coarse" form with hyphae of 5 - 10 μm. The fine endophyte was most common on the highest (550 m) *Nardus* dominated peaty gley soil. Mixed infections with the coarse and fine endophyte together were frequent and in older roots infections with non-mycorrhizal fungi, especially *Rhizoctonia,* were commonly observed. The fertiliser treatments did not cause any detectable morphological changes in the endophytes.

Spores

Despite the high levels of infection the total number of spores was low, frequently well below 100 per litre of soil. Sporocarps were very rarely observed. Although spores were rare, seedlings planted in these soils rapidly became infected with mycorrhizal endophytes.

Pot experiments

Responses to phosphate began to appear after eight months growth and were clearly apparent by twelve months. The phosphate response was greatest in the Malham series soil (Table 4) and a mycorrhizal response was obtained with all five grass species after twelve months growth in this soil. The mycorrhizal response increased further by fourteen months increasing the cumulative dry weight of the tops by 16% (average for all spp), but by 65% by the phosphate treatment. No mycorrhizal response was detected in the Roddlesworth or Brickfield series soils after fourteen months, despite heavy mycorrhizal infection and strong responses to phosphate (39% and 29%).

Table 4.   Cumulative weight of tops (gDW) of grasses grown in three Pennine soils under three different treatments.
(Grand means from all species)

Harvest No.	Age Months	Brickfield Soil			Malham Soil			Roddlesworth Soil		
		C	M	P	C	M	P	C	M	P
1	6	0.77	0.73	0.75	0.75	0.42	0.62	0.51	0.51	0.46
2	8	1.26	1.19	1.33	1.11	0.84	1.02	0.95	0.89	0.86
3	12	2.22	2.20	2.60	1.69	1.81	2.38	1.65	1.63	2.06
4	14	2.53	2.53	3.27	1.87	2.16	3.09	1.85	1.87	2.58

C = Control Plants
M = Mycorrhizal Plants
P = With Phosphate

DISCUSSION

The level and depth distribution of mycorrhizal infection in the Pennine grasses agreed broadly with results obtained by Mejstrik (1972) for a *Molinia* community, where a similar decrease in infection at 30 cm was recorded, but in his work infection mostly increased between April and September.   In the Pennines the overall percentage of infected root did not fluctuate greatly with season, which suggests that the level of infection in individual species did not vary either, though no direct results are available.

High levels of mycorrhizal infection with low spore numbers have been reported from various habitats, (Mosse and Bowen, 1968; Redhead, 1971).   In a perennial host where living and infected root is present the whole year there is less selective pressure for spores than when roots die back at the end of the growing season, or when annual hosts die.   It is likely that the mycorrhizal endophytes in permanent grassland are non or poorly sporing forms, and new root growth will probably become infected from the already existing mycorrhizal mycelium rather than by spore germination.

The "fine" endophyte has only been reported from semi-natural areas (Ali, 1969: Crush, 1973a, b; Greenall, 1963) as was found here.

Neither the nitrogen applications nor the pH change caused by the higher lime dressing had any effect on mycorrhizal infection, in contrast to other reports, (Hayman, 1970; Mosse, 1972 a, b:).   Differences in nutrient levels of the soils and the pH tolerance of the hosts may account for the discrepancies.   Decreased infection caused by phosphate dressings parallels that reported in pots (Daft and Nicolson, 1969: Mosse, 1973b;) and in the field, (Khan, 1972; Hayman, 1970; Ross, 1971).   The amount of phosphate used was high compared to a normal dressing (ca. 60 kg P/ha), but despite this the

Olsen phosphate value remained low.   The two forms
of phosphate fertiliser affected the growth and infection of the grasses to a similar extent, which was
not unexpected in these acid soils.
    Although a mycorrhizal response was finally
obtained only in the Malham series soil, there are
reports of mycorrhizal responses by grasses in other
low fertility soils (Crush, 1973a, b;   Cooper, 1974).
Pennine soils are normally of very low fertility but
soil sterilisation caused marked growth stimulation
(100 - 200%) which persisted for many months.   This
may have interfered with a possible mycorrhizal
response although the Olsen phosphate level was not
altered significantly by sterilisation.   Read and
also Mosse and Hayman (personal communication) have
observed similar sterilisation effects in heath
soils.   Under similar conditions grasses responded
much more rapidly to applied phosphates in the unsterile soil than the sterilised soil.   The results
with irradiated soil may therefore not be a valid
indication of the mycorrhizal effect in the field.
    The fact that the grasses gave large growth
responses to phosphate long before they showed any
mycorrhizal responses may possibly be explained by
root morphology.   All the pot cultures had a very
large root length, and the calculated mean distance
between roots was 0.4 - 1.5 mm.   Since many roots
also carried root hairs of length up to 1 mm the mean
distance between absorbing surfaces would be even
less.   This large root length also gave a very low
phosphate inflow of about 1.2×10^{-16} moles $cm^{-1}sec^{-1}$,
some 400 times lower than that obtained for onions by
Sanders and Tinker (1971).   The roots would probably
be able to meet this demand without the need for any
additional supply via mycorrhizal hyphae.   Further-
more the distance between roots and root hairs is so
small that the individual depletion zones will overlap largely and there will be little or no undepleted
soil for mycorrhizal hyphae to exploit.   However, as
uptake proceeds, the removal of phosphate will slowly

diminish the soil solution concentration and in consequence (Vaidyanathan & Nye, 1970) the diffusion
coefficient (D) will decrease. The individual
depletion zones which are roughly $2\sqrt{Dt}$ in diameter
(where t is time) will become progressively narrower,
(Baldwin & Nye, 1974) and a situation may be reached
where there is once again an appreciable volume of
relatively undepleted soil between the roots and
hyphae can then be beneficial in exploiting these
zones. Mycorrhizal hyphae may possibly have a higher absorbing power than roots and consequently could
carry on absorbing phosphorus at a useful rate when
roots were unable to do so.   (Mosse et al., 1973).
    A similar situation may exist in the field since
the overall root density is similar.   (Table 5).
The age structure of the root mass is unknown however, and the root hair frequency is much less than
in pots.   Even so, in the top 5 cm of soil if only
10% of the root is absorbing, there is still on
average less than 2 mm between absorbing surfaces.
    Although mycorrhizal infection is widespread in
Pennine grassland, the importance of the association
is unknown.   Monro et al., (1973) found nitrogen to
be the main limiting nutrient on Welsh upland and
this may also be so on Pennine soils as there was a

Table 5.   Mean distance (MM) between roots
at different depths, brown earth grassland.

Depth (cm)	Total Root	Unsuberised Root	Mycorrhizal Root
5	0.39	0.42	0.66
10	0.89	0.97	1.34
15	1.23	1.67	2.14

large growth response to nitrogen.   However, this
may not be the case in the absence of mycorrhizal
infection.   There were clearly large nitrogen-
phosphorus interactions, which implies that mycorrhizas may be particularly important where nitrogen
supply was adequate.   Low temperatures (7°) prevail
for much of the year in the Pennines and the plants
have a low growth rate.   Under similar conditions
in the laboratory, mycorrhizas can be ineffective or
detrimental (Furlan & Fortin, 1973; Hayman, 1974).
During phases of more rapid growth (Morris & Thomas,
1972) Baylis (1970) suggested that root hairs rather
than mycorrhizas would be of greater benefit, because
mycorrhizal infection would lag behind root elongation.
        There are thus various reasons why mycorrhizas
may not be effective under natural conditions.   This
work has established that under glasshouse conditions,
grass species responded to mycorrhizal infection when
grown in Pennine soil, but the effectiveness of the
mycorrhizas under field conditions is, as yet,
unknown.

ACKNOWLEDGMENTS

        The authors wish to thank Messrs. R. A. Jarvis
and D. M. Carroll of the Soil Survey of England and
Wales for advice in selecting the experimental field
sites; and Messrs. S. Butcher, A. and G. Stockdale,
and J. Stephenson for permitting the field experiments to be conducted on their farms.
        G.P.S. acknowledges receipt of a M.A.F.F.
studentship.

REFERENCES

ALI, B. (1969). Occurrence and characteristics of
    versicular-arbuscular endophytes of *Nardus stricta*.

Nova Hedwigia, 17, 409-425.

BALDWIN, J. P. & NYE, P. H. (1974). A model to calculate the uptake by a developing root system or
root hair system of solutes with concentration
variable diffusion coefficients.   *Pl. Soil,*
40, 703-706.

BAYLIS, G. T. S. (1970). Root hairs and Phycomycetous
mycorrhizas in phosphate deficient soil.   *Pl.Soil,*
33, 713-716.

BULLOCK, P. (1964). A study of the origin and devel-
opment of soils over carboniferous limestone in the
Malham district of Yorkshire.   M.Sc. Thesis,
University of Leeds, England.

COOPER, K. (1974). Mycorrhizal associations in N.Z.
ferns.   Ph.D. Thesis, University of Otago, New
Zealand.

CROMPTON, E. (1966). The Soils of the Preston District of Lancashire.   Memoirs of the Soil Survey
of Great Britain, England and Wales.   A.R.C.
Harpenden, England.

CRUSH, J. R. (1973a). Significance of endomycorrhizas
in tussock grassland in Otago, New Zealand.
N.Z. J. Bot., 11, 645-660.

CRUSH, J. R. (1973b). The effect of *Rhizophagus
tenuis* mycorrhizas on rye grass, cocksfoot and
sweet vernal.   *New Phytol.,* 72, 965-973.

DAFT, M. J. & NICOLSON, T. H. (1969). Effect of
Endogone mycorrhiza on Plant growth.   II. Influ-
ence of soluble phosphate on endophyte and host in
Maize.   *New Phytol.,* 68, 945-952.

FURLAN, V. & FORTIN, A. J. (1973). Formation of

558   Sparling, G. P. and Tinker, P. B.

endomycorrhizas by *Endogone calospora* on *Allium
cepa* under three temperature regimes.   *Naturaliste
can.,* <u>100</u>, 467-477.

GERDEMANN, J. W. & NICOLSON, T. H. (1963). Spores of
mycorrhizal *Endogone* species extracted from soil by
wet sieving and decanting.   *Trans. Br. mycol.
Soc.,* <u>46</u>, 235-244.

GREENALL, J. M. (1963). The mycorrhizal endophytes of
Griselenia littoralis (Cornaceae).   *N.Z. J. Bot.,*
<u>1</u>, 389-400.

HAYMAN, D. S. (1970). *Endogone* spore numbers in soil,
and versicular-arbuscular mycorrhiza in wheat as
influenced by season and soil treatment.   *Trans.
Br. mycol. Soc.,* <u>54</u>, 53-63.

HAYMAN, D. S. (1974). Plant growth responses to
versicular-arbuscular mycorrhiza.   VI.   Effect of
light and temperature.   *New Phytol.,* <u>73</u>, 71-80.

KHAN, A. G. (1972). The effect of vesicular-arbuscular mycorrhizal associations on growth of cereals.
I.   Effects on maize growth.   *New Phytol.,* <u>71</u>,
613-619.

MANLEY, G. (1955). The Climate of Malham Tarn.   Rep.
of the Council for the promotion of Field Studies.
43-56.

MEJSTRIK, V. K. (1972). Vesicular-arbuscular mycorrhizas of the species of a Molinietum coeruleae.
L. I. association: the ecology.   *New Phytol.,*
<u>71</u>, 883-890.

MONRO, J. M. M., DAVIES, D. A. & THOMAS, T. A. (1973).
Potential pasture production in the uplands of
Wales.   3. Soil nutrient resources and limitations.
J. Br. Grassland Soc., <u>28</u>, 247.

MORRIS, R. M. & THOMAS, J. G. (1972).   The seasonal
pattern of dry matter production of grasses in the
North Pennines.   *J. Br. Grassld. Soc.,* 27, 163.

MOSSE, B. (1972a). Effects of different *Endogone*
strains on the growth of *Paspalum notatum*.   *Nature,
London,* 239, 221-223.

MOSSE, B. (1972b). The influence of soil type and
Endogone strain on the growth of mycorrhizal
plants in phosphate deficient soils.   *Rev. Ecol.
Biol. Sol.,* 9, 529-537.

MOSSE, B. (1973a). Advances in the study of versicular-arbuscular mycorrhizas.   *A. Rev. Phytopath.,*
11, 171-196.

MOSSE, B. (1973b). Plant growth responses to
vesicular-arbuscular mycorrhizas IV.   In soil given
additional phosphate.   *New Phytol.,* 72, 127-136.

MOSSE, B. & BOWEN, G. D. (1968). The distribution of
Endogone in some Australian and New Zealand soils
and in an experimental field soil at Rothamsted.
Trans. Br. mycol. Soc., 51, 485-492.

MOSSE, B., HAYMAN, D. S., & ARNOLD, D. J. (1973).
Plant growth responses and versicular-arbuscular
mycorrhizas V. Phosphate uptake by three plant
species from P deficient soils labelled with 32P.
New Phytol., 72, 809-815.

NEWMAN, E. I. (1966). A method of estimating the
total length of root in a sample.   *J. Appl. Ecol.,*
3, 139-145.

NICOLSON, T. H. (1958). Vesicular-arbuscular mycorrhiza in the Gramineae.   *Nature, London.,* 181,
718-719.

NICOLSON, T. H. (1959). Mycorrhiza in the Gramineae. I. Vesicular-arbuscular and endophytes with special reference to the external phase. *Trans. Br. mycol. Soc.*, 42, 421-438.

NICOLSON, T. H. (1960). Mycorrhiza in the Gramineae. II. Development in different habitats, particularly sand dunes. *Trans. Br. mycol. Soc.*, 43, 132-145.

OLSEN, S. R., COLE, C. V. *et al.*, (1954). Estimation of available phosphorus in soils by extraction with sodium bicarbonate. *Circ. U.S.D.A.*, 939.

PHILLIPS, J. M. & HAYMAN, D. S. (1970). Improved procedure for clearing roots and staining parasitic and vesicular-arbuscular mycorrhizal fungi for rapid assessment of infection. *Trans. Br. mycol. Soc.*, 55, 158-161.

REDHEAD, J. F. (1971). *Endogone* and endotrophic mycorrhizae in Nigeria. *XV. I.U.F.R.O. Congr. Sec.* 24.

ROSS, J. P. (1971). Effect of phosphate fertilisation on yield of mycorrhizal and non-mycorrhizal soyabean. *Phytopathology,* 61, 1400-03.

SANDERS, F. E. & TINKER, P. B. (1971). Mechanism of absorption of phosphate from soil by *Endogone* mycorrhizas. *Nature, London,* 233, 278-279.

VAIDYANATHAN, L. V. & NYE, P. H. (1970). The measurement and mechanism of ion diffusion in soils. The effect of concentration and moisture content on the counter diffusion of soil phosphate against chloride ion. *J. Soil. Sci.*, 21, 15-27.

ARBUSCULAR MYCORRHIZAS IN PLANTS COLONISING COAL SPOILS IN SCOTLAND AND PENNSYLVANIA

M. J. DAFT[1] E. HACSKAYLO[2] AND T. H. NICOLSON[3]

[1, 3] *Department of Biological Sciences, University of Dundee, U.K.*
[2] *U.S.D.A. Forest Service, Beltsville, Maryland, U.S.A.*

INTRODUCTION

It has been suggested that plants with mycor-rhizal infections could be of importance in plants colonising industrial wastes (Nicolson, 1967). Nutrient-deficient sites are more readily colonised by plants with symbiotic associations which fix atmospheric nitrogen and which can extract phosphate efficiently (Harley, 1970). Coal wastes are gener-ally deficient in nutrients and plant cover is often difficult to establish (Bradshaw, 1970; Schramm, 1966). The landscape in certain parts of Britain is scarred by coal tips showing various stages in plant colonisation. In the anthracite region of Pennsyl-vania some 80940 hectares are disturbed by strip and deep mining. Of this acreage 76 per cent is the result of strip mining and less than half is under-going revegetation (Frank, 1964).

Many of the trees which are used in reafforest-ation in Pennsylvania are ectomycorrhizal and Schramm (1966) showed the importance of this association. Neither there, nor on plants in the South Wales coalfield (Gadgil, 1967) were endomycorrhizas noted, but in a recent study in Scotland, the majority of herbaceous plants colonising coal wastes were observed to contain root endophytes (Daft & Nicolson, 1974).

This paper compares the endomycorrhizas found on plants colonising the coal wastes in Pennsylvania, U.S.A. and in Fife, Scotland and describes experiments to assess their potential importance.

MATERIALS AND METHODS

Sampling of plants and spoil

Three tips in both Pennsylvania and Scotland were surveyed for general plant cover and root specimens of the major groups of plants collected with spoil from around their roots. Only 6-8 specimens of each species were taken to keep disturbance of the tip surface to a minimum. Subsequent treatment of the plants and identification and assessment of the fungal endophytes were as described in Daft and Nicolson (1974). The pH values of samples of spoil material were measured on a Pye model 292 pH meter. Total and extractable phosphate levels of samples from the Scottish tips were determined using the molybdenum blue and calcium chloride methods respectively. In the calcium chloride extraction, 5 g tip material was used with 0.1 M $CaCl_2$ and shaken for 18 hours.

Fungal endophytes

Two members of the *Endogonaceae, Gigaspora gigantea* (Nicol. & Gerd.) Gerdemann & Trappe, comb. nov. and *Glomus macrocarpus* var. *geosporus* (Nicol. & Gerd.) Gerdemann & Trappe, comb. nov. were used and both were originally extracted from coal spoil. The nomenclature used for fungal endophytes in this paper is after Gerdemann and Trappe (1974).

Experimental plants

Three hosts were selected; a sweet corn hybrid, Golden Queen (chosen for its short life cycle), a hybrid alfalfa var. Du Puits (chosen because it is easy to infect with *Rhizobium* and *Glomus*) and a wild strawberry taken from the tip at the Devon Colliery, Fife, Scotland (chosen for being a natural coloniser of the coal spoil). All the seeds were germinated on sterile sand; the seedlings were selected for uniformity, transferred to the growth medium and inoculated with the fungal endophyte (Daft & Nicolson, 1966). In the case of the alfalfa plants the abraded seeds were dipped into an effective *Rhizobium* suspension before germinating.

Growth media

Coal spoil and horticultural sand were sterilised by autoclaving (120° for 1 h) and where necessary the pH of the spoil was adjusted with calcium oxide. In some experiments bonemeal was mixed in with the medium and this was the sole source of phosphate.

Growth measurements

Leaf lengths, numbers, fresh and dry weights and reproductive organs were used to measure plant growth.

Chemical analyses

Complete chemical analyses of the experimental plants from experiments carried out in the U.S.A. were done at the University of Georgia, U.S.A.

RESULTS

Sites

The main characteristics of the six sites studied are given in Table 1. The two anthracite sites

Table 1. Characteristics of the spoil sites, plants present and overall levels of mycorrhizal infections.

Site and date mining ceased	Spoil type	pH	Plant cover (%)	Monocotyledons		Dicotyledons	
				Number of species sampled	Average infection (%)	Number of species sampled	Average infection (%)
Pennsylvania							
Cumbola 1927	Anthracite	4.1	15	1	33	7	46
Tamaqua –	Anthracite	2.7-4.0	30	1	100	3	61
Nanty Glo 1944	Bituminous	2.7	40	12	30	14	50
Scotland							
Devon 1960	Bituminous	4.3-4.9	45	7	51	10	34
Glen Ochil 1962	Bituminous	7.4	85	3	64	0	–
Balgonie 1960	Bituminous	5.7	30	3	31	0	–

in Pennsylvania were both classified by Czapowskyj
and McQuilkin (1966) as type 1 and the refuse was
composed of small pieces of black, highly pyritic
carbonaceous shales, slates, silt, stones and coal
fragments.   At Nanty Glo the bituminous spoil was
from the deep mining of low voltal coal.   Each
Scottish tip was made up of 'run of the mine' dirt,
and washer discard (blaes, sandstone and fireclay)
in various proportions.   The pH values were in the
main low and this may be a significant factor, as
maximum plant cover was found at Glen Ochil, which
had the highest pH.   Generally, spoil material had
low nitrogen and phosphorus content.   The Scottish
spoil material had total phosphate contents of 675,
766 and 1087 µg/g and extractable phosphorus levels
of 2.31, 1.37 and 0.35 µg/g for the Devon, Glen
Ochil and Balgonie collieries respectively.   No
data was obtained for the N and P contents of the
spoil from the Pennsylvanian tips.

Herbaceous plants and grasses were the main
primary colonisers in Scotland (Plate XXIII, 1-2))
and at Glen Ochil, where the tip had been levelled
and graded, there was a complete ground cover, but
all three sites in Pennsylvania had a poor ground
cover.   At Tamaqua some experimental areas had been
planted and this in some part explains why the plant
cover was greater than at the similar anthracite
site at Cumbola.   All the plants growing on the
three Scottish tips were natural revegetation by
local components of the flora.

Mycorrhizal infections in colonising plants

A summary of the number of species and average
levels of infection over all species found on the six
sites are shown in Table 1.   There was a greater
number of species colonising the anthracite spoil
at Cumbola compared with Tamaqua but levels of
infection were much higher at the latter.   The
crown vetch and couch grass at Tamaqua were so very

heavily infected that several hundred spores of
G. gigantea were easily collected from their roots.
The large azygospores, and characteristic echinulate
structure of *G. gigantea* are shown in Plate XXIV, 5 &
6   .   It was noticeable that the colonisation by
grasses was still very slow at Nanty Glo (Plate XXIII,
3) although the trees were surviving.   On the
Devon tip (Plate XXIV, 1&2) natural colonisation was
taking place with the plants well established on the
lower slopes and the upper unstable slopes were being
colonised by the Rosebay Willowherb (*Chamaenerion
angustifolium*).   Most of the plants at Glen Ochil
and Balgonie were grasses and the infection levels
were much higher at the former site.

Some of the plants growing at four of the six
sites, together with their levels of infection and
type of endophytes are given in Table 2.   The
recorded levels of infection at Cumbola were high
except in *Prunus* and this may be due to the very dark
roots that make assessment difficult.   An experimental plot at Nanty Glo was laid out in 1970 and all
plants sampled were infected with a *Glomus* sp.,
probably *G. fasciculatus* (Thaxter sensu Gerdemann)
Gerdemann & Trappe comb. nov.. *Lonicera* and *Malus* were
two of the genera used and with such introduced
plants it is difficult to know if they were infected
before, or became infected after, planting.   Schramm
(1966) commented on the success of the nodulated
plants growing on the anthracite waste and he considered that their success was due to their ability to
compensate for the low levels of nitrogen in the
spoil.   We made a particular point to look for nodulated plants as it has been shown that mycorrhizal
and nodulated plants fix atmospheric nitrogen more
efficiently than nodulated only plants (Daft and
El-Giahmi, 1974, 1975).   The introduced *Coronilla
varia* at Tamaqua was well nodulated and heavily
infected with *Gigaspora gigantea* (Table II).   *Alnus
glutinosa* at Nanty Glo had large nodules and was
infected with two different fungal endophytes and

Table 2. Host plants, root infection levels and types of endophytes*
from Pennsylvania and Scotland.

Host	Percentage infection	Types of endophytes identified
Cumbola (Penn.)		
Apocynum canbincum	62	*Gigaspora gigantea*
Echium vulgare	29	-
Hieracium sp.	78	Chlamydosporic type
Prunus alleghaniensis	10	-
Robinia hispida	49	External mycelium only (*Glomus*)
Rubus sp.	51	*G. gigantea* plus *Glomus* type
Tamaqua (Penn.)		
Coronilla varia	85	*G. gigantea*
Nanty Glo (Penn.)		
Eleagnus umbellata	38	*Glomus fasciculatus*
Robinia hispidia	38	*G. fasciculatus*
R. pseudoacacia	66	*G. gigantea*
Festuca arundinacea	46	*G. gigantea*
F. arundinacea	77	*G. gigantea* + *G. fasciculatus*
Lolium perenne	46	*G. gigantea*
Poa compressa	13	-
Yucca baccata	45	*G. gigantea* + chlamydosporic type
Lonicera sp.	99	*G. fasciculatus*
Malus baccata	56	*G. fasciculatus*
Rubus sp.	54	Chlamydosporic type (*Glomus* sp.)
Alnus alnus	63	*G. gigantea* + chlamydosporic type (*Glomus* sp.)

Table 2. (continued)

Host	Percentage infection	Types of endophytes identified
Devon (Scot.)		
Agrostis tenuis	41	Mostly narrow endophyte plus some *G. fasciculatus*
Dactylis glomerata	38	*G. calospora*
Festuca ovina	41	*G. fasciculatus*
Holcus lanatus	84	*G. fasciculatus*
Chamaenerion angustifolium	41	Mostly narrow endophyte plus some *G. fasciculatus*
Cirsium arvense	34	*Gigaspora* and *Glomus* spp. predominating but some narrow endophyte
Fragaria vesca	63	*G. fasciculatus*
Rubus fruticosus	20	Both endophytes
Senecio viscosus	40	*Glomus* and *Gigaspora* spp. predominating but some narrow endophyte

* Nomenclature for endophytes after Gerdemann & Trappe (1974).

several vesicles of one of the species is shown in
Plate XXIV, 8.   The two species of *Robinia*, *R. pseu-
docacia* and *R. hispida*, had only 1 type of fungal
endophyte (Plate XXIV,4&9) as did *Eleagnus umbellata*.
Each of these plants was a successful colonizer.   As
mycorrhizal infection enhances the uptake of phosphate and the different endophytes within the nodules
can fix atmospheric nitrogen, dual infected plants
are probably very suitable for revegetating the spoil
heaps.   There was however, a surprising lack of nodulated plants colonising the Scottish sites.

Plants growing on the Devon tip had infection
levels ranging between 20-84% (Table 2).   From
direct examination of roots and soil sievings and by
growing field material in sterile sand (Gilmore,
1971), three main types of endophytes were identified.
Firstly, the majority of infections were caused by
chlamydosporic endophytes of the genus *Glomus* which
formed typical arbuscules and vesicles.   The most
frequent species was *G. fasciculatus* which could be
recognized by its extensive spore formation in roots
(Plate XXIV, 8).   Also *G. macrocarpus* var. *geosporus* was obtained from plants of *Holcus lanatus*.
Secondly, *Gigaspora calospora* (Nicol. & Gerd.)
Gerdemann & Trappe, comb nov. was identified from
roots of *Dactylis glomerata* by means of its azygospores (Nicolson & Gerdemann, 1968).   Azygospores
with typical bulbous attachments were abundant and
also the irregular accessory vesicles from sieved
material (Plate XXIV, 7)..   Thirdly, a less usual
type with very narrow hyphae (1 μm) was often present on our pioneer colonising species.   This endophyte, which would appear to have no affinity with
the *Endogonaceae* group, has been reported previously
from New Zealand (Greenall, 1963;   Baylis, 1969;
Crush, 1973) and in the U.K. (Ali, 1969).

Indigenous endophytes and plant growth

   Effectiveness of Gigaspora gigantea inoculum

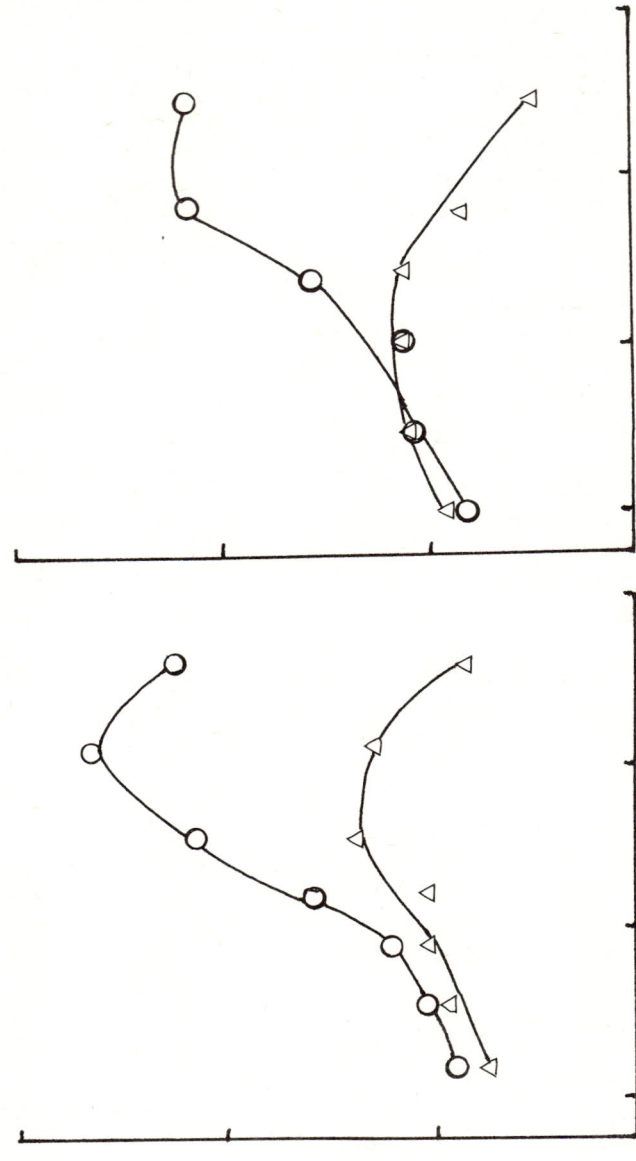

Fig. 1a. Growth of maize in sand, infected with *Gigaspora gigantea* from washed crown vetch o – o, non-mycorrhizal control Δ – Δ .

Fig. 1b. Growth of maize in sterile anthracite (pH adjusted to 6.4) and infected with *Gigaspora gigantea* o – o, non-mycorrhizal control Δ – Δ.

direct from the field.   The azygospores of *G. gigantea* collected from the crown vetch growing at Tamaqua provided an inoculum to test the effectiveness of this indigenous species in promoting plant growth. Maize plants were grown in sand culture containing bonemeal (level 0.5, Daft & Nicolson, 1966).   They were inoculated with the spores of *G. gigantea* and given 30 ml, twice weekly, of a Long Ashton nutrient solution (Hewitt, 1952), minus the phosphate ion. The growth of the 8 infected and control plants was estimated by measuring the length of each leaf, at intervals over a period of 72 days.   The results are shown in Figure 1a.   Mycorrhizal infection produced a significant increase in plant size after 38 days ($P<0.05$).   Maximum leaf lengths were found after 61 days for both treatments, and then declined due to senescence.   When the plants were harvested the control plants had fewer leaves, had not produced any cobs (female inflorescences) and the tassels (male inflorescences) were very small and contained no pollen (Table 3).   The mycorrhizal plants had produced 4 cobs from the 8 replicates and pollen had

Table 3.   Effect of *Gigaspora gigantea,* sieved from anthracite waste, on the growth and reproduction of maize in sand culture. Plants grown for 78 days.

	Mean[a] number of leaves	Total number of:		Mean[a] dry weight (g)		Percentage infection
		Cobs	Tassels	Shoot	Root	
Control	10.1	0	3	1.6	0.9	–
Mycorrhizal	12.3	4	6	4.2*	2.1*	66

[a] Mean of 8 plants   * Significant at 5% level

Table 4. Chemical analysis of maize plants, grown in sand, infected with *Gigaspora gigantea*.

Major elements (%)

		N	P	K	Ca	Mg	S
Root	Control	1.20	0.10	1.91	0.18	0.02	0.21
	mycorrhizal	1.10	0.12	1.27	0.56	0.05	0.13
Shoot	Control	1.40	0.09	2.92	0.46	0.31	0.15
	Mycorrhizal	1.10	0.10	1.67	0.34	0.29	0.10

Minor elements ($\mu g\ g^{-1}$)

		Mn	Fe	B	Cu	Zn	Al	Mo	Sr	Ba	Na
Root	Control	72	133	40	14	138	354	2.6	10	11	943
	Mycorrhizal	73	518	9	25	82	476	2.5	13	8	1175
Shoot	Control	110	44	37	1	86	49	1.1	13	0	162
	Mycorrhizal	59	49	23	7	67	51	0.7	8	1	146

formed in the tassels. Dry weights of the shoots and roots were much greater in the mycorrhizal plants in which the infection level was high.

Chemical analyses of the roots and shoots of the plants from each treatment are given in Table 4. The concentration of phosphorus, Fe, Cu, and Al were all higher in the infected plants and K, S, N and Zn higher in the control plants. These results are expressed as percentages for the major elements and in ppm for the minor elements. If the results are calculated on a total dry weight basis the mycorrhizal plants contained more of every element analysed. These results show that nodulation by *Gigaspora gigantea* obtained directly from the coal spoil is effective in promoting growth and reproduction in maize plants grown under these conditions. While this is a well-known endophyte and the first of the genus to be described (Gerdemann, 1955), it has only once been used in growth experiments before (Clark, 1969).

Effectiveness of Glomus macrocarpus var. geosporus. Other experiments used *G. macrocarpus* var. *geosporus* as the endophyte. This inoculum was obtained from a "pot culture" which originally came from *Holcus lanatus* grown on the Devon tip and as such is an indigenous endophyte. Wild strawberry plants colonising the tip at Devon were collected and the roots completely removed along with the old leaf bases. The rootless crowns were then dipped in hypochlorite solution as a further attempt to reduce the natural mycorrhizal fungi. After washing in water the crowns were rooted in sterile sand. When the crowns had produced sufficient roots they were selected for uniformity and then potted into sterilised coal spoil taken from the tip. Half of the plants were inoculated (with *G. macrocarpus* var. *geosporus*) and given water only for 11 months. From the results given in Table 5, it can be seen that the plants responded to infection with the endophyte. Some of the control plants had become infected by contamination from the inoculated plants or from an original infection whilst growing on the tip. It is possible that the pretreatments of the plants did not remove all of the natural infection, never-the-less a much higher level of infection was produced in the plants inoculated with the *Glomus*. This preliminary experiment indicated that plants which colonise the tip respond to high levels of mycorrhizal infection produced by an indigenous endophyte even when grown in coal spoil.

Daft, M. J. *et al.*

Table 5. Effect of mycorrhizal infection
on the growth of *Fragaria vesca* collected
from the Devon tip, grown in pots containing
sterilised spoil material. Plants grown
for 12 months and infected with *Glomus
macrocarpus* var. *geosporus* (ex. coal spoil).

(a)	Mean fresh weight (g)	Mean number of leaves	Mean infection (%)
Control	1.73	5.4	17.6
Mycorrhizal	2.08**	7.3	82.1

(a) Mean of 20 - 24 plants
** Significant at 1% level

Coal spoil as a growth medium

Anthracite and bituminous spoil were collected
from Cumbola and Nanty Glo. After sterilisation
the pH of the material was considered to be too low
for normal plant growth and so lime was added to
raise the pH values of the spoil material. A comparison was made between the two types of spoil,
before and after adjustment of the pH, for the
growth of mycorrhizal maize plants. This gave 8
treatments, anthracite spoil at pH 4.1 and 6.4, bituminous spoil at pH 2.7 and 6.0, each treatment with
mycorrhizal and non-mycorrhizal plants. The plants
were inoculated with *G. macrocarpus* var. *geosporus* in
the usual way and given only distilled water. The
plants grown in the bituminous spoil without pH
adjustment died within 2 weeks, after 5 weeks no
growth of the plants had taken place in the pH
adjusted bituminous spoil and there were marked signs
of toxicity. In the unadjusted anthracite spoil
there was very little growth of the plants after 5

weeks and an examination of the roots showed no
signs of infection.   Mycorrhizal and non-mycorrhizal plants grew in the adjusted anthracite material.
The results of this part of the experiment are shown
in Figure 1b.   Both treatments grew at a similar
rate until 40 days from inoculation and then the
mycorrhizal plants grew much faster.   At the end of
the experiment some of the mycorrhizal plants were
showing signs of tassel development.   It was apparent that there were considerable differences in the
two types of coal spoil, and only after adjustment
of the pH was normal plant growth possible in the
anthracite material.

The growth of dual infected plants in the anthracite material was investigated in the next experiment.   Nodulated and nodulated plus mycorrhizal
alfalfa plants (*Rhizobium* and *G. macrocarpus* var.
geosporus infected) were grown in anthracite at pH
4.1 and 6.4 and given distilled water only.   Growth
of the plants is shown in Figure 2 as the total
number of leaves present in each treatment.   It can
be seen that even with dual inoculated plants,
unadjusted anthracite material is unsuitable for
normal alfalfa growth.   At the higher pH level the
plants grew well for 30 days and then the nodulated,
non-mycorrhizal plants began to drop their leaves
prematurely whereas the nodulated and mycorrhizal
plants continued to grow much faster.   This experiment again showed that plants can grow normally in
anthracite spoil after pH adjustment and that there
are sufficient nutrients in the spoil material for at
least a limited growth period.

DISCUSSION

The problems in revegatating coal spoils are
many and varied.   Each of the three types considered
in this investigation show the complexity of the
situation.   All of the plants examined except a

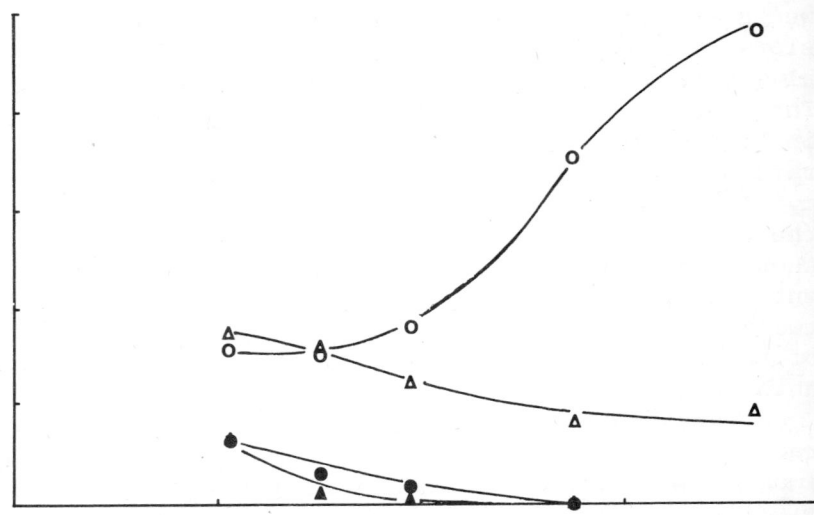

Figure 2.   Growth of nodulated and mycorrhizal alfalfa plants in anthracite waste
at two pH values.

o - o   pH adjusted to 6.4 and infected with
        Rhizobium and *Gigaspora gigantea*
Δ - Δ   pH adjusted to 6.4 and infected with
        Rhizobium
● - ●   pH 4.1 and infected with *Rhizobium*
        and *Gigaspora gigantea*
▲ - ▲   pH 4.1 and infected with *Rhizobium*

Carex species and *Reseda luteola* were infected with
arbuscular mycorrhizas.   Gerdemann (1968) lists the
Cyperaceae and the *Resedaceae* as being non-mycorrhizal.   The endophyte flora in Pennsylvania was rather
more diverse than on the Scottish tips.   In the
former *Gigaspora gigantea* and a number of chlamydosporic *Glomus* spp. were the most frequent.   In the
Scottish material two *Glomus* spp. (*G. fasciculatus*
and *G. macrocarpus* var. *geosporus*) were the most
abundant endophytes.   Also present was *Gigaspora*

calospora and the endophyte with narrow hyphae which was more restricted to pioneer colonising plants. The latter two forms were noted in the U.S.A. material but were of much less frequent occurrence (Table 2).

Our preliminary examination of some of the factors concerned in the growth of plants in coal waste suggests that mycorrhizal infection may be essential for the survival of most of the herbaceous plants. The field collected *Gigaspora gigantea* enhanced the growth and reproduction of maize plants. After adjustment of the pH there were sufficient nutrients in the anthracite waste to support plant growth when the plants were mycorrhizal or mycorrhizal and nodulated. The indigenous strawberry plants taken from the tip at the Devon colliery and grown in sterilized spoil material responded to infection with an isolate of *G. macrocarpus* var. *geosporus* obtained from the same tip. Hence, both groups of the *Endogonaceae* found in the tip material can stimulate plant growth and it would have been interesting to find whether narrow endophyte acted similarly. This last endophyte has been found abundantly in high altitude soils in N. Zealand (Crush, 1973) and could stimulate plant growth under very low phosphate conditions. For this reason its presence in pioneer plants on coal tips could be of considerable significance.

Mycorrhizal and nodulated plants appear to offer certain added advantages as colonisers of coal spoil. In this symposium the effects of mycorrhizal infection in legumes have been discussed. One of the major contributions of the fungal endophyte to the legume host is an increase in the uptake of phosphorus and the efficiency of the nodules is increased (Daft and El-Giahmi, 1974). Five nodulated and mycorrhizal plant species were found growing on the spoils in Pennsylvania. The *Robinia* species were particularly well adapted to growth on the spoil. Their habit and vegetative reproduction are factors

that made these plants suitable as colonisers.　The
accumulation of leaf litter around the stems helped
in the germination of seeds and reduced the temperature of the black surface.

Schramm (1966) was the first to demonstrate the
importance of ectotrophic mycorrhizas for trees
colonising the coal spoil.　Our preliminary work
suggests that arbuscular mycorrhizas play a similar
function in herbaceous plants.　The next step would
be to conduct tests under field conditions.　Rein-
oculation of plants or inoculation of spoil may
assist in the accelerated colonization of unsightly
industrial waste and in this way return the earth's
green mantle to areas devastated by the acts of man.

REFERENCES

ALI, B. (1969). Occurrence and characteristics of
the vesicular-arbuscular endophyte of *Nardus
stricta*.　*Nova Hedwigia*, <u>17</u>, 409-425.

BAYLIS, G. T. S. (1969). Host treatment and spore
production by *Endogone*.　*N.Z.J. Bot.*, <u>7</u>, 173-174.

BRADSHAW, A. D. (1970). Plants and industrial waste.
Trans. Proc. bot. Soc. Edinb., <u>41</u>, 71.

CLARK, F. B. (1969). Endotrophic mycorrhizal infection of tree seedlings with *Endogone* spores.
For. Sci., <u>15</u>, 134-137.

CRUSH, J. R. (1973). Significance of endomycorrhizas
in tussock grassland in Otago, New Zealand.
N.Z.J.Bot., <u>11</u>, 645-660.

CZAPOWSKYJ, M. M. & McQUILKIN, W. E. (1966).
Survival and early growth of planted forest trees
on strip mine spoils in the anthracite region.
U.S.D.A. Forest Serv. Res. Paper NE-46.

DAFT, M. J. & EL-GIAHMI, A. A. (1974). Effect of
   Endogone mycorrhiza on plant growth.
   VII. Influence of infection on the growth and
   nodulation in French bean *(Phaseolus vulgaris)*.
   New Phytol., 73, 1139.

DAFT, M. J. & EL-GIAHMI, A. A. (1975). This
   symposium, p. 581.

DAFT, M. J. & NICOLSON, T. H. (1966). Effect of
   Endogone mycorrhiza on plant growth.   1.   *New
   Phytol.*, 65, 343-350.

DAFT, M. J. & NICOLSON, T. H. (1974). Arbuscular
   mycorrhizas in plants colonizing coal wastes in
   Scotland.   *New Phytol.*, 73, 1129.

FRANK, R. M. (1964). A guide for screen and cover
   planting of trees on anthracite mine spoil areas.
   U.S.D.A. Forest Serv. Res. Paper NE - 22.

GADGIL, P. D. (1967). Atmospheric pollution in the
   Lower Swansea Valley.   In *The Lower Swansea
   Valley Project,* ed. Hilton, K. J. (Longmans),p.
   301.

GERDEMANN, J. W. (1955). Relation of a large soil-
   borne spore to phycomycetous mycorrhizal infections.   *Mycologia,* 47, 619-632.

GERDEMANN, J. W. (1968). Vesicular-arbuscular
   mycorrhiza and plant growth.   *A. Rev. Phytopathol.*, 6, 397-418.

GERDEMANN, J. W. & TRAPPE, J. M. (1974). The
   Endogonaceae in the Pacific Northwest.   *Mycologia
   Memoir* No. 5.

GILMORE, A. E. (1971). The influence of endotrophic
   mycorrhiza on the growth of peach seedlings.
   J. Am. Soc. hort. Sci., 96, 35-38.

Daft, M. J. *et al.*

GREENALL, J. M. (1963). The mycorrhizal endophytes of *Griselina littoralis* (Cornaceae), *N.Z.J. Bot.*, <u>1</u>, 389-400.

HARLEY, J. L. (1970). The importance of micro-organisms to colonizing plants. *Trans. Proc. bot. Soc. Edinb.*, <u>41</u>, 65-70.

HEWITT, E. J. (1952). Sand and water culture methods used in the study of plant nutrition. *Commonw. Brit. Hort. Plantation Crops Tech. Comm.*, <u>22</u>, p.189.

MOSSE, B. & HAYMAN, D. S. (1971). Plant growth responses to vesicular-arbuscular mycorrhiza. II. In unsterilized field soils. *New Phytol.*, <u>70</u>, 29-34.

NICOLSON, T. H. (1967). Vesicular-arbuscular mycor-rhiza - a universal plant symbiosis, *Sci. Prog., Oxf.*, <u>55</u>, 561.

NICOLSON, T. H. & GERDEMANN, J. W. (1968). Mycor-rhizal *Endogone* species. *Mycologia,* <u>60</u>, 313-325.

SCHRAMM, J. R. (1966). Plant colonization studies on black wastes from anthracite mining in Pennsylvania. *Trans. Am. phil. Soc.*, <u>47</u>, 331.

THAXTER, R. (1922). A revision of the *Endogonaceae*. *Proc. Am. Acad. Arts Sci.*, <u>57</u>, 291-351.

EFFECTS OF <u>GLOMUS</u> INFECTION
ON THREE LEGUMES

M. J. DAFT AND A. A. EL-GIAHMI

Department of Biological Sciences,
University of Dundee, U.K.

INTRODUCTION

The uptake of inorganic nutrients by plants is influenced by microorganisms in the rhizosphere, on the surface and within the tissues of the roots. Symbiotic endophytes such as *Rhizobium* and *Glomus* are examples of microorganisms that are involved in the uptake of two vital elements, nitrogen and phosphorus. Some legume crops are both nodulated and mycorrhizal (Asai, 1944). Ross and Harper (1970) showed that the growth and yield of nodulated soybeans was increased after inoculation with *Glomus mosseae*. Later Ross (1971) suggested that mycorrhizal infection enhanced the uptake of phosphorus in soybean roots. Phosphorus (Daft & Nicolson, 1966; Mosse, 1973), zinc (Gilmore, 1971) and copper (Ross & Harper, 1970; Daft & Hacskaylo, 1975) are found in higher concentrations in mycorrhizal plants, and these elements are known to influence nodulation and nitrogen fixation (van Schreven, 1958; Demeterio, *et al.*, 1972; Hewitt, 1958). In this paper we show some of the effects of mycorrhizal infection on the growth and physiology of two nodulated and one non-nodulated legume.

MATERIALS AND METHODS

The hosts used were French bean (*Phaseolus vul-garis* var. Canadian Wonder), alfalfa (*Medicago sativa* var. Du Puits) and peanut (*Arachis hypogea* var. Tripoli No. 4).   Selected uniform bean and alfalfa seedlings were grown in sand with bonemeal providing 30 mg P/kg sand.   Inoculation with *Glomus* spores and with effective *Rhizobium* strains, growth conditions, and feeding with Long Ashton solution minus N and P were as reported by Daft and El-Giahmi (1974). Peanut seedlings were not inoculated with *Rhizobium* and were therefore fed with nutrient solution cont-aining nitrogen but no phosphate.   Plants were weighed, nodules from each plant were weighed and counted, and nitrogenase activity was measured (Sprent, 1969).   Leghaemoglobin content was deter-mined spectrophotometrically (Keilin and Hartree, 1951).

RESULTS

Effect on growth and development

The dry weights of mycorrhizal bean, alfalfa and peanut are given in Table 1.   Dual-infection of bean and alfalfa (RG) produced larger plants than the nodulated only treatments (R) and mycorrhizal peanuts were larger than the non-mycorrhizal controls (Fig. 1).   Fig. 2 shows the rate of growth of the bean plants inoculated with *Rhizobium* (R), and *Rhizobium* and *Glomus* (RG).   After 21 days plants with *Glomus* weighed significantly more (P < 0.05). At each harvest the shoot/root ratio of the RG plants was lower than that of the plants from the R treat-ment.   RG plants also had significantly more leaves at each harvest (P < 0.01).   In the R plants the number of leaves decreased in the latter part of the experiment and this is typical of non-mycorrhizal

Table 1. Dry weight of shoot from three hosts infected with *Rhizobium* and *Glomus*. All plants received nutrient solution minus phosphorus, and for the bean and alfalfa plants also minus nitrogen.

Host	Number of plants/ treatment	Duration of experiment (days)	Mean dry weight (g)			
			Rhizobium (R)	*Rhizobium* plus *Glomus* (RG)	*Glomus* (G)	Non-infected (C)
Bean	8	63	1.4	2.3**	–	–
Alfalfa	20	90	0.04	0.08**	–	–
Peanut	25	125	–	–	3.21**	1.49

** significantly larger at 1% level

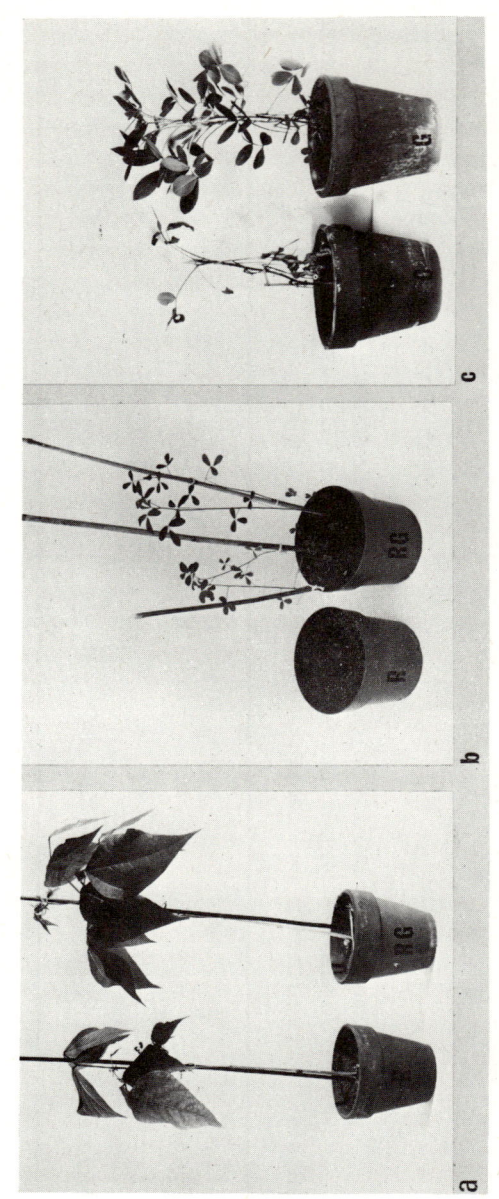

Figure 1. Plants of a) bean, b) alfalfa, c) peanut (63, 90 and
125 days old respectively) inoculated with *Rhizobium* (R), *Glomus* (G),
Rhizobium and *Glomus* (RG) and uninoculated (C).

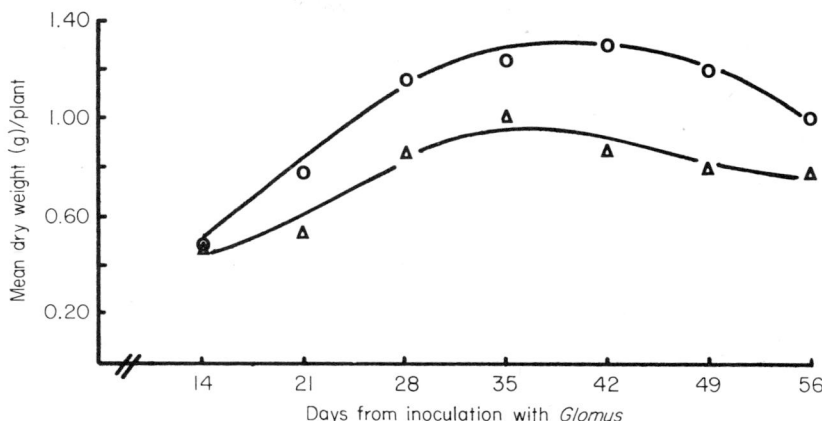

Figure 2.   Mean dry weight of bean plants
infected with *Rhizobium* (R) and *Rhizobium*
plus *Glomus* (RG), grown in sand containing
bonemeal and fed with nutrient solution
minus nitrogen and phosphorus.
(o - o *Rhizobium* and *Glomus*; Δ - Δ *Rhizobium*)

plants growing in poor nutrient conditions.   Similar
results have been reported before for tomatoes (Daft
and Nicolson, 1969b) and for maize (Daft and
Hacskaylo, 1975).   The RG plants also had more
flowers and fruits.

Effect on nodulation

The mycorrhizal infection, fresh weight, number
and leghaemoglobin content of the bean nodules at 4
harvests are shown in Table 2.   Mycorrhizal infection increased continuously, reaching a maximum of
58% after 56 days.   Fresh weight, number and leghaemoglobin content of the nodules reached a maximum
between the second and third harvests and then
declined.   In all but one harvest each of these

Table 2. Development of nodules and mycorrhiza in bean plants inoculated with Rhizobium (R) and Rhizobium plus Glomus (RG)

Days from inoculation	Mean mycorrhizal infection(%)		Mean fresh weight of nodules (g)/plant		Mean number of nodules/ plant		Leghaemoglobin (ng/mg) fresh weight of nodule	
	R	RG	R	RG	R	RG	R	RG
14	0	4	15	60	38	38	250	170
18	0	28	120	270	54	73	620	790
42	0	44	110	280	40	74	350	490
56	0	58	50	275	40	57	170	210

parameters was greater in the RG plants.   The mycorrhizal infection (at a comparatively low level of
20%) had enhanced nodulation after 28 days.

Table 3.   Mean nitrogenase activity[1] of two
hosts infected with *Rhizobium* and *Glomus*.

Host	Number of plants/ treatment	Days from inoculation	*Rhizobium* (R)	*Rhizobium* and *Glomus* (RG)
Bean	8	35	19	23
		49	13	22*
		63	22	29
Alfalfa	20	90	460	2420**

[1] Nitrogenase activity expressed as p moles/
min/mg nodule fresh weight for bean;pmoles/min
/plant for alfalfa.

Difference between R and RG significant at
* 5% and ** 1% level.

Nitrogenase activity of the nodules from the RG
plants was higher at each harvest (Table 3), indicating more efficient fixation of atmospheric nitrog-
en.   Chemical analyses of the bean shoot tissues
gave 19 and 25% total protein content and 0.13 and
0.29% of phosphorus in the R and RG plants respectively.   Statistical analyses of data from the
alfalfa experiment indicated that percentage mycorrhizal infection in RG plants was significantly
correlated (P = 0.01) with shoot dry weight (r =
0.810) and with acetylene reduction rate (r = 0.895),

but in non-mycorrhizal (R) plants there was no significant correlation (r = 0.543) between shoot dry
weight and acetylene reduction rate.

Effect of Glomus inoculum concentration

The number of *Glomus* spores used in the inocula
influenced the results in a similar set of experiments.   A low inoculum concentration (5 spores per
plant) gave increased plant size, weight of nodules
and nitrogenase activity, and these increases were
even larger in plants given an inoculum containing
50 spores/plant compared with the nodulated only
control plants.   After 49 days the mycorrhizal
infection levels for the 5 and 50 spores/plant
treatments were 27 and 46% respectively (Daft &
El-Giahmi, 1974).

Effect of soluble phosphate and nitrate

Daft and Nicolson (1969a) have shown that the
addition of soluble phosphate depresses the development of mycorrhizal infection in maize.   Table 4
shows that nodule numbers and weight were greater in
the RG (-N-P) treatment than in the R(-N-P) and
R(+N+P) treatments (P < 0.05).   Addition of nitrate
markedly suppressed nitrogen fixation.   The suppression of nitrogen fixation by nitrate is usual and
the greater nodule weight is presumably attributable
to the phosphorus which also stimulated fixation
compared to the R(-N-P) treatment.   Plants with
mycorrhiza had slightly heavier nodules and a marg-
inally greater fixation rate than those given phosphate and this may indicate that, apart from supply-
ing phosphate, the mycorrhiza could have some other
synergistic interaction with rhizobia.

DISCUSSION

From these experiments it appears that mycor-

Table 4.   Effect of *Rhizobium* and *Glomus* and different nitrogen (N) and phosphorus (P) treatments on the production of nodules and their rate of acetylene reduction.

Bean plants grown for 49 days (mean of 9 replicates).

	Rhizobium plus *Glomus*, without nitrogen and phosphorus RG (−N−P)	*Rhizobium*, without nitrogen and phosphorus R (−N−P)	*Rhizobium*, without nitrogen; with phosphorus R (−N+P)	*Rhizobium*, with both nitrogen and phosphorus R (+N+P)
Number of nodules/plant	104 ± 7.1	79 ± 7.5	96 ± 4.9	81 ± 5.0
Fresh weight (g) of nodules/plant	460 ± 5.0	280 ± 3.0	420 ± 3.0	330 ± 2.0
p moles C_2H_4/min/mg fresh weight nodule	28 ± 3.6	16 ± 2.1	22 ± 3.8	9 ± 1.8

rhiza can greatly assist nodulation and nitrogen
fixation in plants also inoculated with rhizobia and
that this is probably related to the better phosphorus supply of the mycorrhizal plants.   Crush (1974)
obtained a similar result with *Centrosema, Stylo-
santhes* and *Trifolium* grown in a P-deficient soil.

It is common practice to grow nodulated plants
on poor agricultural soils to increase their fertility and effective strains of the nodulating bacterium
are often sown along with the plant seeds.   Our
results show that an effective *Glomus* strain may
contribute to the efficiency of such a system.
Harley (1970) considered that plants with dual symbiotic associations were successful as primary colonisers due to their ability to compensate for the
infertility of the habitat.   On coal wastes in
Pennsylvania, species of *Robinia, Coranilia* and
Eleagnus were successful and these plants were both
nodulated and mycorrhizal (Daft & Hacskaylo, 1975).

REFERENCES

ASAI, T. (1944). Über die Mykorrhizenbildung der
    Leguminosen Pflanzen.   *Jap. J. Bot.,* 13, 463–485.

CRUSH, J. R. (1974). Plant growth responses to
    vesicular-arbuscular mycorrhiza.   VII.   Growth
    and nodulation of some herbage legumes.   *New
    Phytol.,* 73, 743.

DAFT, M. J. & NICOLSON, T. H. (1969b). Effect of
    Endogone mycorrhiza on plant growth.   III.   Influ-
    ence of inoculum concentration on growth and
    infection in tomato.   *New Phytol.,* 68, 953–963.

DAFT, M. J. & EL-GIAHMI, A. A. (1974). Effect of
    Endogone mycorrhiza on plant growth.   VII.   Influ-
    ence of infection on the growth and nodulation in
    French Bean (*Phaseolus vulgaris*).   *New Phytol.,*

73, 1139.

DAFT, M. J. & HACSKAYLO, E. (1975). Vesicular-arbuscular mycorrhiza in the anthracite and bituminous coal wastes of Pennsylvania. (In preparation).

DAFT, M. J. & NICOLSON, T. H. (1966). Effect of *Endogone* mycorrhiza on plant growth. *New Phytol.*, 65, 343-350.

DAFT, M. J. & NICOLSON, T. H. (1969a). Effect of *Endogone* mycorrhiza on plant growth. II. Influence of soluble phosphate on endophyte and host in maize. *New Phytol.*, 68, 945-952.

DEMETERIO, J. L., ELLIS, R. & PAULSEN, G. M. (1972). Nodulation and nitrogen fixation by two soybean varieties as affected by phosphorus and zinc nutrition. *Agron. J.*, 64, 566.

GILMORE, A. E. (1971). The influence of endotrophic mycorrhizae on the growth of peach seedlings. *J. Am. Soc. hort. Sci.*, 96, 35-38.

HARLEY, J. L. (1970). The importance of micro-organisms to colonising plants. *Trans. Proc. bot. Soc. Edinb.*, 41, 65-70.

HEWITT, E. J. (1958). Some aspects of mineral nutrition in legumes. In *Nutrition of the legumes* (ed. Hallsworth, E. G.), Butterworths, London. p15.

KEILIN, D. & HARTREE, E. F. (1951). Purification of horse radish peroxidase and comparison of its properties with those of catalase and methaemoglobin. *Biochem. J.*, 49, 88-104.

MOSSE, B. (1973). Advances in the study of vesicular-arbuscular mycorrhiza. *A. Rev. Phytopath.*, 11, 171-196.

ROSS, J. P. (1971). Effect of phosphate fertilisation on yield of mycorrhizal and non-mycorrhizal soybeans. *Phytopathology*, 61, 1400 - 1403.

ROSS, J. P. & HARPER, J. A. (1970) Effect of *Endogone* mycorrhiza on soybean yields. *Phytopathology*, 60, 1552 - 1556.

VAN SCHREVEN, D. A. (1958). Some factors affecting the uptake of nitrogen by legumes. In *Nutrition of the legumes*. (ed. Hallsworth, E. G.). Butterworths, London. p. 137.

SPRENT, J. I. (1969). Prolonged reduction of acetylene by detached soybean nodules. *Planta*, 88, 372.

THE EFFECT OF <u>CYLINDROCARPON</u> ON PLANT GROWTH RESPONSES TO VESICULAR-ARBUSCULAR MYCORRHIZA

D. K. PAGET

Rothamsted Experimental Station, Harpenden, Herts., U.K.

INTRODUCTION

For a wide range of plants, phosphorus uptake and growth are increased by vesicular-arbuscular (VA) mycorrhiza (Mosse, 1973). In soil from Meathop Wood, a mixed deciduous woodland in Lancashire, such increases are much greater in irradiated than in unsterile soil. One year old seedlings of the dominant woody species *Fraxinus excelsior* showed a 95% increase in phosphorus uptake and 82% increase in dry weight. Even greater increases were found in herbaceous species, e.g. for *Viola riviniana* a 158% increase in phosphorus uptake and 102% increase in dry weight, and for *Fragaria vesca* a 206% increase in phosphorus uptake and 105% increase in dry weight were recorded (Mosse & Hayman, unpublished data). Possible explanations for the much better performance of VA mycorrhiza in the irradiated soil are :
1) altered nutrient status of the soil; 2) elimination of pathogens; 3) greater VA infection and phosphorus uptake because of lack of competition with other soil microorganisms.

1) Phosphorus is the nutrient limiting plant growth in Meathop Wood. The soluble phosphorus

content, measured by extraction with 0.5M NaHCO₃, (Olsen, *et al.*, 1954), increased on irradiation from 8.0 to 8.4 mg/kg soil, and measured by extraction with 0.01M CaCl₂ (Aslyng, 1954) from 0.4 to 0.5 μmol/l (Mosse, unpublished data). These increases are not large enough to account for the differences in growth.

Meathop Wood soil contains on average 0.34% total nitrogen. Whilst it is likely that irradiation affects the form of nitrogen in the soil, i.e. by increasing the ammonium nitrogen content, *Fragaria* seedlings initially grew equally well in irradiated and in unsterile soil, and after three weeks weighed 22 mg in both soils. After five weeks the plants in the unsterile soil had become mycorrhizal (30% infected roots) and weighed 35 mg, whereas the non-mycorrhizal plants in the irradiated soil weighed only 28 mg. The improved growth of the mycorrhizal plants is attributed to their greater phosphorus uptake (Mosse & Hayman, unpublished data).

2) It seemed unlikely that the smaller weight and correspondingly lower phosphorus uptake of mycorrhizal plants could be attributed to pathogens in the unsterile soil since no obviously diseased roots were observed. It is also improbable that the indigenous pathogens should cause such similar growth depressions in widely differing plant species.

3) Plants growing in irradiated soil had less mycorrhizal infection (e.g. *Fraxinus* 58% infected roots) than plants in unsterile soil (71% infected roots), so that the better growth in irradiated soil cannot be attributed to a greater mycorrhizal infection. Alternatively there may be competition for available phosphorus between mycorrhizal fungi and other soil microorganisms in the unsterile soil. This possible explanation has been investigated further using a soil saprophyte, *Cylindrocarpon destructans*, as a representative of the soil microflora. *Cylindrocarpon* is common to woodland situations both in the soil and on plant roots (Kubíková, 1963;

Matturi & Stenton, 1964).   This paper presents the
results of preliminary experiments designed to
investigate interactions between *Cylindrocarpon* and
VA endophytes.

MATERIALS AND METHODS

Meathop Wood soil was irradiated with a gamma
dose of 0.8 Mrads. *Cylindrocarpon destructans*
(Zins) Scholten was isolated from Meathop Wood soil
and initially cultured on Czapek Dox Agar.
Endogone. The inoculum consisted of spores, external mycelium and small adhering root fragments
obtained by wet sieving soil from pot cultures inoculated with type E_3 endophyte (Mosse, 1972).

Experiment 1

Cylindrocarpon was cultured on Czapek Dox Agar
at $20^{\circ}C$ for 14 days.   One eighth of a petri dish of
agar ($\equiv$ 2½ mls of agar) with or without the fungus
was added to pots containing 200 g of irradiated
Meathop Wood soil.   After 14 days to allow establishment of *Cylindrocarpon, Fragaria vesca* seedlings,
previously germinated in a John Innes compost/sand
mixture, were transplanted into the pots and E_3 inoculum (sievings) added.   There were three replicates
for each treatment.   The experiment was harvested
after 45 days.

Experiment 2

Cylindrocarpon was cultured in 50 ml 0.4% Malt
Extract liquid medium at $20^{\circ}C$ for 14 days.   The
culture medium was then filtered off and the fungus
washed five times with tap water.   The treatments
(each with five replicates) were control (no inoculum), with *Cylindrocarpon,* with E_3, and with
Cylindrocarpon and E_3.   The irradiated soil was

inoculated with the washed *Cylindrocarpon* and left
to stand for 14 days.   *Fragaria* sp. var. Baron
Solemacher seedlings, previously germinated in sterilised sand, were then transplanted and E_3 inoculum
(sievings) added.   The experiment was harvested
after 65 days.

Experiment 3

Experiment 2 was repeated with two extra treatments: a) autoclaved *Cylindrocarpon,* added to the
soil 14 days before transplanting seedlings to
investigate whether nutrients, particularly phosphorus, released on fungal autolysis, could significantly
stimulate plant growth.   b) a double inoculation of
Cylindrocarpon 14 days prior to and at the time of
transplanting the seedlings, to investigate the
effect of placement of the fungus on plant growth.
The experiment was harvested after 62 days.

In all experiments fresh and dry weights of
plants were recorded.   The phosphorus content of the
dried plants, ground up and ignited with magnesium
acetate at 450°C and dissolved in HCl, was determined
colorimetrically by the molybdenum blue method (Fogg
& Wilkinson, 1958), using a Technicon Auto Analyzer.

A small portion of the root system of all
plants was cleared in KOH and stained in lactophenol-
cotton blue (Phillips & Hayman, 1970).   The degree
of mycorrhizal infection was assessed by microscopic
examination.

At the harvesting of experiments 2 and 3, the
soil and root systems of the plants were examined
for establishment of *Cylindrocarpon*.   For the treatments with *Cylindrocarpon* four soil samples were
taken per pot and three or four small aggregates per
sample plated out in 2% water agar.   From the carefully washed root systems ten to twelve 1 cm root
lengths were taken and shaken in 10 ml sterile tap
water for 2, 5 and 15 minutes, transferring the roots
to a fresh vial after each washing.   After the final

washing all roots were plated out on 2% water agar.
For treatments without *Cylindrocarpon,* small pieces
of root with adhering soil particles and small soil
aggregates were plated out on 2% water agar.   All
plates were examined regularly for colonisation by
Cylindrocarpon and other microorganisms.

Solubilisation of soil phosphate by Cylindrocarpon

Irradiated Meathop Wood soil was inoculated with
Cylindrocarpon previously cultured as in Experiment 2.
There were three replicates.   After 40 days the soil
was examined for establishment of the fungus, and the
soluble phosphate content of the soil, with and without *Cylindrocarpon* inoculum, was measured by extraction with 0.5M $NaHCO_3$ and 0.01M $CaCl_2$.

RESULTS

Experiment 1

Soil inoculation with *Cylindrocarpon* nearly
doubled plant dry weight ($P < 0.01$) and increased
phosphorus content, but not significantly, in spite
of a slight reduction in mycorrhizal infection
(Table 1).

Table 1.   Mean dry weight and phosphorus
content of *Fragaria* seedlings inoculated with
Endogone (E_3) and *Cylindrocarpon.*

	Dry weight (mg)	μg P/ plant	% Mycorrhizal infection*
E_3 + Cylindrocarpon	850	1.643	73
E_3	490	1.229	90

* one quarter of the root system per plant examined

There was thus no evidence that *Cylindrocarpon* reduced the beneficial effects of mycorrhizal infection; on the contrary it seemed to have a stimulatory action on plant growth.

Experiment 2

Cylindrocarpon again had no adverse effects on phosphorus uptake or dry weight of mycorrhizal plants (Table 2); in this experiment, however, neither phosphorus uptake nor dry weight was increased by its presence.

The mean dry weight of plants inoculated with *Cylindrocarpon* alone was twice that of controls, and phosphorus uptake one and a half times greater (Table 2).   However this result was not significant because the plants in soil inoculated with *Cylindrocarpon* were extremely variable (Fig. 1), ranging from 31 mg dry weight (165 µg P) to 281 mg dry weight (753 µg P).   The uninoculated control plants were more uniform, weighing from 51-108 mg (154-232 µg P). *Cylindrocarpon* grew readily from soil particles but only rarely from washed roots (Table 2).   It was not found in roots stained with lactophenol-cotton blue. Its recovery from soil particles or root surfaces was little affected by E_3.   Other microorganisms which grew from soil particles and roots included *Penicillium* sp., *Trichoderma* sp., species of Actinomycetes, a non-spore forming rod-shaped bacterium, and other unidentified fungi.

Experiment 3

The results for Experiment 3 are presented in Table 3.   As in Experiment 2, plants grown in soil with a single inoculation of *Cylindrocarpon* had a greater mean phosphorus uptake and dry weight than uninoculated control plants.   However, the former were again much more variable (12-148 µg P, 11-113 mg dry weight) than the control plants

Table 2. Mean dry weight and phosphorus uptake of *Fragaria* seedlings inoculated with *Endogone* (E_3) and/or *Cylindrocarpon* and establishment of *Cylindrocarpon* in soil and on roots (Experiment 2).

Treatment	Dry weight (mg)	P uptake (µg/plant)	% Mycorrhizal infection*	*Cylindrocarpon* % soil aggregates colonised+	% roots colonised‡
Control	68	185	0	0	0
Cylindrocarpon	140	336	0	94	4
E_3	232	747	96	0	0
E_3 + *Cylindrocarpon*	239	718	92	76	7

* mean infected length of fifty 1 cm root lengths.

+ out of 60 soil aggregates.

‡ out of fifty five 1 cm root lengths.

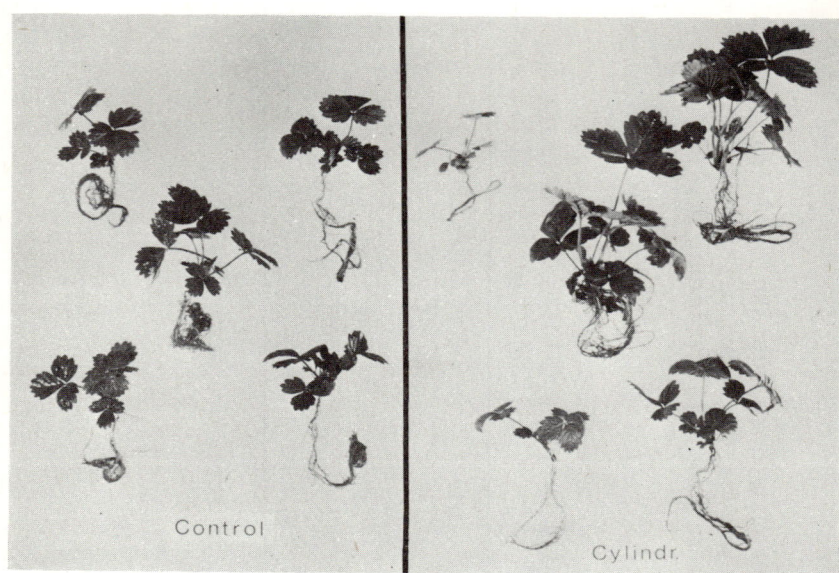

Fig. 1.   *Fragaria* seedlings grown in
irradiated Meathop Wood soil with and
without *Cylindrocarpon*.

(8-22 µg P, 12-26 mg dry weight).   Consequently the
differences were not significant.

   Double inoculation of the soil with *Cylindrocarpon* significantly increased the phosphorus uptake
(v. control P < 0.001, v. autoclaved *Cylindrocarpon*
P < 0.01, v. *Cylindrocarpon* (1 inoc.) P < 0.01), and
dry weight (v. control P < 0.001, v. autoclaved *Cylindrocarpon* P < 0.01, v. *Cylindrocarpon* (1 inoc.)
P < 0.01) of plants over all non-mycorrhizal treatments.

Table 3. Mean dry weight, % P, and phosphorus uptake of *Fragaria* seedlings inoculated with *Endogone* (E$_3$) and/or *Cylindrocarpon* and establishment of *Cylindrocarpon* in soil and on roots (Experiment 3).

Treatment	D.Wt. (mg)	% P Shoot	% P Root	P uptake (μg/plant)	% Myc. Inf.*	*Cylindrocarpon* % soil aggregates colonised+	*Cylindrocarpon* % roots colonised‡
Control	19	0.19	0.29	13	NE	0	0
Autoclaved Cylindrocarpon	39			42	0	0	0
Cylindrocarpon (1 inoc.)	52			62	0	55	67
Cylindrocarpon (2 inoc.)	99			118	0	84	81
E$_3$	146			294	90	0	0
E$_3$ + Cylindrocarpon (1 inoc.)	149	0.36	0.46	408	86	59	33

*   mean infected length of fifty 1 cm root lengths.

+   out of 80 soil aggregates.

‡   out of fifty 1 cm root lengths.

NE not examined for mycorrhizal infection as plants so small.

Plants inoculated with E_3 and *Cylindrocarpon* contained a very much higher % P in both shoot and root and had a greater phosphorus uptake ($P < 0.10$) than those inoculated with E_3 only but they were similar in weight. It would appear that the extra phosphorus represents luxury uptake without any corresponding growth response. In his studies of the mineral nutrition of *Fragaria* sp. var. Cambridge Favourite, Holevas (1966) grew mycorrhizal and non-mycorrhizal plants in soil with added phosphate. Good growth of both sets of plants was obtained with a maximum phosphorus concentration in the leaves of 0.29% recorded.

In this experiment, unlike Experiment 2, *Cylindrocarpon* grew readily from washed root segments, though not as appreciably from soil particles (compare Tables 2 and 3). Other microorganisms recorded in Experiment 2 were also found.

Solubilisation of soil phosphate by Cylindrocarpon

The presence of *Cylindrocarpon* in the soil did not increase the available soil phosphorus after 40 days (Table 4).

Table 4. Soil phosphorus levels following inoculation with *Cylindrocarpon*.

	% soil aggregates colonised*	Soil Phosphorus mg/kg soil in $NaHCO_3$ +	Soil Phosphorus μ moles/l in $CaCl_2$ ‡
Control	0	7.3	0.69
Cylindrocarpon	67	7.3	0.27
Original soil	-	8.0	0.32

* out of 36 soil aggregates

\+ Olsen *et al.* (1954)

‡ Aslyng (1954)

DISCUSSION

It had previously been recorded that the effect of VA mycorrhiza on plant growth was greater in irradiated than in unsterile soil from a mixed deciduous woodland. I examined whether this could be attributed to competition for soil phosphorus between a VA endophyte and other soil microorganisms, *Cylindrocarpon destructans* being used as a representative of the indigenous fungal population. In these experiments *Cylindrocarpon* established by inoculation in irradiated soil did not apparently compete with *Endogone* for available phosphorus or reduce its beneficial effects on plant growth; on the contrary there appeared to be some synergistic effect. In Experiment 1 inoculation with E_3 and *Cylindrocarpon* caused a significant increase in dry weight and a small increase in phosphorus uptake. In Experiment 3 phosphorus uptake, but not growth, was greatly increased. It seemed that Cylindrocarpon might actually increase available phosphorus in the soil. Sergeeva (1961) has demonstrated phosphate solubilisation by *Cylindrocarpon in vitro* and Agnihotri (1970) has shown this for a number of other fungi. However, in soil kept in the greenhouse *Cylindrocarpon* did not increase available phosphorus.

Some of the plants with a single inoculation and those with a double inoculation of *Cylindrocarpon* grew considerably better than uninoculated control plants, more than can be explained by the release of nutrients following fungal autolysis. These results, together with the interaction sometimes found with the endophyte, were not expected. The stimulatory action may be hormonal.

The effect on plant growth of *Cylindrocarpon* alone sometimes appears to be pathogenic. The fungus is thought to be a mild pathogen on e.g. *Lupinus arboreus* (Matturi & Stenton, 1964), black oak and to a lesser extent scarlet oak (Hart, 1965), and Evans, Cartwright & White (1967) report that a phyto-

toxin, nectrolide (identical with Brefeldin A),
produced by the fungus causes stunting and inhibition
of blackbutt seedlings.   As in the present experiments, they found no evidence of root infection by
Cylindrocarpon and suggested that the phytotoxin
acted pathogenically at the root surface.   The
smaller plants (e.g. Fig. 1) obtained in soil with a
single inoculation of the fungus in these experiments
may have been due to the presence of such a phytotoxin.   Finally the pathogenic effects of a single
inoculation with *Cylindrocarpon* were not apparent in
mycorrhizal plants.   It is interesting to speculate
whether *Endogone* could similarly reduce the pathogenicity of other organisms.

ACKNOWLEDGMENTS

I wish to thank Dr. Barbara Mosse for her
advice and helpful discussions and Dr. J. Frankland
for the original culture of *Cylindrocarpon*.   The
work was carried out during the tenure of a grant
from the International Biological Programme.

REFERENCES

AGNIHOTRI, V. P. (1970). Solubilisation of insoluble
    phosphates by some soil fungi isolated from nursery
    seedbeds.   *Can.J. Microbiol.* <u>16</u>, 877-880.

ASLYNG, H. C. (1954). The lime and phosphate potent-
    ials of rock;   the solubility and availability of
    phosphates.   *Yb.R.vet.agric.Coll., Copenhagen,*
    1-50.

EVANS, G., CARTWRIGHT, J. B. & WHITE, N. H. (1967).
    The production of a phytotoxin nectrolide by some
    root-surface isolates of *Cylindrocarpon radicicola*
    Wr.. *Pl. Soil,* <u>26</u>, 253-260.

FOGG, D. N. & WILKINSON, N. T. (1958). The colori-
metric determination of phosphorus. *Analyst,
Lond.,* 83, 406–414.

HART, J. H. (1965). Root rot of oak associated with
Cylindrocarpon radicicola. Phytopathology,
55, 1154–1155.

HOLEVAS, C. D. (1966). The effect of a vesicular-
arbuscular mycorrhiza on the uptake of soil phosphorus by strawberry (*Fragaria* sp. var. Cambridge
Favourite). *J. hort. Sci.,* 66, 57–64.

KUBÍKOVÁ, J. (1963). The surface mycoflora of ash
roots. *Trans. Br.mycol.Soc.,* 46, 107–114.

MATTURI, S. T. & STENTON, H. (1964). Distribution and
status in the soil of *Cylindrocarpon* species.
Trans. Br.mycol.Soc., 47, 577–587.

MOSSE, B. (1972). Effects of different *Endogone*
strains on the growth of *Paspalum notatum.
Nature, London,* 239, 221–223.

MOSSE, B. (1973). Advances in the study of vesicular-
arbuscular mycorrhiza. *A. Rev. Phytopath.,*
11, 171–196.

OLSEN, S. R., COLE, C. V., WATANABE, F. S. & DEAN,
L. A. (1954). Estimation of available phosphorus in
soils by extraction with sodium bicarbonate.
Circ. U.S. D.A., 939.

PHILLIPS, J. M. & HAYMAN, D. S. (1970). Improved
procedures for clearing roots and staining parasitic and vesicular-arbuscular mycorrhizal fungi
for rapid assessment of infection. *Trans.Br.
mycol.Soc.,* 55, 158–161.

SERGEEVA, N. V. (1961). The species composition of
   bacteria decomposing tricalcium phosphate and
   their physiological activity.   (Russian).
   Is.Moldavsk.Fil.Akad.Nauk. SSSR, 7, 66-71.

INTERACTION OF ENDOMYCORRHIZAL FUNGI AND ROOT-KNOT NEMATODE ON SOYBEAN

N. C. SCHENCK,[1] R. A. KINLOCH[2] AND D. W. DICKSON[3]

[1]Plant Pathology Department; [2]Agricultural Research Center, Jay, [3]Department of Entomology and Nematology, University of Florida, Gainesville, Florida, U.S.A.

INTRODUCTION

Zak (1964) postulated that ectomycorrhizae may influence the development of other organisms in the rhizosphere. Marx (1972) reviewed the impressive world-wide research that clearly demonstrated that ectomycorrhizae exert a marked effect on other micro-organisms in the rhizosphere, including many plant pathogens.

Although endomycorrhizae are obligate symbionts in plant roots, their hyphae extend into the rhizosphere. Few studies have investigated the effect of endomycorrhizal fungi on other microorganisms in the rhizosphere. Ross (1972) found that an endomycorrhizal fungus, *Endogone* sp., increased the susceptibility of soybean (*Glycine max* (L.) Merrill) to *Phytophthora megasperma* Drechs. var. *sojae* Hildeb. Baltruschat *et al.* (1973) indicated *Endogone mosseae* Nicol. and Gerd. had an antagonistic effect on *Thielavia basicola* (Berk. and Br.) Zopf. and *Meloidogyne incognita* (Kofoid and White) Chitwood on tobacco (*Nicotiana tabaccum* L.). Chou and Schmitthenner

(1973) found that *E. mosseae* did not significantly affect development of *Pythium* or *Phytophthora* on soybean.

In a recent survey of soybeans in Florida, Schenck and Kinloch (1974) found the incidence of endomycorrhizal fungi was related to the incidence of certain nematodes. Low populations of endomycorrhizal fungi were invariably associated with high populations of root-knot *(Meloidogyne incognita)* and cyst nematode *(Heterodera glycines* Ichinohe) but not with other nematode species. The purpose of this study was to evaluate the interrelationships of three common species of endomycorrhizal fungi on soybean in Florida to different population levels of Southern root-knot nematode.

MATERIAL AND METHODS

Greenhouse pot experiments were performed at the Agricultural Research Center, Jay, Florida, and the University of Florida at Gainesville. At Jay the soybean cultivar Ransom (*M. incognita* susceptible) and at Gainesville the cultivars Pickett (susceptible) and Forrest (resistant) were used. Five replications of each treatment were arranged on a greenhouse bench in a split-plot design with nematode levels forming the main treatments and endomycorrhizal species the subtreatments.

Isolates of endomycorrhizal fungi, obtained from soybean roots, were cultured on sorghum (*Sorghum vulgare* Pers.) in a greenhouse. Spores were removed from pot culture soil by wet sieving and decanting. About 30 detritus-free spores were placed in a 25-ml screw cap bottle containing 15 cc of sterile, sandy soil and stored at 6-8 C until needed. The three species of endomycorrhizal fungi were: *Endogone heterogama* Nicol. and Gerd., *E. calospora* Nicol. and Gerd. (a white azygospore-producing isolate with echinulate extramatrical vesicles (Schenck & Hinson,

1973) and *E. macrocarpa* Nicol. and Gerd. (a chlamydospore producing species not forming sporocarps in
pot culture).

At Jay, root-knot juveniles were extracted from
field soil, naturally infested with *M. incognita,* by
centrifugation and the Baermann-funnel technique.
These juveniles were used as inoculum.   At Gainesville, a population of *M. incognita* maintained on
tomato (*Lycopersicon esculentum* Mill.), cultivar
Rutgers, was used to obtain egg inoculum.   Individual eggs were obtained by shaking tomato roots 4 min
in 1.0% solution of sodium hypochlorite to separate
the eggs (Hussey and Barker, 1973).   Nematodes were
added to pots containing 3 kg of steam-sterilised soil
(120°C for 4 h; Orangeburg series, adjusted to pH
6.5 with 4 g/kg lime) at the following levels:
1) zero level = no nematodes added;   2) low level =
at Jay 500 juveniles per pot, at Gainesville 1,000
eggs per pot;   3) high level = at Jay 5,000 juveniles
per pot, at Gainesville 5,000 eggs per pot.

The procedures for infesting the soil with nematodes and mycorrhizal fungi were as follows: at
Gainesville the nematode eggs were distributed in
approximately 10 kg of soil by sprinkling the soil
with a suspension of nematode eggs in 400 cc of water
while mixing the soil by hand.   This amount of soil
was then distributed equally in five pots.   A 2.5-cm
layer of sterile soil was then added to each pot.
Soil containing approximately 30 spores of the desired endomycorrhizal fungus was added to the center
of each pot and covered with an additional 5-cm layer
of sterile soil.   Seed of the desired soybean cultivar were planted near the center of each pot prior
to watering.   At Jay the procedure was similar
except that root-knot nematode juveniles were added
rather than nematode eggs.

All pots were thinned to one soybean seedling
approximately 14 days after emergence.   Data recorded in this experiment included seed weight, dry
weight of roots, number of nematode juveniles per 100

cc of soil, and number of endomycorrhizal spores per
50 cc of soil.   Data were analysed statistically
using an analysis of variance.   Treatment means were
compared using Duncan's multiple range test.

RESULTS

Effect of nematodes - With the root-knot nematode susceptible cultivars (Pickett and Ransom), seed
yield and spores of endomycorrhizal fungi decreased
with increasing nematode inoculum (Tables 1 and 2).
In general, root dry weights were lowest with the
maximum nematode inoculum levels, but low nematode
inoculum levels with cultivar Pickett increased root
weights on mycorrhizal plants.   The highest numbers
of nematode juveniles per 100 cc of soil occurred at
the end of the experiment in pots receiving the high
nematode inoculum levels, except for cultivar Pickett
colonized with *E. heterogama* or *E. calospora*.

Effect of mycorrhiza - With the susceptible
varieties, seed yield and usually root weight of soybeans colonized with *E. macrocarpa* were superior to
those colonized by the other species of endomycorrhiza (Table 1).   Lowest seed yields and root
weights were generally obtained on nonmycorrhizal
plants.   More spores of *E. calospora* were recovered
after the experiment than with the other species of
endomycorrhiza.   In addition, nematode juveniles
were recovered in greater numbers from plants colonized by *E. calospora* than from plants colonized by
other mycorrhizal species.   In general, mycorrhizal
plants had greater numbers of nematodes than non-
mycorrhizal plants (Table 2).
With the resistant cultivar Forrest, highest
seed yields and root weight with low and high levels
of nematodes occurred on plants colonized with
E. heterogama.   At the low nematode inoculum level,
the greatest number of *M. incognita* juveniles were

Table 1. Root weight and seed yield of three soybean cultivars after 110 days exposure to endomycorrhizal fungi and root-knot nematodes in 15-cm pots in a greenhouse.[1]

Endomycorrhiza	Nematode level	Seed yield (g)			Dry root weight (g)		
		Pickett	Ransom	Forrest	Pickett	Ransom	Forrest
Endogone calospora	None	1.85 bc	1.60 b	1.53 ab	0.92 c	30.7 b	0.60 cd
	Low	1.32 c	0.42 e	1.91 a	1.46 bc	15.0 c	1.04 bc
	High	0.60 d	0.22 f	1.03 b	1.53 b	9.5 c	0.99 bc
Endogone macrocarpa	None	2.46 a	2.04 a	1.58 ab	1.24 bc	45.8 a	0.90 cd
	Low	1.90 b	0.62 d	1.45 ab	2.09 a	23.6 bc	1.44 b
	High	0.94 cd	0.48 d	1.95 a	1.22 b	11.8 c	0.94 c
Endogone heterogama	None	1.43 bc	1.06 c	0.77 bc	1.18 bc	21.4 bc	0.60 cd
	Low	1.17 cd	0.38 e	2.06 a	1.18 bc	6.5 c	2.11 a
	High	1.69 d	0.16 f	2.02 a	0.95 c	12.5 c	1.27 bc
None	None	0.50 d	1.02 c	0.35 bc	0.94 c	24.0 bc	0.21 d
	Low	0.65 d	0.38 e	0.64 bc	0.83 cd	14.6 c	0.53 c
	High	0.52 d	0.04 f	0.22 c	0.34 d	7.4 c	0.24 d

[1] Means within the same vertical column followed by same letter do not differ significantly (5% level) by Duncan's multiple range test.

Table 2. Root-knot nematode juveniles and spores of endomycorrhizal fungi in rhizosphere soil of three soybean cultivars after 110 days in 15-cm pots in a greenhouse[1].

Endomycorrhiza	Nematode level	Nematode juveniles/100 cc soil			Spores/50 cc soil		
		Pickett	Ransom	Forrest	Pickett	Ransom	Forrest
Endogone calospora	None	0 c	0 c	1 b	124 a	182 a	80 b
	Low	714 a	139 b	175 b	109 a	28 b	127 a
	High	512 ab	292 a	566 a	83 ab	8 b	56 bc
Endogone macrocarpa	None	1 c	0 c	0 b	62 b	32 b	35 bc
	Low	224 bc	79 bc	49 b	63 b	25 b	32 c
	High	373 b	113 b	735 a	40 bc	5 b	45 bc
Endogone heterogama	None	0 c	0 c	0 b	15 c	12 b	4 c
	Low	463 ab	258 a	94 b	19 bc	21 b	18 c
	High	148 bc	225 ab	430 a	8 c	0 b	10 c
None	None	2 c	0 c	0 b	0 c	7 b	1 c
	Low	63 c	91 bc	52 b	0 c	11 b	0 c
	High	45 c	154 b	131 b	0 c	0 b	0 c

[1] Means within the same vertical column followed by same letter do not differ significantly (5% level) by Duncan's multiple range test.

recovered from *E. calospora* colonized plants, while
at the high nematode inoculum levels most juveniles
were recovered on *E. macrocarpa* colonized plants.

DISCUSSION

In general root-knot nematodes decreased soybean seed yields and endomycorrhizal fungi increased
them.   This was not unexpected.   Root-knot nematodes are parasitic and endomycorrhizal fungi symbiotic on soybean.   Other results of interest were:
1) seed yields of the susceptible cultivars were
increased most by *E. macrocarpa* while yield of the
resistant cultivar was increased most by *E. heterogama;*   2) the number of juvenile nematodes generally
increased with increasing root weight;   3) the
absence of nematodes had the least effect on yields
of the resistant cultivar;   4) yields and root weight
were consistently lowest with high nematode inoculum
levels and no mycorrhiza on the roots;   5) nematodes
were detrimental to the sporulation of endomycorrhizal fungi on soybean roots, thus verifying the field
observations of Schenck and Kinloch (1974) that spore
counts of endomycorrhiza fungi were consistantly low
when associated with high populations of root-knot
nematodes.
However, the most significant result from this
study was the difference in interaction of nematodes
and mycorrhiza on the susceptible and resistant cultivars.   Numbers of nematode juveniles generally
increased with increasing root weight.   However, an
increase in yield and root weight on the susceptible
varieties was concomitantly associated with a decrease in nematode juveniles with *E. macrocarpa*
(Fig. 1).   On the resistant cultivar, a similar
result was obtained with *E. heterogama* (Fig. 2).
These results indicated *E. macrocarpa* on the susceptible cultivars and *E. heterogama* on the resistant
cultivar had an adverse effect on nematode develop-

Schenck, N. C. *et al.*

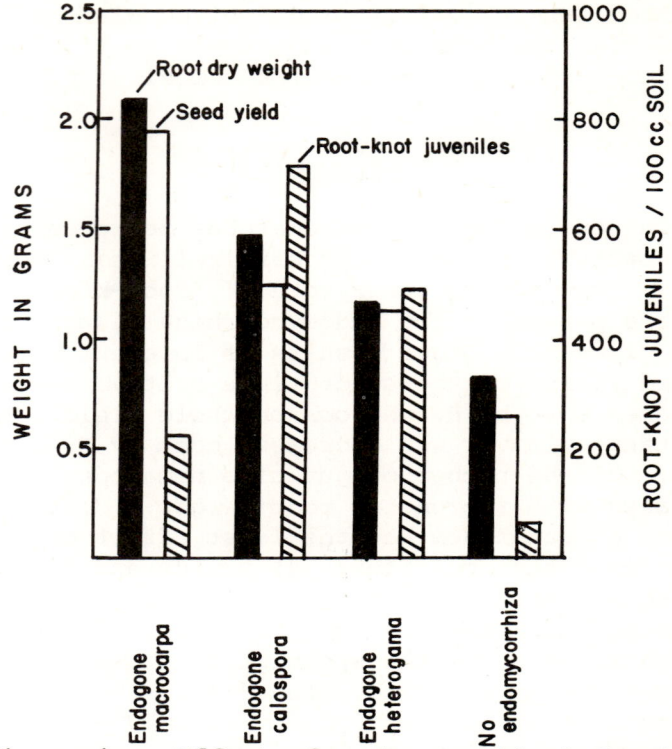

Figure 1. Effect of endomycorrhizal fungi
on root weight and seed yield of Pickett
soybean and root-knot nematode juvenile
populations. (Low nematode inoculum level
- 1,000 nematode eggs per pot).

ment. This antagonistic effect was noted at the
low nematode level for the root-knot susceptible variety and at the high inoculum level on the resistant
variety. Nematode juvenile populations at the term-
ination of the experiment in the low inoculum level
on Pickett were comparable to those on Forrest at the
high nematode inoculum level (Table 2).

These results support those of Baltruschat *et
al.* (1973) who reported a reduction in root-knot

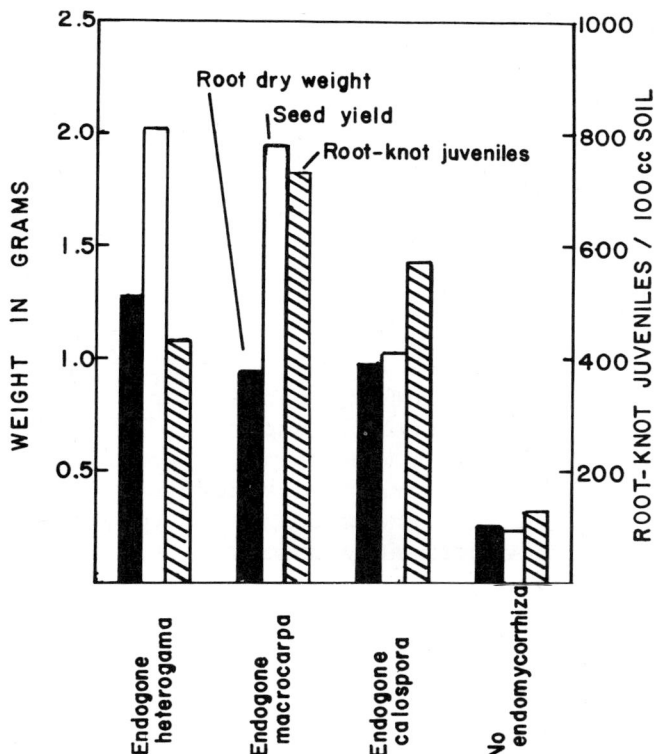

Figure 2.   Effect of endomycorrhizal fungi
on root weight and seed yield of Forrest
soybean and root-knot nematode juvenile
populations.   (High nematode inoculum level
- 5,000 nematode eggs per pot).

nematode juveniles and their effect on tobacco in
association with *E. mosseae*.   In addition to
supporting previous work, the results of our study
indicate that mycorrhiza-nematode interactions differ
with various cultivars.   Mycorrhizal species affect
different cultivars in dissimilar ways, thus necessi-
tating testing individual cultivars with each mycorrhizal species.   The response of one mycorrhizal-
plant-nematode combination may not be extended to

other mycorrhizal-plant-nematode combinations.

Under the conditions of our experiment, nematode juveniles were recovered at the end of the experiment in larger numbers from soybean plants with endomycorrhiza than those without endomycorrhiza. Since nematologists frequently wish to obtain large numbers of nematodes for plant inoculation work, it may be desirable to utilize plants colonized with endomycorrhizal fungi as host plants for increasing nematode populations.

During this experiment, no attempt was made to determine nematode or endomycorrhizal fungi populations or their effects on each other. Thus the mechanism for the apparent suppression of nematodes by *E. macrocarpa* on the susceptible cultivar and *E. heterogama* on the resistant cultivar is unknown. Studies to evaluate the factor(s) involved in this suppression are currently underway.

REFERENCES

BALTRUSCHAT, H., SIKORA, R. A. & SCHONBECK, F. (1973) Effect of V-A mycorrhizae on the establishment of *Thielaviopsis basicola* and *Meloidogyne incognita* on tobacco. *2nd Internation. Cong. Plant Pathol.*, 661 (Abstr.)

CHOU, L. G. & SCHMITTHENNER, A. F. (1973). Effect of *Rhizobium japonicum* and *Endogone mosseae* on virulence of *Pythium ultimum* and *Phytophthora megasperma* var. *sojae* on soybeans. *2nd Internation. Cong. Plant Pathol.*, 662. (Abstr.)

HUSSEY, R. S. & BARKER, K. R. (1973). A comparison of methods of collecting inocula of *Meloidogyne* spp., including a new technique. *Pl. Dis. Reptr.*, 57, 1025-1028.

MARX, D. H. (1972). Ectomycorrhizae as biological

deterrents to pathogenic root infections.   *A. Rev. Phytopathol.*, 10, 429–454.

ROSS, J. P. (1972). Influence of *Endogone mycorrhiza* on *Phytophthora* rot of soybean.   *Phytopathology,* 62, 896–897.

SCHENCK, N. C. & HINSON, K. (1973). Endotrophic vesicular-arbuscular mycorrhizae on soybean in Florida. *Mycologia,* 63, 672–674.

SCHENCK, N. C. & KINLOCH, R. A. (1974). Pathogenic fungi, parasitic nematodes, and endomycorrhizal fungi associated with soybean roots in Florida. *Pl. Dis. Reptr.,* 58, 169–173.

ZAK, B. (1964).Role of mycorrhizas in root disease. *A. Rev. Phytopathol.,* 2, 377–392.

EFFECT OF SYSTEMIC FUNGITOXICANTS ON THE DEVELOPMENT OF ENDOTROPHIC MYCORRHIZA

B. L. JALALI AND K. H. DOMSCH

Institut für Bodenbiologie,
Forschungsanstalt für Landwirtschaft,
Braunschweig, W. Germany

INTRODUCTION

Diverse groups of fungitoxicants are widely used in the control of plant diseases. The application of these toxicants may, however, result in indiscriminate killing of both pathogenic as well as nonpathogenic/beneficial microorganisms (Domsch, 1974). It is becoming increasingly clear that the vesicular-arbuscular mycorrhiza is one of those fungus-root associations with a probiotic influence on higher plants. However, not much attention has been paid to fungicidal impact on the development of the fungal endophyte. Since vesicular-arbuscular infection is a labile condition, environmental factors greatly influence the balanced relationship between plant roots and the endophyte (Mosse, 1963, 1973). This is of particular significance because V-A mycorrhiza occur on a large number of commercial crops, and the wide use of toxic materials in normal agricultural practices may, therefore, upset this balance.

Harmful effects of soil fumigants have been reported on the growth of ectotrophic fungi on forest trees (Iyer & Wilde, 1965; Persidsky & Wilde, 1960; Wilde & Persidsky, 1956), and on endo-

trophic fungi on corn (Nesheim & Linn, 1969), and
citrus seedlings (Kleinschmidt & Gerdemann, 1972).
However, no studies have been reported on seed and
foliar applications with fungitoxicants on the
formation of V-A mycorrhiza.   The present work was,
therefore, undertaken in an attempt to study the
effect of some systemic fungitoxicants on the
mycorrhizal development on wheat (*Triticum aestivum*)
roots, and on root hair development.

MATERIALS AND METHODS

'Jubilar' variety of wheat (*Triticum aestivum*)
was used as the test plant, since most of the fungitoxicants employed in these studies are recommended
for the control of various diseases affecting small
grains.   Stock cultures of *Endogone* sp. (yellow
vacuolate type) were maintained on onion (*Allium
cepa*) seedling roots in order to minimize the
possibility of contamination by subsequent cultures
with wheat root pathogens.   Mycorrhizal roots of
onion were washed free from adhering sand and cut
into small segments.   Each experimental pot, containing 900 g sand, maintained at 30% moisture
content, was then mixed with 2 g mycorrhizal inoculum.
     In the seed treatment experiments, three
systemic fungitoxicants were used: benomyl 1-(N-
butylcarbamamoyl)-2-(methoxycarboxamide)-benzimid-
azole ;   thiabendazole 2-(4-thiazolyl)-benzimidaz-
ole ;   and ethirimol 5-n-butyl-2-ethylamino-4-
hydroxy-6-methyl pyrimidine .
     Five fungitoxicant-treated seeds of wheat were
sown in each experimental pot containing mycorrhizal
inoculum and grown in a light-, temperature- and
moisture controlled growth- chamber.   Each treatment
was replicated three times.   At weekly intervals
each pot received 10 ml Knoop nutrient solution.
     In foliar application experiments nine fungi-

toxicants were used.   They were: triforine   1,4-
bis-(2,2,2-trichlor-1-formamidoethyl)-piperazine ;
chloramformethane   1-(3,4-dichloranilino)-1-formyl-
amino-2,2,2-trichlorethane ;   tridemorph N-tride-
cyl-2,6-dimethylmorpholine ;   thiophanate   1,2-bis-
(3-ethoxycarbonyl-2-thioureide)-benzene ;   triadem-
ifon   1-(4-chloro-10 oxy)-3,3-dimethyl-1-(1,2,4-
triazol-1-yl)-2-butanon ;   benomyl; maneb   Mn(II)-
(N,N'-ethylene-bis-(dithiocarbamate) ;   dichlofluanid   N,N-dimethyl-N-phenyl-(N-fluordichlormethyl-
thio)-sulfamide ;   and captan   n-trichloromethylmer-
capto-4-cyclo-hexene-1,2-dicarboximide .

Each fungicidal material was sprayed at the
recommended dose three weeks after sowing, when the
fungal endophyte had just begun to infect wheat
roots.   The sand surface of each pot was well
covered with polythene sheets in order to avoid
drain-off contamination of the sand with sprayed
material.

The assessment of the mycorrhizal development
was made after six or nine weeks of plant growth,
depending upon the experiment.   Roots were washed
free from sand and cut into 1 cm pieces;   they were
then cleared by heating in 10% KOH for one hour at
90°C.   After rinsing and acidifying with dil.HCl,
the root segments were stained by simmering in
0.05% trypan-blue in lactophenol for 5 minutes
(Phillips & Hayman, 1970), mounted in lactophenol,
and examined for mycorrhizal infection.   Infection
was recorded by grading (grade 0 - 4) mycorrhizal
structures along each root segment.   The root segments were also examined for chlamydospore numbers
and root hair development simultaneously.

RESULTS

Effect of seed treatment

Seed treatment of wheat with the three systemic
fungitoxicants *viz.* benomyl, thiabendazole and

ethirimol, each used at three different concentrations, had an adverse effect on the formation of
mycorrhiza on roots (Table 1).   The effect was more
pronounced on 9 week old seedlings than on 6 week
old seedlings with the exception of the thiabendazole treatment (0.6 mg/100 g seeds).

Table 1.   Effect of seed treatment with
three systemic fungitoxicants on infection
of wheat roots by *Endogone* sp.

Fungitoxicant	Dosage (mg a.i./100 g seed)	Mycorrhizal grading*	
		6th week	9th week
Benomyl**	0.5	2.90	2.70
	2.5	2.42	2.23
	10.0	1.80	1.51
Thiabendazole	0.6	2.92	3.00
	3.0	2.20	1.82
	12.0	1.55	1.12
Ethirimol	2.50	2.00	1.54
	3.50	1.43	1.36
	5.00	1.20	1.04
Control	–	3.08	3.41

* Mean of 3 replications;   each replication consisted
of 5 pots, from which 50 root segments were taken
at random.   All effects were significant at the
5% level, except for benomyl (0.5 mg a.i./100 g
seed) and thiabendazole (0.6 mg a.i./100 g seed)
treatments.
**The recommended dosage for seed treatment with
benomyl is 150 mg a.i./100 g seed.

Of the three systemic fungitoxicants, ethirimol at 5.0 mg a.i./100 g seed was the most harmful for the development of the fungal endophyte.   Benomyl at the two higher concentrations appeared to have a less depressing effect on the mycorrhiza formation than other comparable fungitoxicants.   It should be mentioned that the highest concentration of this treatment used in these studies, although below the recommended dose, still restricted the mycorrhizal development significantly, particularly so at the 9th week stage of plant growth.

Observations on root hair development revealed that thiabendazole treatment at 12 mg a.i./100 g seed had a depressing effect; benomyl and ethirimol treatments did not result in such effects.

Effect of foliar application

The response of foliar treatments with systemic and, for the purpose of comparison, nonsystemic fungitoxicants showed that most of the treatments had a less drastic effect on the establishment of mycorrhiza in root tissues than seed treatments. Triforine and benomyl treatments had relatively the strongest inhibitory effects (Table 2).   However, it was of interest that four of the nine treatments *viz.* triforine, tridemorph, chloramformethan, and the non-systemic, dichlofluanid, reduced the number of chlamydospores by 50% or more.   Of the non-systemic toxicants, maneb and captan had no such pronounced effect on chlamydospore development. None of these nine foliar treatments had any noticeable effect on root hair development.

DISCUSSION

All three fungitoxicants applied as seed dressings had an inhibitory effect on the development of vesicular-arbuscular mycorrhiza on wheat roots, even at the lowest concentration.   From the majority of

Table 2. Effect of foliar application of fungitoxicants on mycorrhizal infection of wheat roots and on chlamydospore numbers (observations taken 9 weeks after sowing).

Fungitoxicant	Dosage (Kg a.i./ ha)	Mycorrhizal Infection grade[1]	Chlamydospore no./40 segments[2]
Triforine	0.2	2.57*	17*
Chloramform- ethane	0.25	2.72*	25*
Tridemorph	0.55	2.84*	21*
Thiophanate	0.38	2.89	32*
Triademifon	0.06	2.84*	35
Benomyl	0.12	2.62*	37
Maneb	0.8	3.17	44
Dichlofluanid	0.5	2.79*	18*
Captan	0.6	3.07	41
Control	-	3.18	48

[1] Mean of 3 replications; each replication consisted of 3 pots, from which 40 root segments were taken at random.

[2] Each segment of 1cm length.

* Significant at 5% level.

reports on the translocation of systemic fungicides it can be concluded that the amount transported within the plant downward into the roots is small and of little significance (Erwin, 1973). Therefore, inhibitory effects could arise from residues of the fungitoxicants taken up by the germinating seed and

translocated passively during root growth, or by
physiologically increasing root "resistance".   In
particular, the possibility of an indirect action on
the balanced fungus-root relationship must be
considered when the formation of chlamydospores was
suppressed by fungitoxicants applied as foliar
spray.   It can be demonstrated that after the tox-
icant application, plants react with changes in
amino acid metabolism (Jalali, unpublished).
Alteration of host metabolism is one of the principles by which systemic fungitoxicants may act on
fungal invaders.   If the concurrent changes in the
plant metabolism persist long enough, conditions may
prevent the fungus from normal development.

Since chlamydospores are the means by which
Endogone spp. survive in time, the ecological significance of this phenomenon should not be over-
looked.   More detailed work is in progress to
elucidate both causes for the inhibition phenomena
as well as ecological and economical implications.

REFERENCES

DOMSCH, K. H. (1964). Soil fungicides.   *A. Rev.
Phytopath.*, 2, 293-320.

ERWIN, D. C. (1973). Systemic fungicides: disease
control, translocation and mode of action.
A. Rev. Phytopath., 11, 389-422.

IYER, J. G. & WILDE, S. A. (1965). Effect of vapam
biocide on the growth of red pine seedlings.
J. For., 63, 703-704.

KLEINSCHMIDT, D. D. & GERDEMANN, J. W. (1972).
Stunting of citrus seedlings in fumigated nursery
soils related to the absence of endomycorrhizae.
Phytopathology, 62, 1447-1452.

MOSSE, B. (1963). Vesicular-arbuscular mycorrhiza: an extreme form of fungal adaptation. In: *Symbiotic Associations.* Ed. Nutman, P. S. & Mosse, B.. *Symp. Soc. gen. Microbiol.,* 13, 146-170.

MOSSE, B. (1973). Advances in the study of vesicular-arbuscular mycorrhiza. *A. Rev. Phytopath.,* 11, 171-196.

NESHEIM, O. N. & LINN, M. B. (1969). Deleterious effects of certain fungitoxicants on the formation of mycorrhiza on corn by *Endogone fasciculata* and on corn root development. *Phytopathology,* 59, 297-300.

PERSIDSKY, D. J. & WILDE, S. A. (1960). The effect of biocides on the survival of mycorrhizal fungi. *J. For.,* 58, 522-524.

PHILLIPS, J. M. & HAYMAN, D. S. (1970). Improved procedures for clearing roots and staining parasitic and vesicular-arbuscular mycorrhizal fungi for rapid assessment of infection. *Trans. Br. mycol. Soc.,* 55, 158-161.

WILDE, S. A. & PERSIDSKY, D. J. (1956). Effect of biocides on the development of ectotrophic mycorrhizae in Monterey pine seedlings. *Proc. Soil Sci. Soc. Am.,* 20, 107-110.